No L.C.

FRESHWATER ECOLOGY

Freshwater Ecology

T. T. Macan M.A. Ph.D.

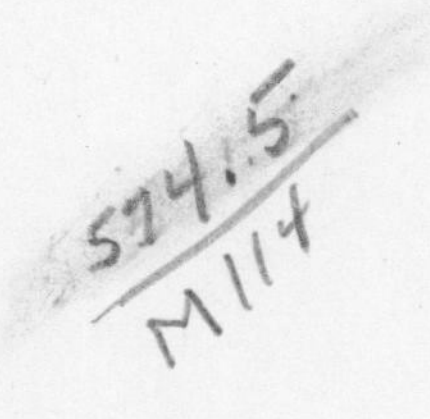

JOHN WILEY & SONS INC
New York, N.Y.

Published throughout the world except
the United States by Longmans, Green & Co. Ltd.

First published 1963

Printed in Great Britain by
The Camelot Press Ltd., London and Southampton

To

ZAIDA MACAN

who has tolerated a rival
in her husband's affections – Science –
with commendable equanimity

Contents

FOREWORD ix

1. *The Background* 1
2. *Physical and Chemical Properties of Water* 11
3. *Communities* 15
4. *Transport* 66
5. *Behaviour* 78
6. *Interrelationships* 91
7. *Physical Factors (1): Water Movement, Desiccation and Miscellaneous* 116
8. *Physical Factors (2): Temperature* 138
9. *Oxygen* 175
10. *Salinity* 207
11. *Calcium* 246
12. *Other Chemical Factors* 254
13. *Production* 267
14. *Methods* 276

REFERENCES 286

INDEX TO SPECIES NAMES 323

INDEX TO ENGLISH NAMES 330

INDEX TO AUTHORS' NAMES 331

SUBJECT INDEX 335

Foreword

A foreword is used in different ways by different authors. Some write it with the fire and fervour of one preaching a new religion, implying freely that those who do not follow their ideas will be cast into outer darkness by the scientific fraternity in this world and probably also by whoever arranges such things in the next. The only doctrine I wish to preach is toleration; the school that is brilliantly leading the world to-day may be nearing the end of its seam, whereas the school that has been plodding along an unprofitable road for years may be about to strike a rich vein. Scientists are not seers. Some believe that all available forces could be concentrated for an attack in one place on any particular problem; to them I would recommend a study of the First World War, in which so many perished because that idea was held so tenaciously. My belief is that an attack in different ways from different sides is likely to prove profitable more quickly.

Other authors write deprecatingly of their own shortcomings and those of the book, which probably does not achieve anything except to confirm continental belief in British hypocrisy.

One of the most popular uses of a foreword is to explain the motives, all of them invariably of the most praiseworthy nature, that led to the writing of the book. I hope that no reader of this book will wonder why it was written, for there could be no more certain criterion of failure; moreover, any reader who does wonder will not be likely to have sufficient interest in the book to want to turn back to the beginning to find out.

This unwillingness to explain why the book was written is not matched by similar feelings towards the question 'How?' In these days when scientific literature is so vast and knowledge is advancing at so rapid a rate, the would-be author, having planned his chapters, does well to look round among his friends to see how many are expert on the subjects treated and can be prevailed upon to bring their expert knowledge to bear on the criticism of a chapter. What has to be judged nicely is how long one can continue to write science in this way and still have any friends. The chapters in this book have been criticized by the following: Professor Kaj Berg 3 and 9, Dr J. H. Mundie 4, Dr H. B. N. Hynes 5 and 7, Dr G. Fryer and Dr. T. B. Reynoldson 6, Professor Gertrud Pleskot 8, Dr J. A. Kitching, F.R.S. 10, Dr K. H. Mann 11, Mr and Mrs F. J. H. Mackereth 12, Mr E. D. Le Cren and Dr J. Talling 13, and Miss C. Kipling 14. My colleagues Dr W. E. Frost and Mr W. J. P. Smyly have permitted me to draw freely on their knowledge of fish and entomostraca respectively. These good friends have headed me off from many a pitfall and indicated many a path that has proved worth following. They have also, to descend to a more mundane level, corrected typists' errors, grammatical faults and constructional infelicities; so has some lady or gentleman in the publisher's office who preserves strict anonymity. My gratitude to them all is immense.

There are many others who should be thanked, persons with whom I have discussed problems of various kinds and persons who have helped in the innumerable ways that are such an agreeable feature of a scientific career. A list would be long and certainly incomplete and therefore none is attempted though I shall take this opportunity to mention gratefully the work of my former assistants, Mrs F. J. H. Mackereth and Mr T. Gledhill. If any reader detects in this book, purporting to be original, an idea that he can remember communicating to me, let him not be too quick to see in the purloining of it a piece of gamesmanship; rather may I plead that the memory is a selective mechanism retaining and rejecting with an alarming independence of that part of the brain from which ethical principles spring.

My present assistant, Miss Rachel Maudsley, has given much help with the long literature list and with the index. The first draft of this was compiled by Miss Susan Wright. Miss Rachel Cruttwell has assisted in these and other activities that intervene between a typescript and a book. It is a pleasure to record my gratitude for this assistance.

CHAPTER 1

The Background

'Ecology may be defined broadly as the science of the interrelation between living organisms and their environment. . . .' (Allee *et al.* 1949). This definition provokes the question: what is not ecology? The answer seems to be: not much. The morphologist studying an organ will want to know how the living organism in its environment used it, the physiologist studying function must ultimately relate his findings to the animal's way of life, taxonomy in its later stages and genetics must take account of natural populations. The author who embarks on a general work on ecology must, therefore, cover so vast a field that he finds it difficult to give coherence to the innumerable and diverse facts and ideas which he assembles. Many, in consequence, prefer to restrict themselves to one part of the field, taking a single theme that can be developed without being overwhelmed by a mass of factual information. It is profitable, before making a choice, to discover how other authors have restricted themselves in their approach to the subject.

Tansley's large book, *The British Islands and Their Vegetation* (1939), starts with an account of physiography, climate, soil, and biotic factors, which include the activities of man in the past and to-day, the activities of other animals, and the interrelationships between the plants themselves. Then he defines the communities, a word he uses in a general sense, by lists of species and an indication of the abundance of each, and discusses their occurrence in relation to the factors already mentioned. In so far as it is possible to summarize 900 pages in nine words, Tansley's approach may be said to be a description of community composition and discussion of limiting factors.

Tansley's ideas were familiar to Elton when he wrote about animal ecology in 1927. The striking contrast between the two books is the absence of any account of the composition of animal communities. Only a few simple ones have been described, partly because animals are more difficult to catch but mainly because they are far more diverse, so that even a preliminary list of them cannot be made without the collaboration of a panel of specialists. Elton writes that he was forced to abandon ideas of work of this type because of the taxonomic difficulties.

Many of Elton's chapter headings and sub-headings are the same as Tansley's but there are some which, if not different, deal with phenomena of a different order of magnitude. First there is the structure of the community. With a few exceptions that do not seem to be regarded as important, a plant community consists of organisms all striving for a place in the sun from which they derive energy to grow and reproduce. Animals obtain their energy in many different ways. Herbivores graze, mine in leaves or trunks, or live in the soil, feeding on dead matter or on roots. The carnivores which hunt these, and the scavengers, are even more diverse. Elton puts forward the concept of the 'niche', by which he means what an animal does. 'Birds of prey which eat small mammals' is an example which he quotes, and another illustrative sentence, selected by Bates (1949) as a quotation to embellish a chapter heading is: 'When an ecologist says "there goes a badger" he should include in his thoughts some definite idea of the animal's place in the community to which it belongs, just as if he had said "there goes the vicar".' Later (Elton and Miller 1954) Elton admits that it is an elusive concept, for animals are versatile and adaptable, and the bird of prey of the original definition will eat something else if small mammals are hard to find.

Elton describes the food-chain, to which expression later writers have preferred the term food-web, and points out that the animal community tends to consist of increasingly small numbers of increasingly large animals, a concept that has since been named the 'Eltonian pyramid'.

Plants are all active at the same time but animals may come out by night or by day, and different times of activity may mean that close neighbours never meet. Moreover, migration causes changes in composition of a community according to the time of year.

Elton devotes two chapters to the numbers of animals, a subject scarcely mentioned by Tansley. The number of oaks in an oak-wood, to take an extreme example, varies little from year to year. If the wood be bounded by terrain which is unsuitable because it is too steep, or too rocky, or too damp, each year's seedlings are doomed to perish on unfavourable ground or beneath the shade of their parents, and have no hope of survival except in a clearing caused by the death or fall of a mature tree. Moreover, an oak is a difficult organism to kill; caterpillars sometimes devour all its leaves but then the caterpillars die, their food supply exhausted, and the tree in due course puts forth another crop of leaves. The ecologist will probably wish to know how many trees there are per acre, but fluctuations in numbers concern him not at all; they are likely to be infinitesimal in his lifetime and for longer than that.

The oak forest utilizes only a small proportion of the energy falling on it from the sun; if it could use more, a stage would come when scarcity of mineral nutrients would limit growth.

Fluctuation in numbers is obviously greater in other plants, notably annuals, but not enough to excite much interest among plant ecologists. There is, however, one important exception, and that is the student of freshwater algae. These plants are not fixed, one individual cannot by casting shade keep others out of an area sufficient for its own needs and their populations wane suddenly when the mineral supply is exhausted (see for example Lund 1950).

Some animal species limit their numbers in the same sort of way as trees, though the comparison cannot be applied too rigidly because the life span is so much shorter. Kalleberg (1958) has described how young trout establish a territory from which they drive out invaders, and how those unsuccessful in this struggle die. The size of the territory depends on the size of the fish and on the configuration of the substratum, and presumably the population of a stream or river depends on the nature of the bottom, though this is a hypothesis that has not been demonstrated. Other animals, particularly herbivores, are more like algae in that their numbers will increase till checked by lack of food. A world carpeted by vegetation but uninhabited by animals is conceivable; all species would have to be wind-pollinated and soil conditions would probably not be as favourable in the absence of burrowing animals but, provided that there were bacteria to decompose plant remains, a stable cover of vegetation could exist. A world of plants and herbivores would be an unstable place. The animals would increase till they had devoured almost every plant and then most would starve, after which regeneration would start from a few survivors and the cycle would repeat itself. A comparable sequence of events probably takes place in small ponds (Pennington 1941). A world with plants, herbivores, carnivores and parasites is more stable, and how they interact to produce stability has been one of the main themes of ecological study in the last few decades. The books of Lack (1954) and Andrewartha and Birch (1954) may be mentioned in addition to the others referred to in this chapter. Andrewartha and Birch maintain that abundance is the key to distribution. They picture species continually struggling to extend their range and limited along the line where conditions are sufficiently adverse to kill more than are born. This is doubtless true of a great many species but it is not true of all. Some require particular conditions at the breeding time and will not breed where they do not find those conditions, even though there is no obvious reason why breeding should not be successful.

A similar apparently capricious selection may be exhibited by some species at all times.

Odum (1959) writes about the ecosystem, that is the sources of energy, the plants and the animals. He dwells on the study of the efficiency of production, which involves a knowledge of the flow of energy through the ecosystem, more than Elton does. Macfadyen (1957) devotes the main part of his book to population dynamics, production, the flow of energy, and the regulation of numbers. He attaches less importance to limiting factors. He quotes seven definitions of the community, though his pages are noteworthy, as are Elton's, for the absence of any account of the composition of a community. He is well aware of the theoretical nature of the definitions, but, to underline the point, I quote here the words of a field naturalist discussing the relation of insects to plant communities: 'Our real knowledge of it is so limited and patchy that it is almost impossible to appreciate the extent of the gaps' (Diver 1944).

A line of investigation that has been popular and productive among freshwater biologists has been to take a group whose systematics can be mastered and attempt to define the habitat of each species.

In a joint paper (Elton and Miller 1954), Elton sets out the results of many years' meditation on the problems of animal ecology. If, as is commonly supposed, all members of a community interact, all must be studied. The authors describe their plan for doing this. It is a long-term project, involving many collaborators, an elaborate record system, and a permanent institute.

Thienemann (1950) is concerned mainly with how historical events and the abilities and requirements of species have reacted to confine them within a given geographical area to-day, rather than with the factors that determine occurrence within that range, but he devotes some hundred pages to this question.

Which of these approaches is most appropriate to fresh water? It would appear to provide an environment particularly suited to the study of community composition, as it consists of isolated and often small pieces. The main hindrance to advance in this field is taxonomic. Elton (1927) writes that, as a result of the discoveries of Darwin: 'Half the zoological world thereupon drifted into museums and spent the next fifty years doing the work of description and classification which was to lay the foundations for the scientific ecology of the twentieth century.' Unfortunately this work did not include the immature stages, and larvae and nymphs of insects are frequently the most numerous components of freshwater communities. Much has been accomplished in recent years but there are still some large gaps, and it is still impossible to put a reliable name to many Trichoptera larvae. The species lists on which the study of communities must be based cannot therefore be complete, though some authors, by breeding out adults or collecting adults in the neighbourhood of the water, have nearly succeeded in making them so. Some of this work is noticed briefly in chapter 3, but obviously the time has not yet come to make it the subject of a book.

Production and its efficiency are of practical importance, particularly in some tropical countries where much of the supply of proteins comes from

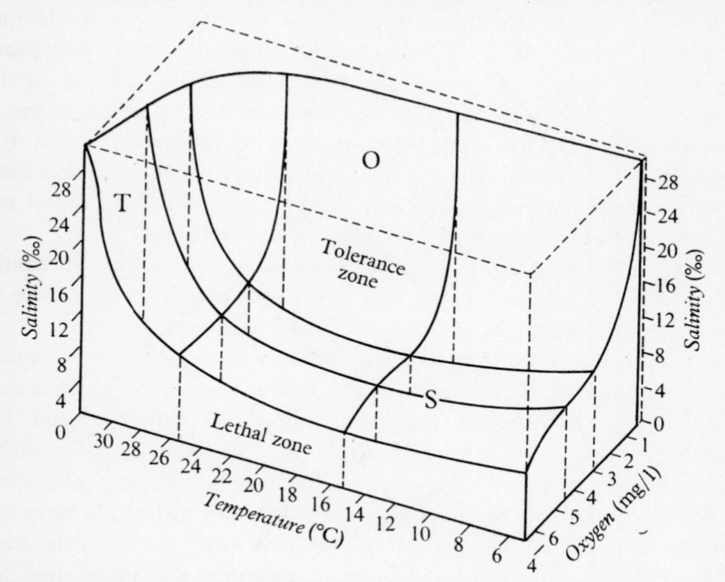

1. Diagram of the boundary of lethal conditions for the American lobster for various combinations of temperature, salinity and oxygen (McLeese 1956) (*J. Fisheries Research Board of Canada*, p. 269, fig. 7).

fish cultivated in ponds. However, the practical application is still largely empirical, and theoretical calculations must still be based on a number of assumptions.

No original line of approach having occurred to me, there remain two possible themes: the popular one of how the numbers of animals are controlled and the less popular one of how the ecological ranges of species are limited. The numbers of freshwater animals are probably controlled by mechanisms of the same type as those that control terrestrial and marine populations, and it is not logical to restrict a discussion of the topic to animals confined to freshwater. Indeed, to do so would necessitate omitting some of the most notable examples. On the other hand freshwater animals are limited by several factors that do not operate elsewhere, and a study of such factors is a logical entity. It is the line I have chosen.

This book then sets out to examine why species are present in some places and absent from others. That at least is the theme which runs

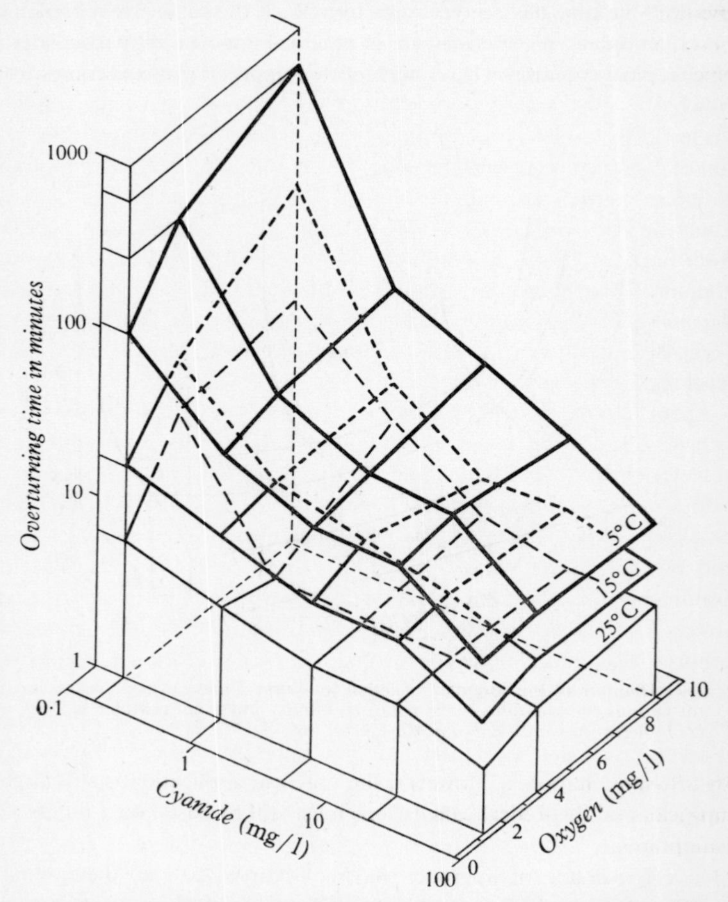

2. Survival time of minnows (*Phoxinus laevis*) plotted against the concentration of cyanide and oxygen at three different temperatures (Wuhrmann and Woker 1955) (*Verh. int. Ver. Limnol.* **12**, p. 800, fig. 1).

through it. Natural phenomena do not fall into categories in real life nearly as neatly as they do in books, and the choice of one line of approach does not mean excluding many ideas and facts that somebody taking a different line would use. It is much more a question of how to marshal and present them. Another statement that must be made at the outset is that the result of a study of limiting factors is not a set of exact figures; no species is confined within such precise limits as say 3-22°C, pH 5·5-8·3, 0·1-1·0 per mille salinity and so on. There are three reasons for this: 1. the threshold for any one factor varies according to the intensity of others. Fig. 1 shows how the tolerance of a lobster varies with temperature, oxygen, and salinity; at the highest salinity and oxygen concentration shown, the lethal temperature is about 32°C but at lower values of either it is less. Fig. 2 shows how the time in which minnows turn over in a solution of cyanide varies with the concentration of cyanide, with the concentration of oxygen, and with the temperature. 2. Tolerance varies with age. Adult *Ostraea madrasensis* can live in water of 20 per mille salinity but the lower limit for eggs is 22 per mille and for developing larvae 28 per mille (Remane and Schlieper 1958). On the other hand Kinne (1954) finds young *Gammarus duebeni* more tolerant of salinity changes than old ones. 3. Competition, or some other interrelationship, often prevents a species reaching a limit which it can attain when alone.

Absence may be due to the complex interaction of a number of factors, or it may be due to the species in question having failed to reach the piece of water being studied. This simple possibility must obviously be examined before more involved ones are investigated. After that the factors could be taken in almost any order. They can be divided into biological, chemical, and physical, and this is the sort of exercise that commends itself to examiners and is thought highly of in some research quarters. There is always the danger, however, that, once ideas have been separated into cohorts, each comes to assume an identity and distinctness that it does not really possess; temperature is a physical, oxygen a chemical factor, but, as will be seen later, in operation the two are inextricably tangled up together. It seems a sound general principle to start with the simple and proceed to the more difficult. If the occurrence of a mosquito depends on the habits of the female when she is laying eggs, to take an example about which more is written later, those habits will repay study before an elaborate analysis of the phytoplankton and the chemistry of water is embarked upon. This is a statement that would not be universally accepted, to judge from many of the papers that have been written. The reasons, I think, are two. First, in an age when science is blessed or cursed, according to one's point of view, with professionalism, the worker who sets out merely to observe fears that he may find himself, when his grant runs out, with rather little material to justify his expenditure of time and effort, whereas he who measures is certain to have figures from which graphs and histograms can be made. Also, in an epoch when the physiologists have been setting the pace, there has been a certain scientific respectability about measurements of physical and chemical variables. These are reasons of expediency and fashion, not logic. This, in my view, demands that the animal should be studied first and accordingly biological factors will be considered before chemical and physical ones.

Factors are taken one by one, which may seem illogical in view of the statement that distribution of any one species depends on the interaction of several factors. The day will come when it will be possible to present a series of neat examples in which the habitats of selected species are described and the reasons why they are confined to them expounded. It has not come yet, and most research hitherto has taken the form of investigating how a single factor affects a particular species, generally one with which nobody else has worked. I believe, therefore, that the arrangement I have used is the best method of presenting the information available at the present moment.

Certain words used in these pages need definition. It is essential that the exact meaning of any word should be clear, and desirable that it should coincide with that of somebody else.

Habitat, which has already been employed several times, is here applied to the sort of place in which a given species is found.

Biotope refers to a given set of conditions. Many authors use the word habitat where biotope is used here. An example will make the usage clearer. The habitat of *Heptagenia lateralis* (Ephemeroptera) is stony bottoms in both running and still water; the habitat of *Rhithrogena semicolorata* is stony bottoms in running water only. Stony bottoms in running water and in still water are two different biotopes.

A *biocoenosis* is the community that inhabits a biotope. Biocoenoses and biotopes are no doubt susceptible to splitting into smaller, and grouping into larger categories, but it will be time enough to do that when something is really known about some of them.

Migration is movement of a species from one place to another and, if nothing untoward prevents it, back again at a later date.

Emigration is used here to apply to any movement away from a place if return is unlikely.

To decide how to use certain words has always been an author's task; to decide which scientific names he should employ is a more recent one. It is difficult to write about it in temperate language. The writer who sits down to produce a general work demands no more of a name than that it shall convey to his readers what he is referring to. In other words, once bestowed it should not change. Unfortunately all too often it does. Some changes cannot be avoided, as when one species proves to be two. A species name may also change because somebody has unearthed an earlier one, which may be exasperating, but is the price to be paid for a system recognized all over the world.

Changes in generic names are entirely different. Each species is an entity, a growing tip on the evolutionary tree, and, provided that the rest of the twig has died out and no fossils have been found, it is distinct. At what level species should be grouped into higher taxa is an arbitrary human decision, and changes are matters of opinion. It seems sometimes as if a taxonomist splits a familiar genus into a number of new ones, or reverses the process, for no better reason than to make his mark, but generally his motive is a sincere belief that the step is a contribution to science. Such changes, however, are based on morphology and possibly on a difference of opinion whether the shape of the nose is more important than the shape of the ear. Frequently little is known about the physiology of the group, nothing whatever about the fossil history, and next to nothing about the genetics. Without these it is impossible to know the real affinities of species, yet the only justification for an unstable system is that its purpose is not merely to label but to indicate affinity.

I have written a more factual account elsewhere (Macan 1955a) using the Corixidae as an illustration. One species has been in six different genera in twenty-five years and has returned once to its starting point. Here, on the threshold of a book, I am less concerned with polemics than with the problem of how to make myself comprehensible to readers of all ages. The use of every latest generic name would no doubt leave the modern generation impressed with my contemporary progressive outlook, but older readers would not always know what animal I was writing about. To please them by using the old names would puzzle the others. Obviously both must be used. I have preferred, however, to give alternative names in an appendix to each chapter rather than to encumber the text with such names as *Potamopyrgus* (=*Hydrobia*, *Paludestrina*) *jenkinsi* (=*crystallinus carinatus*). In the text I have been conservative, deliberately and provocatively conservative. *Planaria alpina*, for example, has been a celebrated animal in ecology for decades and many papers have been written on it under that name. Some important physiological work has been done on it too. I remain unconvinced that the taxonomists have made out a good case for changing the name to *Crenobia alpina*. Such changes can be kept in sight by the specialist but they add unnecessary difficulty to the work of any zoologist who ranges widely over the animal kingdom. If those not primarily concerned with systematics insist on the long-established names, as applied workers did on '*Anopheles*', some stability will eventually be achieved. If not, generic names will continue to change as taxonomists succeed, or fall out with each other.

CHAPTER 2

Physical and Chemical Properties of Water

The physical properties of water and the origin, concentration and circulation of the substances dissolved in it have been studied with some thoroughness, and Ruttner (1952) devotes nearly half of his well-known *Grundriss der Limnologie* (English version *Fundamentals of Limnology*, translated by Frey and Fry 1953) to them. Such studies must be fundamental but many of the findings have, in the present state of knowledge, no immediate relevance to faunistic and floristic problems. There is, therefore, no need for a long account here. Information that seems relevant is included at the beginning of each chapter in which the effect of a physical or chemical factor is discussed. The purpose of the present one is to draw attention to those properties of inland waters which present their inhabitants with problems of a different nature or on a different scale from any confronting creatures dwelling on land or in the sea.

Much fresh water flows constantly in one direction, a property which brings certain benefits to successful colonists. Some strain the water by means of nets which they spin or by special developments of the mouth-parts, and are able to obtain food with little expenditure of energy. Water flowing swiftly is nearly always well oxygenated. In order to exploit these advantages, stream-dwellers have had to develop some means of clinging to the substratum or a behaviour pattern that keeps them in the substratum, behind or under stones, where the current is not strong enough to wash them away. Even then occasional displacement downstream is inevitable because the substratum is rarely quite stable, and this demands that there must be a reaction to move upstream at some stage in the life history. Little is known about the way in which species maintain an even population from top to bottom of suitable stretches.

Air does not absorb much of the sun's radiant energy, which therefore falls on to and warms the ground. Continual warming or cooling keeps the air just above the ground in circulation and the concentration of gases does not vary enough to affect living organisms. Water not only requires more energy to move it, but absorbs the sun's rays rapidly so that the bottom of even a comparatively shallow lake is not warmed at all, nor inhabited by plants, since there is no energy for photosynthesis either. In most lakes in summer the upper layers, heated by the sun, soon become so much warmer than the lower layers that a marked discontinuity zone is established, and the lower layers are sealed off till temperature at the surface falls in the autumn. The same may happen in a pond, though the system is less stable, and a cycle of stratification and turnover which takes a year in a lake may be completed in twenty-four hours in a small piece of water. The bodies of animals and plants living in the productive upper layer, together with dead leaves and debris of all kinds washed in by the inflowing rivers, fall down into the lower layer and decompose. This process requires oxygen and, if there is much to decompose and not a great deal of oxygen owing to the small volume of the lower layer, all will be used up. Some animals dwelling at the bottoms of lakes may, therefore, have to live without any oxygen for a period of several months, and any inhabitant of stagnant water may be subjected periodically to low concentration or even complete absence, though not for a long period if the piece of water is a small one. Violent fluctuation in the concentration of oxygen is peculiar to the freshwater environment, though absence is a condition to which certain intestinal parasites have also had to adapt themselves.

Water is heaviest at 4°C and thereafter lighter till it freezes, with the result that ice forms at the top and rarely extends very far down. This may also lead to oxygen deficiency, but the important consequence is that most freshwater animals never experience a temperature below freezing-point. Land animals do, and the amount of frost that different species can stand is an important factor governing the range of many.

The factors governing heat uptake of water are complex. The cooling effect of evaporation becomes more marked with rising temperature but it depends on the humidity of the air and the speed of the wind, which plays a part also by stirring the water. It is, however, probably safe to make the generalization that a piece of water does not get as warm as an adjoining piece of land. The ecologist studying land animals must constantly bear in mind the effect not only of high temperature but of the reduced humidity that often goes with it. Obviously this is of no moment to the freshwater biologist, but, if his studies of the effect of temperature are relieved of this complication, balance is restored by another unknown to his land colleague, because, as the temperature of water rises, its saturation concentration of oxygen falls.

Whereas the sea is chemically the most constant of all environments, fresh water is extremely variable. A certain number of marine animals can tolerate some dilution, such as invaders of estuaries are exposed to, or some concentration, which may occur in lagoons that are entered by the sea only at irregular intervals, and accordingly the marine biologist may consider that such places come within his province. Strongly concentrated seawater and places with a high salinity due to salts in different proportion to that in the sea, are inhabited by species whose relatives are found in fresh water, and the field of the freshwater biologist ranges from places that are almost pure water to places saturated with some of the salts present. These places are few and far between but even among the commonly encountered ponds, streams, and lakes, whose waters have a very low salt content compared with the sea, the range of variation is immense. Moreover, values may change quite markedly with climatic conditions, heavy rain diluting and dry weather concentrating.

The content of calcium is one of the biggest variables in fresh water, and also one on which can be based a division with which faunistic differences correspond, it being frequently possible to distinguish 'hard-water species' and 'soft-water species'. Nobody has found faunistic or floristic differences to correspond to the groupings based on other ions and there is no point in mentioning them here. Tables showing the composition of different waters are given in Taylor (1958) pp. 314 to 383.

The oceans of the world are all connected; most of the land is concentrated in the great land-masses; but all pieces of fresh water are tiny in comparison and none is connected with more than a relatively small number of others. The smaller a piece of water the more likely it is to be isolated and also the shorter its life.

All animals and plants that inhabit water are less subject to violent temperature changes than those that are terrestrial, but apart from this the freshwater environment is probably the one that has confronted invaders with most problems. Adaptation must be made to conditions which include sudden fluctuations in oxygen concentration in an environment which, highly variable in its content of dissolved substances, consists of a lot of small pieces, many quite isolated, and with a very short existence compared with land or sea.

Some freshwater animals originated in the sea and have adapted themselves to life in more dilute water. As will be stressed in later chapters, it is misleading to think in terms of sea water, brackish water and fresh water in this context, convenient though those terms may be in others. To an adaptable animal it is a continuum ranging from sea water at one end to pure water at the other. Different species have proceeded further towards extremely dilute conditions and this may explain the distributions of some to-day within what is commonly referred to as fresh water. The less there is in solution, therefore, the more vacant ecological niches. Insects, reverting to an aquatic life after successful adaptation to a terrestrial one, have filled many of these. Some have retained the impervious cuticle, developed to withstand desiccation on land, finding it useful to prevent loss of salts from the body fluid in a dilute medium. They come to the surface to take air into the spiracular system, which was another adaptation to life on land. Others have become more truly aquatic and, taking in oxygen from the surface, have found other means of maintaining the concentration of salts in the body fluid in face of the increased permeability which the new mode of respiration demands. Their ability to fly must have facilitated colonization of a medium distributed in small isolated pieces. An inherent adaptability has helped insects to benefit from these initial advantages. No other group, for example, has developed the ability to live without oxygen to the same extent.

CHAPTER 3

Communities

I remarked in chapter 1 that the time is not yet ripe for a comprehensive book on the communities of fresh water. Whoever one day feels it his duty to make the attempt will be hard put to it to prevent the thin stream of his narrative getting lost in a vast reed-swamp of lists of names of species.

It is my belief that the species is the basic unit in ecology and that a list must be reasonably complete to be of value for the present purpose. For this chapter I have selected those communities which conform to this standard and have attempted a preliminary charting of the reed-swamp. It is unfortunately necessary to except the Chironomidae, a family with some 400 species most of which cannot be distinguished as larvae. Space being limited, I have also omitted the communities of the open water and of the mud, and it has seemed preferable to describe the fauna of saline water in chapter 10 along with the ecological and physiological findings.

The definition of plant communities and examination of the conditions under which each is found has been a rewarding line of investigation. The study of animal communities is likely to proceed along different lines because there is a fundamental difference between them and plant communities. From one point to another along the gradient of an ecological factor, one or a few species of plants establish dominance, and they become a potent, perhaps the most potent, influence on the other members of the community. At the points mentioned it or they can no longer compete with some rival, which, accordingly, assumes dominance and determines the pattern of another community, delimited fairly sharply from the first. The most striking example I have ever seen was in the fens that have filled up many bays of the Danish Lake Lyngby. The gradient here was an actual one—the slope of the land. Arising abruptly from the water was a dense bed of *Phragmites*, bounded presumably by water too deep for it to colonize. It extended beyond the water's edge at normal level. Then, along a line presumably determined by the frequency of flooding, *Alnus* wood started, and the *Phragmites* was reduced from a vigorous stand of plants growing close together and reaching far above the head of any human trying to penetrate it to scattered specimens scarce a third as high as those not shaded by the trees. There is nothing in the animal kingdom to correspond exactly with a dominant plant, and there is no reason why animals should be affected by one species of plant to the same extent as other plants are. Each one then may have a different range along any particular gradient with the result that lines between communities must be so arbitrary as to have little value.

Anyone reading Macfadyen (1957), chapter 16, the fruit of a painstaking study of what others have written about communities, will conclude that freshwater biologists are a long way behind, and that much more must be known about the communities of other media, about the constituent species, their life histories, and their relations one to another; otherwise writers would not have ventured on so many definitions or hazarded such downright statements as that the community is more than the sum of its parts. I believe that the conclusion would be false, and that knowledge about other communities, except for a few simple ones, is not more complete than it is about those in fresh water. The ideas put forward, except in so far as some are mainly botanical, are hypothetical. There will always be those who like to predict how something operates, how something is made up, or what something will become, from pure reasoning. Their concepts may lead to important research; sometimes they lead a generation into the wilderness. The rival school prefers to gather its facts first and to build its theories on them. A certain amount of abuse is exchanged between extremist partisans, but I think that most scientists would allow that, though science could not advance without the fact-gatherers, it would advance more slowly than it does without the theoreticians. My intention now is to stick closely to the second method, presenting facts first and then trying to draw conclusions from them.

RUNNING WATER

Ford Wood Beck is a small stony stream flowing through pasture in a low-lying valley in the English Lake District. It is about 1 km long from mouth to main source though further to the sources of the two tributaries, and nowhere wider than 2 m. One tributary originates in a cut in an old peat bog but everywhere else the current is swift enough to keep the substratum stony, though only just in the bottom of the valley (stations 1 and 2, fig. 3). At station 3, where the stream comes down the side of the valley, it is more torrential and small waterfalls or steep faces of rock alternate with pools.

Table 1

FORD WOOD BECK

Numbers of animals per square metre (approx.)

The figures show the number of specimens taken with a coarse net at station 1 in December 1951 and January, February, March, and April 1952. Each collection lasted ten minutes. The figures are approximately numbers per square metre. Appropriate approximations have been made for summer growers. A dash indicates that the species were taken elsewhere in the stream or at some other time.

The Hydrachnellae were collected by means of a special technique by Mr. T. Gledhill and numbers are not comparable. The + sign indicates the species that were fairly numerous. The list of chironomids is based on the captures in an emergence trap. Nearly 6,000 specimens were taken in 1957 and they were identified by Dr J. H. Mundie, who found some sixty species. Only the commoner ones are included in the table and the figures show for each its percentage of the total number of all chironomids. *Simulium hirtipes* was the dominant species in that genus, for the identification of which I am indebted to Professor D. M. Davies. Miss R. M. Badcock kindly named adults and larvae of *Hydropsyche*. The names of Trichoptera in the family Limnephilidae were obtained by trapping the emerging adults.

		Species	Numbers /m²
Platyhelminthes	Tricladida	*Planaria* (*Crenobia*) *alpina* Dana	5
		Polycelis felina (Dalyell)	43
Mollusca	Pulmonata	*Ancylus fluviatilis* (Müller)	42
Crustacea	Amphipoda	*Gammarus pulex* Linn.	1432
Arachnida	Hydrachnellae	*Hydrovolzia placophora* (Monti)	
		Protzia eximia Protz	
		Panisus torrenticolus (Piersig)	+
		Sperchon clupeifer (Piersig)	+
		S. glandulosus (Koenike)	
		S. brevirostris (Koenike)	
		S. setiger (Thor)	
		Sperchonopsis verrucosa (Protz)	+
		Lebertia fimbriata Thor	
		L. glabra Thor	+
		L. insignis (Neuman)	

		Species	Numbers /m²
Arachnida	Hydrachnellae	*Torenticola anomala* (Koch)	
		Hygrobates fluviatilis (Ström)	
		Atractides gibberripalpis Piersig	
		A. nodipalpis (Thor)	+
		A. tener (Thor)	+
		A. octoporus (Piersig)	
		Feltria romijni Besseling	
		Wettina podagrica (Koch)	
		Vietsaxona lundbladi (Motas and Tanasachi)	
		Ljania bipapillata Thor	+
		Aturus scaber Kramer	+
		Kongsbergia materna Thor	
Insecta	Ephemeroptera	*Baetis rhodani* (Pictet)	792
		Rhithrogena semicolorata (Curtis)	308
		Ecdyonurus torrentis Kimmins	196
		Baetis pumilus (Burm.)	129
		Heptagenia lateralis (Curtis)	24
		Ephemerella ignita (Poda)	350
		Paraleptophlebia submarginata (Steph.)	4
		Habrophlebia fusca (Curtis)	–
		Ecdyonurus venosus (Fabr.)	8
	Plecoptera	*Nemoura cambrica* (Steph.)	166
		Isoperla grammatica (Poda)	50
		Leuctra hippopus (Kempny)	27
		Amphinemura sulcicollis (Steph.)	20
		Leuctra inermis Kempny	8
		Leuctra fusca (Linn.)	75
		Perla bipunctata Pictet	16
		Chloroperla torrentium (Pictet)	12
		Protonemura meyeri (Pictet)	25
		Perlodes microcephala (Pictet)	5
		Brachyptera risi (Morton)	8
		Leuctra nigra (Oliv.)	15
		Chloroperla tripunctata (Scop.)	5
		Protonemura praecox (Morton)	5

		Species	Numbers /m²
Insecta	Heteroptera	*Velia caprai* Tamanini	–
	Coleoptera	*Hydraena gracilis* Germar	4
		Elmis maugei Bedel	4
		Latelmis volckmari (Panzer)	–
		Helodid larvae	9
	Neuroptera	*Sialis fuliginosa* Pictet	
	Trichoptera	*Agapetus fuscipes* Curtis	123
		Silo pallipes (Fabr.)	60
		Plectrocnemia conspersa Curtis	1
		Wormaldia occipitalis (Pictet)	5
		Hydropsyche instabilis (Curtis)	2
		Rhyacophila dorsalis (Curtis)	13
		Diplectrona felix McL.	–
		Hydropsyche fulvipes (Curtis)	6
		Sericostoma personatum (Spence)	3
		Odontocerum albicorne (Scop.)	1
		Glossosoma boltoni Curtis	–
		Rhyacophila obliterata McL.	–
		Philopotamus montanus (Don.)	4
		Chaetopteryx villosa (Fabr.)	} 86 (bracketed: *Chaetopteryx villosa* to *S. stellatus*)
		Halesus radiatus (Curtis)	
		Drusus annulatus (Stephens)	
		Stenophylax latipennis (Curtis)	
		S. stellatus (Curtis)	
	Diptera Tipulidae	*Pedicia rivosa* (Linn.)	–
		Dicranota sp.	–
		Other tipulids	–
	Muscidae	*Limnophora*	–
	Simuliidae	*Simulium variegatum* Mg.	} 40 (bracketed: *Simulium variegatum* to *S. hirtipes*)
		S. ornatum Mg.	
		S. latipes (Mg.)	
		S. hirtipes Fries	
	Chironomidae		
		Orthocladius semivirens (Kieffer)	57%
		O. veralli (Edwards)	10%

		Species	Numbers /m²
Insecta	Chironomidae		
		Thienemanniella clavicornis (Kieffer)	8%
		Corynoneura lobata Edwards	7%
		Micropsectra brunnipes (Zetterstedt)	4%
		M. subviridis (Goetghebuer)	4%
Vertebrata	Pisces	*Salmo trutta* Linn.	–

Table 1 shows the number of all species caught at the lowermost station in five ten-minute collections with a coarse net, which is roughly equivalent to number per square metre. The basing of a total on catches in five successive months introduces an error, because recruitment and death of all species will not be the same during that period. It is, however, but a small one in comparison with the variation from sample to sample that is probably attributable to the irregular nature of the bottom.

Once a list of this kind has been prepared, the next step is to find out what is known of factors that could modify it. Some Ephemeroptera are confined to the lower reaches of Ford Wood Beck, though whether on account of the slightly higher summer temperature or the richer food supply is not known (Macan 1957). Other species and some Trichoptera (Mackereth 1960) are confined to or commoner in the upper reaches and in the tributaries, probably because the water is cooler there than it is lower down at the height of summer. A striking example of a cold-water species of restricted range may be seen in fig. 3. Of the two large stoneflies, *Perla bipunctata* (*carlukiana*) is confined to the main stream, but *Perlodes microcephala* (*mortoni*) is most abundant in Sykeside Beck (Mackereth 1957). This tributary dries up in summer droughts but probably never before most nymphs of *Perlodes*, which completes development in one year, have emerged. The eggs are not harmed by the disappearance of water. This is true of many species but few can tolerate desiccation in an active stage (Hynes 1958b). Nymphs of *P. bipunctata* cannot and, as they take three years to complete development, they are unable to inhabit temporary streams.

Altitude and changes in fauna have long been known to be associated. Insects emerging from station 1, Ford Wood Beck, which is 44·5 m above sea-level, have been trapped for many years, and in 1958 Gledhill (1960) installed a trap in Whelpside Ghyll at a height of 624 m. The numbers of species taken in the high and the low trap respectively were: Plecoptera 12 : 8, Ephemeroptera 4 : 6, and Trichoptera 6 : 7. There were not only more species of Plecoptera but a big preponderance in numbers as well. Other features of Whelpside Ghyll in contrast with Ford Wood Beck were: more baetids relative to ecdyonurids; the presence of *Ameletus inopinatus*,

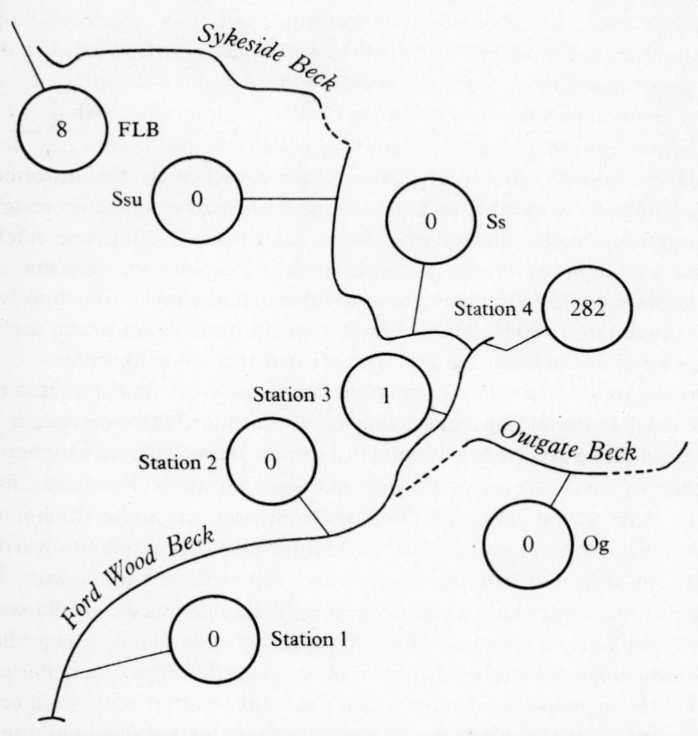

3. Ford Wood Beck and its tributaries and the position of the collecting stations. The numbers in the circles are the numbers of *Planaria alpina* caught in a fine net in five minutes in July 1961. Broken lines show where the stream runs underground (Original).

Protonemura montana, and *Diura bicaudata*; the absence of *Leuctra fusca* and *Agapetus fuscipes*. The first two are typical members of a group of animals found only at high altitudes; *Leuctra fusca* is one of the few Plecoptera that grow during the warmest part of the year.

It is obvious at a glance that Whelpside Ghyll is steeper than Ford Wood Beck, and it was simple to establish by means of a maximum and minimum thermometer that it is colder; the greatest summer temperature is about 5°C lower. Presumably there is less in it for the fauna to eat since it flows off a barren mountain top that supports nothing but a few sheep, whereas nearly all Ford Wood Beck flows through pasture on which cows feed and to which the farmer adds fertilizer now and then. Moreover, dead leaves blow into it from near-by woods, whereas Whelpside Ghyll is well above the present tree-line. It seems likely that all the presences and absences noted are related to temperature, but it is impossible to be certain that slope and food supply do not also play a part. The distribution of *Diura bicaudata* is a product of temperature and competition, and this statement will be substantiated in the chapter on interrelationships.

The part played by food in these two English streams has not been elucidated, but the effect that it can have has been convincingly demonstrated in Lappland. The work of Illies (1956) is the best for present purposes, since he took more pains than anyone else to name every species. Immediately below a lake, net-spinning Trichoptera and *Simulium* spp. were the most abundant animals in a stream, and all others were scarce. The abundance is attributed to the rich supply of food in the form of plankton and detritus washed out of the lake, and it is thought that the net-spinners also caught other stream-dwellers. Two hundred metres lower down the filtering species were much fewer and such animals as *Baetis* sp. and *Heptagenia dalecarlica*, which might be expected to abound in a stream of this type, were numerous.

After the main survey of Ford Wood Beck had been completed, mains water came to the village of Outgate. Four houses, provided with their own wells and hand pumps, had had baths and indoor sanitation before 1951, but after that date the number rose to fourteen. No alteration was made to the septic tank when more houses were connected to the sewer. There was also an increased flow into a second septic tank, which served the farm, and much more washing of byres and milk-bottles was done after water had become available from taps. It is believed that, as a result of the extra load on the septic tanks, more organic matter both dead and living in the form of ciliates, bacteria and fungi, was washed down the beck, and that this provided food. Every species seemed to benefit to judge from the emergence-trap collections but the flatworm *Polycelis felina* benefited more than any; at least its numbers increased much more than those of any other species. Numbers taken in ten minutes collecting at each of five stations in March were six in 1951 and over 1,000 in 1959. After 1955 the numbers of *Ecdyonurus torrentis*, *Perla bipunctata*, *Rhithrogena semicolorata* and *Baetis rhodani* began to decline and the first two eventually disappeared. Macan (1962b) puts forward the suggestion that this decline was due to predation by *Polycelis*, and adumbrates the possibility that *Rhithrogena* would have been eliminated also but for repopulation of the main stream from Sykeside Beck, where *Polycelis* has not become numerous. The four species affected live on the larger stones, whereas the others tend more to inhabit the gravel and smaller stones beneath the larger ones.

An instructive comparison can be drawn between Ford Wood Beck and the Afon Hirnant (Hynes 1961b), a Welsh mountain stream that rises at 580 m and is colder. It is probably also poorer, at least in the upper reaches, but there are no quantitative data, which would be extremely difficult to obtain. It is longer than the stream I studied and eventually flows into the River Dee. Hynes could not name to species all the Trichoptera and did not attempt to identify water-mites, chironomids and *Simulium*. He records some Crustacea which I probably missed because my net was too coarse to catch them. There remain the Platyhelminthes, Mollusca, larger Crustacea, Plecoptera, Ephemeroptera and Coleoptera as the basis for comparison. Between the two lists differences are few and it is satisfactory to find that many can be explained in terms of what is already known, but it must be stressed that there are differences, notably of the abundance of certain species in the two streams, which cannot be accounted for at the present time.

Polycelis felina did not occur in that part of the Hirnant where Hynes did his main work but it did come in lower down. This is in accord with what is known elsewhere and the relations between this species, temperature, rate of flow, and *Planaria alpina* is discussed in later chapters. Also related to its higher altitude is the occurrence in the Hirnant but not in Ford Wood Beck of *Diura bicaudata* and *Baetis tenax* (Macan 1961).

The most striking difference is the absence of *Gammarus pulex*, which is abundant in Ford Wood Beck and also in the spring where Whelpside Ghyll originates. It is generally absent from Welsh mountain streams, and the most likely explanation is that the dissolved substances in the water are not right for it, though no particular deficiency has yet been identified.

Ecdyonurus venosus occurs in both streams, *E. torrentis* in Ford Wood Beck only. All the other differences are species recorded in the Hirnant but not in the Lake District stream. *Amphinemura standfussi* is a rare species in Britain; *Nemoura cinerea* is stated by Hynes (1958a) to be a species of slow water, though Illies records it from fast currents as well. He finds it to be unusually tolerant of adverse physical and chemical conditions but unsuccessful in competition with other species (Illies 1955a). *Siphlonurus lacustris* and two beetles probably occur because parts of the Hirnant are slower than any stretch of Ford Wood Beck.

The preparation of lists, such as those that have just been discussed, is but an early stage on the way to understanding what a community is and how its components are related; it also involves time and hard work. Further advance demands knowledge about the relations of each species to the physical environment, to its food, and to any species that may prey on it; at least this should be known for the important species which, according to Hynes (1961b) are those herbivores which are numerous, but carnivores whose numbers are much smaller. Hynes, himself, has worked out the life histories of many Hirnant species in addition to investigating their food, and Jones (1950) has made a particularly important study of the food of invertebrates in streams. Hynes believes that in his Welsh mountain stream dead leaves from deciduous trees are the most important source of food and that their remains are extensively eaten in the winter and spring. Animals that grow in the summer feed more on algae and other living plants in the water. Incidentally, while recording what they find inside animals, these authors point out that they know little about digestion and the sources of nourishment. Hynes believes that a diet of dead leaves would be useless to many animals but for the bacteria and fungi which are taken in at the same time.

Hynes's analysis of the community in the Hirnant is expressed in fig. 4. The vertical axis indicates size, the horizontal one time. Each line represents the development of a group of animals and the steeper the slope the quicker the development. The little vertical lines below these growth curves cover the period when eggs are hatching. The figure includes two years merely because it is convenient to label the F groups on the left and the S groups on the right.

Distinction between winter-growers and summer-growers is commonplace in stream studies. Hynes's objective is to distinguish different categories of life history more precisely and to work out how much the animals of each category contribute to production. When accurate data about production are not available, recourse is generally had to 'standing crop', which is the number of each species present at any particular moment, and this is basically what Hynes has done to obtain his indices in fig. 4. The 'index' is the figure shown against each name in the figure, for example 10 *Crenobia alpina*, 1 Nematoda, and so on in the extreme right-hand one. It is not, however, based upon the actual numbers but is obtained by adding together percentages, after the numbers of each species in each catch have been converted to a percentage of the total number of animals in the catch. Data were obtained by various means for four stations in various months. It is obviously necessary to ensure that the proportion of catches in stony

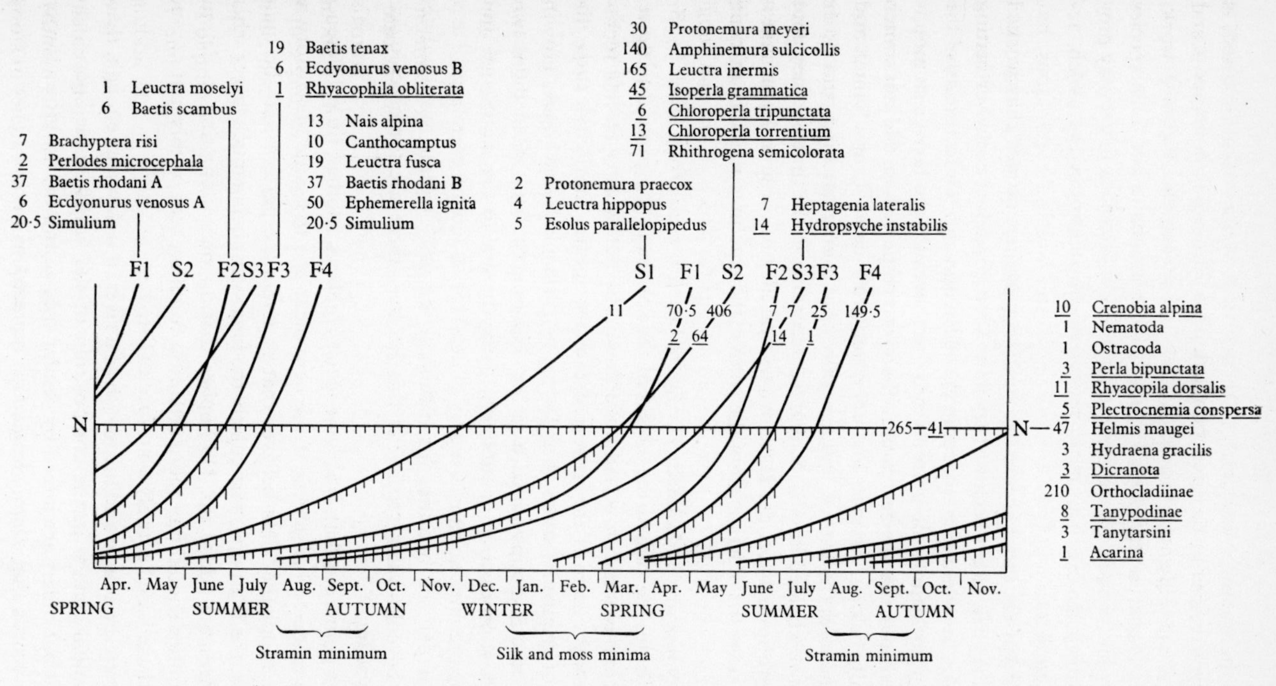

4. Analysis of a stream fauna according to life history and an index based on relative abundance with a correction for size (explanation in text) (Hynes 1961) (*Arch. Hydrobiol.* **57**, p. 374, fig. 7).

parts of the stream to catches in moss is similar to the proportion of stony to mossy stretches in the area sampled. This Hynes has done and in addition, by including more coarse-net than fine-net samples, he has increased the index of species represented by many large specimens and thereby made a correction for their greater weight. In this way the index figures are given a bias which makes them more representative of production and less of actual numbers.

'N' designates those species whose size groupings do not change much during the year, either because they have a long breeding period or because they take several years to complete development. Unfortunately it is necessary to include the chironomids here because the larvae cannot be identified and the species cannot therefore be placed in the class where they really belong. Animals in the S group grow slowly over a long period, which includes the winter, and generally have a long hatching period; the group is subdivided into S_1, S_2 and S_3, according to the time of emergence. The F group includes the fast growers and all but those in F_1 remain in the egg stage for much or all of the winter; the hatching period is generally short.

Carnivores are underlined. The figure beneath each group label, F_1, S_1, etc., indicates the sum of the indices of the species in the group, herbivores on top, carnivores below and underlined. The noteworthy feature of this small stony stream is the importance of the species that hatch early, the indices of groups F_1 and S_2 being some two and a half times that of all the other F and S groups added together. Hynes speculates about the reason for the low production in summer. It may be due to the scarcity of food then, the previous autumn's supply of dead leaves having been largely used up, or it may be because, for historical reasons, streams are inhabited largely by cold-water species. If this be true, there are some ecological niches vacant.

Hynes points out that, when one species replaces another, little difference may be made to his scheme if the newcomer has the same life-history as the one supplanted. He believes that in rivers the pattern may be quite different, the F groups, particularly the later ones, being much larger when the indices are added up. He implies that anyone attempting to delimit communities should pay more attention to differences of this nature, and less to those based entirely on a species list.

Between the source and the mouth of a river, if it flows far enough, there is a gradual change in the composition of the fauna; some species are replaced but others are joined by similar species and the list grows longer. Hynes records that *Diura bicaudata* is replaced by *Perlodes microcephala*,

Planaria alpina by *Polycelis felina*, *Lumbriculus variegatus* by *Stylodrilus heringianus*, *Heptagenia lateralis* by *H. sulphurea*, and *Ephemerella ignita* by *E. notata*. *Rhithrogena semicolorata* is joined by *R. haarupi*, *Leuctra fusca* by *L. geniculata*, *Perlodes microcephala* by *Isogenus nubecula*, and various Elmidae by *Limnius tuberculatus*. I would add that *Baetis rhodani* and *B. pumilus* are joined by four or five more species (Macan 1957). Hynes (1961b) writes that these and other replacements and additions take place at different points and doubtless for different reasons, and further on: 'How far, therefore, the elaborate subdivisions of streams into named zones, as advocated by Illies and Schmitz, are of value, except in the most general terms, is a matter of opinion.' I have expressed my view that we still have far too few fundamental facts to justify any scheme of classification (Macan 1961).

Space does not permit comparison with the River Rheidol (Jones 1950) but I would like to mention the German River Mölle (Illies 1952a) because it illustrates a point. It rises from a number of springs in a porous calcareous soil at an altitude of 150-250 m. The temperature of the springs lay between 8 and 10°C throughout the year, and range increased with distance from the source being about 16·5°C at the lowest point where it was measured.

Polycelis felina was not found in the Mölle and there is some evidence that it is not favoured by calcium, particularly when it has to compete with *P. alpina* and *P. gonocephala*. Mollusca and Crustacea were similar, if spring species such as *Niphargus* are omitted. Illies identified fourteen species of water mite in the Mölle and we found twelve in Ford Wood Beck; seven of these were common to the two streams and three represented by close relatives. It must be pointed out, however, that when Mr Gledhill started looking specially for mites, he quickly doubled the number of species taken during the original survey. As did Hynes, Illies found *Ecdyonurus venosus* and not *E. torrentis*, and he also records *Habrophlebia lauta*, which perhaps takes the place of our *H. fusca*. Otherwise he found all the mayflies which I believe to be typical of a small stony stream except *Heptagenia lateralis*, whose absence is hard to account for. In addition he collected *Ephemera danica*, which indicates the existence of sandy stretches in the Mölle, and two species not recorded in Britain. The resemblance between the Plecoptera, in contrast, was much less, and there were only seven common species out of a total of fifteen in the English and fourteen in the German stream. This is a group in which the wings are not used much and in which, as a result, there is thought to be little travelling. Isolation since the Ice Age has led to speciation (Illies 1953, 1955b) and lists from Scandinavia, the Alpine Region, and Britain, to mention three, have many differences. The ecologist, therefore, must compare species whose life histories and habitats are the same rather than species whose names are the same.

The beetles of the two streams are not by any means the same, but it is not possible to say why this should be so.

The only west European river that has been surveyed thoroughly is the Susaa, which Kaj Berg (1943, 1948) and his Danish colleagues have studied in their usual painstaking way. It is nearly 90 km long and in that distance drops 90 m, the fall being 0·31, 0·06 and 0.03 per cent in the upper, middle, and lower reaches respectively. None the less, parts are quite swift with a stony bottom, though much of the river is sluggish with extensive beds of rooted vegetation. It has a hard water rich in ions, it flows through agricultural land, and it is polluted by sewage at several points.

The authors set out to obtain a complete species list and to discover something about the life history and way of life of all common species. Quantitative collections were made at selected points, and figures for production are given.

Berg (1948, p. 286) concludes with a section on 'Some biotopes of the Susaa and their characteristic and numerous species'. A characteristic species must be fairly numerous at least, and distinctly more numerous than in any other biotope. A few species have been treated as characteristic of two biotopes. Species which are not abundant are omitted from Berg's lists and there is no means of compiling complete lists from the data published. The system has the merit of simplicity and economy, but, if a species be characteristic of one biotope yet just not numerous enough to qualify for inclusion in the next, it makes the differences between the two appear to be greater than they really are.

Berg recognizes eleven biotopes altogether, but five of these are sufficient for the present theme, which is not concerned with the fauna of mud, open water or brackish water.

Current speeds are described as follows:

Very swift	over 100 cm/sec
Swift	50–100 cm/sec
Moderate	25–50 ,,
Slight	10–25 ,,
Very slight	under 10 ,,

A swift current, according to this terminology, would move nothing larger than stones about the size of a pigeon's egg, and a slight current would move particles ranging from mud to sand.

Table 2

FIVE BIOTOPES IN THE RIVER SUSAA (Berg 1948)

SUSAA 1

Stones and sand with accumulations of dead leaves in very slight to moderate or occasionally strong current

		Characteristic species	Other numerous species
Annelida	Hirudinea		*Glossiphonia complanata* (Linn.) *Erpobdella octoculata* (Linn.)
Crustacea	Ostracoda	*Candona parallela* Müller	*Candona albicans* Brady
Arachnida	Hydracarina	*Spherchon glandulosus* Koen.	
Insecta	Plecoptera	*Isoperla grammatica* (Scop.) *Leuctra hippopus* (Kempny)	*Nemoura variegata* Oliv.
	Ephemeroptera	*Ephemera danica* Müll.	
	Hemiptera	*Velia* sp.	
	Trichoptera	*Rhyacophila septentrionis* McL. *Agapetus fuscipes* Curtis *Tinodes pallidula* McL. *Plectrocnemia conspersa* (Curtis) *Crunoecia irrorata* (Curtis) *Stenophylax stellatus* Curtis *Silo pallipes* Fabr.	*Halesus radiatus* (Curtis) *Anabolia laevis* Zett.
	Coleoptera	*Helmis maugei* Bedel *Helodes minuta* (Linn.)	
	Diptera		*Simulium latipes* (Meigen)

SUSAA 2

Stones bare or with moss and algae, in a moderate to strong current

		Characteristic species	Other numerous species
Platyhelminthes	Tricladida	*Bdellocephala punctata* (Pallas) *Planaria lugubris* O. Schmidt *Polycelis nigra* Ehrenb.	
Porifera			*Ephydatia fluviatilis* Linn.
Annelida	Oligochaeta	*Chaetogaster limnaei* V. Baer *Eiseniella tetraedra* (Sav.)	
	Hirudinea		*Glossiphonia complanata* *Erpobdella octoculata*
Crustacea	Copepoda	*Nitocra hibernica* Schmeil	
	Ostracoda	*Cypridopsis vidua* (O.F.M.)	
Insecta	Neuroptera		*Sisyra fuscata* (Fabr.)
	Ephemeroptera	*Baetis tenax* Eaton *scambus* Eaton	
	Trichoptera	*Hydropsyche angustipennis* (Curtis) *Goera pilosa* (Fabr.) *Leptocerus cinereus* Curtis	*Halesus radiatus* *Anabolia laevis*
	Coleoptera	*Limnius tuberculatus* Müll. *Helmis maugei*	
	Diptera	*Simulium venustum* Say *Calliophrys riparia* (Fallén)	
Mollusca	Operculata	*Neritina fluviatilis* (Linn.)	
	Pulmonata	*Ancylus fluviatilis* O.F.M.	*Planorbis albus* O.F.M.

SUSAA 3

Gravel and sand with occasional stones with slight to moderate current

		Characteristic species	Other numerous species
Porifera			*Spongilla lacustris* (Linn.) *Ephydatia fluviatilis*
Platyhelminthes	Rhabdocoela	*Gyratrix hermaphroditus* (For. and du. P.)	
Annelida	Oligochaeta	*Peloscolex ferox* (Eisen)	
	Hirudinea		*Erpobdella octoculata*
Crustacea	Copepoda	*Cyclops fimbriatus* Fischer	
Arachnida	Hydracarina		*Hygrobates fluviatilis* (Strøm)
Insecta	Plecoptera		*Nemoura variegata*
	Ephemeroptera		*Caenis horaria* (Linn.)
	Neuroptera		*Sisyra fuscata*
	Hemiptera	*Aphelocheirus aestivalis* Westw.	
	Trichoptera	*Hydropsyche angustipennis* *Goëra pilosa* *Polycentropus flavomaculatus* (Pictet) *Tinodes waeneri* (Linn.) *Molanna angustata* Curtis *Leptocerus senilis* Hagen *annulicornis* Stephens *cinereus*	*Anabolia laevis*
	Coleoptera	*Limnius tuberculatus* *Helmis maugei*	

		Characteristic species	Other numerous species
Mollusca	Operculata	*Valvata piscinalis* (O.F.M.)	*Bithynia tentaculata* (Linn.)
	Pulmonata		*Planorbis albus*
	Lamelli-branchiata	*Dreissena polymorpha* (Pallas)	
Pisces			*Cobitis taenia* Linn. *Acerina cernua* (Linn).

SUSAA 4

Submerged vegetation and some stones in slight to moderate current

		Characteristic species	Other numerous species
Coelenter-ata			*Pelmatohydra oligactis* (Pallas) *Hydra vulgaris f. attenuata* Pallas
Porifera			*Spongilla lacustris*
Annelida	Hirudinea		*Glossiphonia complanata* *Erpobdella octoculata* *Piscicola geometra* (Linn.)
Crustacea	Ostracoda	*Cypridopsis vidua*	
	Amphipoda	*Gammarus pulex* (Linn.)	
Arachnida	Hydracarina		*Lebertia insignis* Neum. *Hygrobates fluviatilis*
Insecta	Ephemerop-tera	*Baetis tenax* *scambus*	
	Odonata	*Agrion splendens* (Harris)	
	Trichoptera	*Neureclipsis bimaculatus* (Linn.)	*Hydropsyche angustipennis*
	Coleoptera	*Limnius tuberculatus*	
	Neuroptera		*Sisyra fuscata*
Mollusca	Pulmonata		*Planorbis albus*

SUSAA 5

Emergent and other marginal vegetation in slight to very slight currents

		Characteristic species	Other numerous species
Porifera			*Spongilla lacustris*
Coelenterata			*Pelmatohydra oligactis* *Hydra vulgaris f. attenuata*
Annelida	Oligochaeta	*Stylaria lacustris* (Linn.)	*Nais obtusa* (Gervais) *Tubifex albicola* (Mich.)
	Hirudinea	*Piscicola geometra*	*Helobdella stagnalis* (Linn.) *Erpobdella octoculata* *Glossiphonia complanata*
Crustacea	Copepoda	*Cyclops albidus* (Jurine)	*Cyclops strenuus* (Fischer)
	Ostracoda	*Cyclocypris laevis* (O.F.M.)	
	Isopoda		*Asellus aquaticus* (Linn.)
Arachnida	Hydracarina	*Limnesia maculata* (O.F.M.) *Hygrobates longipalpis* (Herm.) *Brachypoda versicolor* (O.F.M.)	*Lebertia insignis* *Hygrobates fluviatilis*
Insecta	Ephemeroptera	*Cloeon simile* Eaton	*Caenis horaria moesta* Bengtss. *Cloeon dipterum* (Linn.)
	Odonata		*Erythromma najas* (Hansem.) *Ischnura elegans* (v. d. Lind) *Agrion pulchellum* v. d. Lind

		Characteristic species	Other numerous species
Insecta	Hemiptera		*Corixa falleni* (Fieb.) *striata* (Linn.)
	Trichoptera	*Orthotrichia tetensii* Kolbe	*Anabolia laevis*
	Coleoptera	*Haliplus fluviatilis* Aubé *Laccophilus hyalinus* (De Geer)	
	Diptera		*Simulium argyreatum* (Meigen)
Mollusca	Operculata	*Bithynia leachi* (Sheppard) *Bithynella steini* Mart. *Valvata piscinalis*	
	Pulmonata	*Planorbis planorbis* (Linn.) *corneus* (Linn.) *crista* (Linn.) *contortus* (Linn.) *Acroloxus lacustris* (Linn.)	

Biotope 1 (table 2) is near the source, where the Susaa is a small stream running through beechwoods. Nearly all the species listed occur also in stony streams but there are fewer of them: seven Plecoptera for example in the Susaa, which is half the total in Ford Wood Beck. *Rhithrogena semicolorata* and *Heptagenia lateralis* are not recorded, though in the text it is stated that *H. sulphurea* occurred. Neither *Planaria alpina* nor *Polycelis felina* lived in the Danish stream. This paucity is no doubt related in some way to the main difference between the streams: Ford Wood Beck runs in the open, the Susaa through woodland sufficiently slowly for packets of dead leaves to accumulate. Hynes (1960) has pointed out that under such circumstances conditions indistinguishable from pollution by human agency may arise.

Biotope 2, stones in a moderate to strong current, offers a substratum similar to that in a stony stream, but the fauna is distinctly different. The absence of Plecoptera and Ecdyonuridae is perhaps the most striking feature at first glance. The species of *Baetis* (if correctly identified, which they may not have been as the two species named are among the most difficult) are not those typical of stony conditions. *Hydropsyche angustipennis* was present in great numbers, which, in view of what Illies (1956) found in Lappland, is likely to be due to a rich supply of food. *Simulium venustum* is a lowland species. The three flatworms are more typical of lakes than of running waters, and their abundance may be tentatively attributed to a suitable substratum and a rich supply of food. *Planorbis albus* is a typical lake species and the other two molluscs occur in both still and running water, *Ancylus* being found almost wherever there is a hard substratum.

Too low a concentration of oxygen at certain times, or possibly too high a temperature may keep stony-stream species out of this biotope, or they may be eaten or otherwise exterminated by species which appear to be favoured by the good food supply. Current does not seem to be of outstanding importance.

Biotope 3 is in slower water than biotope 2 and consequently the bottom deposits are finer. There is frequently a layer of organic matter lying on the sand and gravel. In spite of the slight flow, rooted plants are still absent. How far the differences between the two biotopes are due directly to current and how far to substratum, which depends on current, remains to be discovered. The restricted range of *Aphelocheirus* puzzled Berg and his co-workers and that of *Dreissena* is attributed by them to its recent arrival.

Biotope 4 is the first with rooted vegetation. *Agrion* (*Calopteryx*) *splendens* is a dragonfly whose nymphs occur under these conditions and rarely extend to still or swift water. In weed-beds in English rivers I have found a much richer fauna of Ephemeroptera (Macan 1957) and often a considerable abundance of *Simulium* spp.

Biotope 5 harbours an assemblage of animals which could easily be found in a lake. *Simulium*, which depends on flowing water to bring it food, is one of the few species that indicate that this is a river fauna.

CONCLUSIONS ABOUT RUNNING WATER

The purpose of this chapter is to give a necessarily superficial sketch of the freshwater fauna, offering tentative explanations of differences noted as an introduction to the rest of the book, in which the various factors are discussed in detail. It is also an experiment, a first attempt to discover whether the definition of communities is likely to help freshwater biologists. It does appear from what has already been presented that a useful basic species list for a small stony stream can be drawn up and that variations can be explained. I am less sanguine about the value of trying to delimit biocoenoses in rivers. There is more to vary, and it may well turn out that there are so many biocoenoses merging almost imperceptibly one into another that their definition will create a basis too complicated to have any value as a springboard for further ideas. However, that cannot be settled till comparisons have been made with Berg's important pioneer records. In the meantime, I believe that the experiment is worth making.

LAKES AND PONDS

There is no more comprehensive report on the fauna of a lake than that of Berg (1938), who studied Esrom Lake in Denmark. The main sampling was done at various depths along a line at right-angles to the shore with a Birge-Ekman grab, but collections of all kinds in other places were made in order to obtain a reasonably complete fauna list. The resulting volume is about as large as this book and, after all the manipulation of figures that has been necessary in order to attempt a comprehensible précis in a few pages, I almost feel that I have done a piece of research on the lake myself.

The lake, 8 km long, 2-3 km wide and 22 m deep, lies in a moraine landscape and is surrounded by farmland and forest. It is a rich lake with some 40 mg/l Ca. The stations of which it is proposed to take note here were as follows:

c. 0·4 m	large wave-washed stones
0·8-1·2 m	sand, gravel and some stones
2 m	coarse sand and some small stones
5 m	sand and gravel
8 m	sand and shell fragments
11 m	sand and shell fragments
14 m	sand and organic debris

At greater depths the bottom was muddy. There are 114 pages devoted to the distribution, life histories, and sometimes the taxonomy of the animals found, and 221 names are mentioned. Many are rare and would contribute little but obscurity to a list. For each station Berg gives a plate showing the relative abundance of the numerous species and beneath each he quotes the relevant figures. The lists, however, are short and some species are grouped together. They are based on primary tables in which there are sixty-eight names, excluding chironomids, and these are more suitable for the present purpose. They refer, however, only to the survey line, which seems too narrow a basis for the fauna list of a large lake. Accordingly I have modified them in two ways in order to arrive at table 3: where totals for a whole group are given by Berg, I have listed the species composing

Table 3

FAUNA OF ESROM LAKE

Numbers per square metre (approx.) at successive depths in m.

The figures in the first two columns are the totals of twelve double samples taken between April and October 1937 and multiplied by two (forty-eight samples); those for *Chara* are the average of two double samples multiplied by twenty-two (forty-four samples); those for the rest are the totals of eleven double samples taken at roughly monthly intervals between February 1933 and January 1934 and multiplied by two (forty-four samples). The samples covered 225 cm^2 and therefore 44·4 samples cover 1 m^2.

			0·1–0·4	0·8–1·2	*Chara*	2	5	8	11	14
Coelenterata		*Pelmatohydra oligactis* (Pallas)	–	–	+	–	–	–	–	–
		Hydra vulgaris Pallas	+	–	–	–	–	–	–	–
Platyhelminthes	Allocoela	*Plagiostomum lemani* (Forel and du Plessis)	+	10	–	+	+	+	–	–
	Tricladida	*Planaria torva* (O. F. Müller)	0	0	22	14	6	20	8	8
		P. lugubris Schmidt	+	–	–	–	–	–	–	–
		Polycelis tenuis (Ijima)	2	34	88	88	2	36	74	26
		P. nigra (Müller)	–	–	–	–	–	–	–	–
		Bdellocephala punctata (Pallas)	0	0	0	4	6	6	6	0
		Dendrocoelum lacteum (O.F.M.)	+	–	–	–	–	–	–	–

			0·1–0·4	0·8–1·2	*Chara*	2	5	8	11	14
Nemertini		*Prostoma graecense* Böhmig	+	–	–	–	–	–	–	–
Annelida	Oligochaeta	Total	304	6008	814	1974	1166	552	900	1324
		Aelosoma hemprichi Ehrenb.								
		Chaetogaster diaphanus (Gruthuisen)								
		Stylaria lacustris (Linn.)								
		Nais pseudoobtusa Piguet								
		N. obtusa (Gervais)								
		Limnodrilus udekemianus (Claparède)								
		L. hoffmeisteri (Claparède)								
		L. claparèdeanus Ratzel								
		L. austrostriatus Southern								
		Tubifex tubifex (O.F.M.)								
		T. nerthus Michaelsen								
		T. barbatus (Grube)								
		T. ignotus (Štolc)								
		Ilyodrilus hammoniensis Michaelsen								
Annelida	Hirudinea	*Piscicola geometra* (Linn.)	2	0	0	0	6	2	2	8
		Hemiclepsis marginata (O.F.M.)	0	1	0	0	0	0	0	0

		Glossiphonia complanata (Linn.)	48	36	66	56	30	24	22	4
		G. heteroclita (Linn.)	0	2	0	4	4	4	0	0
		Helobdella stagnalis (Linn.)	44	102	22	114	24	16	8	2
		Erpobdella octoculata (Linn.) } *E. testacea* (Savigny)	360	536	660	374	44	26	28	6
Polyzoa		*Cristatella mucedo* Cuvier	0	0	0	8	18	14	6	4
Crustacea	Cladocera	*Eurycercus lamellatus* (O.F.M.)	4	322	44	126	2	0	0	2
		Alona affinis Leydig	0	22	0	0	0	0	0	0
	Ostracoda	*Herpetocypris reptans* (Baird)	0	8	0	6	0	4	0	2
		Cypridopsis vidua (O.F.M.)	0	0	110	24	0	4	0	0
		C. newtoni (Brady and Robertson)	0	0	0	36	0	0	0	0
		Candona candida (O.F.M.) } *C. neglecta* Sars	12	6	1650	274	74	14	60	10
	Copepoda	*Cyclops viridis* (Jurine)	12	286	110	34	12	10	24	14
		Canthocamptus staphylinus (Jurine)	0	184	44	90	8	4	0	0
	Isopoda	*Asellus aquaticus* Linn.	548	312	5214	1488	464	1214	1042	126
	Amphipoda	*Gammarus pulex* (Linn.)	768	112	22	0	0	0	0	0
		Pallasea quadrispinosa Sars	0	0	22	0	30	4	128	146

			0·1 – 0·4	0·8 – 1·2	*Chara*	2	5	8	11	14
Insecta	Ephemeroptera	*Caenis moesta* Bengtsson	210	540	1430	942	664	518	40	22
		Centroptilum luteolum (O.F.M.)	20	58	22	22	0	0	0	0
	Plecoptera	*Nemoura avicularis* (Morton)	76	4	0	0	0	0	0	0
	Coleoptera	*Haliplus* total	72	36	0	12	2	0	2	0
		H. confinis Steph. *H. ruficollis* (De Geer) *H. obliquus* (Fabr.) *H. lineatocollis* (Marsh) *H. fulvus* (Fabr.)								
		Limnius tuberculatus (Ph. Müller)	198	36	0	0	0	0	0	0
	Trichoptera	*Agraylea multipunctata* Curt.	66	38	88	12	4	0	2	0
		Hydroptila femoralis (Eaton)	24	32	0	0	0	0	0	0
		Oxyethira costalis (Curt.)	20	72	374	52	42	14	10	2
		Polycentropidae total	124	30	44	4	6	14	8	4
		Polycentropus flavomaculatus (Pict.) *Cyrnus flavidus* McLachlan								

		Tinodes waeneri (Linn.)	24	0	0	0	0	0	0	0
		Leptoceridae total	52	26	0	24	8	24	20	0
		Leptocerus cinereus Curt. *Mystacides nigra* (Linn.)								
		Goera pilosa (Fabr.)	28	10	0	0	0	0	0	0
	Lepidoptera	*Acentropus niveus* (Oliv.)	12	12	0	0	0	0	0	0
	Hemiptera	*Micronecta poweri* (D. and S.)	164	0	0	68	0	0	0	0
Arachnida	Hydracarina	*Diplodontus despiciens* (O.F.M.)	8	2	0	12	18	8	28	8
		Limnesia undulata (O.F.M.)	0	34	0	152	144	80	186	656
		L. maculata (O.F.M.)	4	8	0	28	40	16	18	38
		Hygrobates longipalpis (Hermann)	0	10	0	6	8	18	20	4
		H. nigromaculatus Lebert	0	0	0	8	0	0	2	0
		Neumania vernalis (O.F.M.)	0	0	0	4	8	8	0	26
		N. callosa (Koenike)	0	0	0	0	2	8	28	68
		N. spinipes (O.F.M.)	0	0	0	0	0	0	0	2
		Piona rotunda (Kramer)	0	0	0	0	0	0	2	0
		P. coccinea C. L. Koch	0	0	0	4	2	0	0	0
		Forelia liliacea (O.F.M.)	0	48	0	12	14	16	22	24

			0·1 – 0·4	0·8 – 1·2	*Chara*	2	5	8	11	14
Arachnida	Hydracarina	*Brachypoda versicolor* (O.F.M.)	0	4	0	10	0	0	0	2
		Midea orbiculata (O.F.M.)	0	2	0	58	6	6	6	2
		Mideopsis orbicularis (O.F.M.)	0	4	0	0	2	4	0	2
		Arrenurus nobilis Neumann	0	0	0	0	2	0	0	0
		A. adnatus Koenike	0	0	0	0	0	0	2	0
		A. perforatus George	0	0	0	4	0	0	0	2
Mollusca	Gastropoda	*Limnaea auricularia* (Linn.)	0	2	0	0	0	0	0	0
		L. pereger (O.F.M.)	4	230	44	108	18	8	4	0
		L. palustris (O.F.M.)	2	0	0	0	0	0	0	0
		Physa fontinalis (Linn.)	14	4	0	0	2	2	0	0
		Planorbis planorbis (Linn.)	0	0	22	46	0	0	0	0
		P. carinatus O.F.M.	2	18	66	8	0	0	0	0
		P. contortus (Linn.)	2	2	0	14	2	8	0	0
		P. albus O.F.M.	4	30	1254	106	4	8	2	2
		P. crista (Linn.)	6	52	1188	70	0	0	0	0

Operculata	*Neritina fluviatilis* (Linn.)	1252	96	0	0	12	18	0	0
	Valvata piscinalis (O.F.M.)	26	104	3564	1036	34	38	128	170
	V. cristata O.F.M.	38	50	110	52	38	6	4	2
	Bithynia tentaculata (Linn.)	48	80	902	582	198	38	34	6
	B. leachii (Sheppard)	12	10	110	104	12	4	4	0
Lamellibranchiata	*Dreissena polymorpha* (Pallas)	42	34	198	176	910	5216	3240	384

the group; and I have added a few species not taken on the sampling line but mentioned in the text as being abundant elsewhere. *Planaria lugubris*, for example, 'is very common on the stones in certain parts of the shore . . .' (p. 41). I have not added abundant species that were evidently chiefly inhabitants of mud, weeds, or reeds.

The first general impression given by table 3 is that the lists of triclads, oligochaetes, leeches, and snails are long, and that insects are not nearly as numerically predominant as they were in Ford Wood Beck. It may be significant that most members of the groups named are typical water animals whose ancestors probably entered fresh water from the sea. The insects in contrast are essentially terrestrial animals that have secondarily invaded fresh water after a period of life on land, when they had to develop a relatively hard cuticle in order to avoid drying up. Possibly the creatures of marine origin have had physiological difficulties in colonizing waters poor in ions and have left unoccupied a series of niches which insects with their hard external covering have been better fitted to fill. However, waters rich in ions are generally rich in food of all kinds and this secondary difference between productive and unproductive waters may be important. It appeared to be the enriched supply of food that caused the increase of *Polycelis felina* in Ford Wood Beck, an increase that led to the decrease of certain insects that were preyed upon, and this kind of interrelationship may be of great importance. Presumably in a productive piece of water one of the main requirements for success is ability to avoid predators, especially in the vulnerable egg stage and during the period just after hatching, and it may be that insects are less good at this than other groups.

On the wave-beaten stones (column 1) *Neritina* was the commonest species, and Berg attributes its success to its ability to cling tightly. The leeches can cling too but, though they do not occur where there is no suitable substratum (Boisen Bennike 1943), a plant suffices, and they are common in still water. Other common animals such as *Gammarus* and *Asellus* presumably seek shelter from waves beneath the stones. The caddis *Tinodes*, whose larva lives in long tubes attached to a hard substratum, is the only species confined to the stony zone. The absence of other specialists is noteworthy; *Ancylus fluviatilis* and the flattened nymphs of the Ecdyonuridae are not recorded in Esrom Lake.

At a depth of 2 m (column 4) there is both the longest species list and the greatest number of animals, some 10,000/m^2. *Dreissena*, the most abundant species at the next three depths, reaches its greatest abundance at 8 m and must have a marked effect on the rest of the fauna because they can find refuge between its shells, and a smooth surface on them. At 14 m there is a typical deep-water mud-fauna of chironomids, oligochaetes and *Pisidium*, together with a large number of mites and a few representatives of many of the species abundant higher up. With increasing depth all but the three groups named first disappear or become very scarce.

The mud-dwelling larva of *Caenis* occurs at all depths and in fair numbers except in the last two. All other insects are commonest in the shallow water and some are confined to it. The beetles, and *Micronecta* during its short adult life, must come to the surface for air, but it is not immediately obvious why some of the larvae should not abound in deep water; *Gammarus pulex* is also confined to the shallows.

The mites, *Limnesia undulata* and *Neumania callosa*, were most abundant at the lowest station as was also the amphipod, *Pallasea quadrispinosa*, which, Berg suggests, finds there the low temperature it requires but not the low oxygen concentration which obtains at greater depths.

The remaining species are at their greatest abundance at some intermediate depth and less numerous in deeper and shallower water. There is considerable difference in the shape of kite histograms showing the numbers of the various species at each depth. The numbers of *Dreissena* are thought by Berg to be related to the hardness of the substratum, but there is no obvious explanation of the wax and wane of most species.

Ehrenberg (1957) investigated the stone-fauna of the German lakes of the same type as Esrom, but she was not interested in this particular problem and does not present her data in such a way that one can find how the numbers of any species increase and decrease with changing conditions in the horizontal plane. There are no outstanding differences between her list of species and that of Berg.

Although a considerable amount must be known about the fauna of the submerged vegetation in Esrom Lake, the information is not available from the published work. Müller-Liebenau (1956) gives an account of the animals found in the weeds, mainly *Potamogeton* spp. in the lakes in Holstein. When her list of species of coelenterates, triclads, nemertines, leeches, molluscs, and larger Crustacea is compared with that of Esrom Lake, the two are almost identical and such species as are absent from one or the other are, almost without exception, rarities. The insect groups are less similar; the presence of two species of *Cloëon* (Ephemeroptera), *Corixa striata* (Hemiptera) and *Erythromma najas* (Odonata) and the absence of *Centroptilum luteolum* (Ephemeroptera) and *Micronecta poweri* (Hemiptera) are to be expected in the more sheltered conditions in a weed-bed. Both authors found six species of *Haliplus* but only two are common, and there is no more similarity in their list of dytiscids. How significant the dissimilarity of the list of Trichoptera may be is hard to say in view of the unsatisfactory taxonomic situation.

Table 4

Number of species in:

	Esrom Lake	German Lakes	both
Bryozoa	3	4	2
Cladocera	21	24	16
Copepoda Cyclopoidea	6	11	6
Harpacticoidea	6	3	2
Ostracoda	10	8	3
Hydracarina	24	32	12

It is among the small animals that the differences, summarized roughly in table 4, are most marked. Unfortunately, not being familiar with any of these groups, I cannot comment on these differences.

As Müller-Liebenau points out, the dying down of the pond weeds in the winter means that the animals that live on them must seek some other substratum, and it is therefore not surprising that the larger inhabitants are found also on weedless shores. Moreover, the algal covering of a leaf and the algal covering of a stone are not greatly different, and both substrata satisfy a herbivore that is not particular about texture. That the smaller animals should correspond less well is to be expected. In the first place, the smaller the size of an animal the greater will be the number of different habitats within a limited area available to different species in its group (Hutchinson 1959). Secondly, some of these organisms have a winter resting stage and are active only when the plants are active.

Windermere may be contrasted with Esrom. It is a glacial lake much of whose drainage area is unproductive mountainside, and much of whose shores are stony and rocky. It has but 5 mg/l Ca, but is not as unproductive as many mountain lakes, because its shores are well populated, and sewage enriches its waters.

Moon (1934, 1936) published the results of a survey which had exactly the same orientation as this chapter. Having selected two lengths of shore along which he could find at one extreme a flat rock face, at the other a reed-bed, and between them all types of intermediate condition, he set out to find how each species was distributed. Unfortunately, knowledge about the taxonomy of freshwater organisms was less then than it is now, and

Moon's lists cannot be taken as they stand. However, since then Hynes has studied the Plecoptera, Mann the Hirudinea, and Reynoldson the triclads, and I have made collections of Ephemeroptera, Mollusca, Odonata, Hemiptera and Coleoptera in Windermere. Moreover, Moon's collection of hydracarina has been submitted to the late Mr W. Williamson and the late Dr Karl Viets, and his Coleoptera to Professor F. Balfour-Browne. It is therefore possible to produce a species list in accord with the latest advances and to assert with confidence to which species most of Moon's names really refer. Only the Trichoptera must remain a blank. I have had the kind assistance of Mr D. E. Kimmins who collected many adult Trichoptera beside Windermere (Kimmins 1943, 1944), but have not succeeded in making out which species Moon encountered. The basis for comparison is also narrowed by Moon's exclusion of the smaller crustacea.

Though reliable names cannot be attached to them, there are obviously many species of Trichoptera in Windermere, and the numbers of some are comparatively high. This, together with the occurrence of the species in other groups listed below, is an immediately striking difference between the two lakes.

Plecoptera:
Diura bicaudata
Leuctra fusca
L. hippopus
Chloroperla torrentium
Nemoura avicularis
Capnia bifrons

Ephemeroptera:
Ecdyonurus dispar
Heptagenia lateralis
Siphlonurus lacustris

Mollusca:
Ancylus fluviatilis

I have seen relatively large numbers of *Heptagenia* in Windermere and of *Ecdyonurus* in Ullswater; and *Diura bicaudata* is always one of the most numerous animals in Ennerdale, one of the least productive of the lakes in the English Lake District.

There is no significant difference between the temperatures of Esrom Lake and Windermere. The former may possibly have a lower oxygen concentration just before dawn owing to its denser phytoplankton, but, if this does take place, it is at a time of year when most of the Plecoptera and Ephemeroptera in Windermere are in the egg stage. There may, of course, be some surprises awaiting armchair theorists when data from continuous oxygen recorders become available, but at present it seems unlikely that the difference between the two lakes is explicable in terms of this factor. This leaves in the field the suggestion already made that most insects succumb to some biotic factor in waters where conditions make a great abundance of other kinds of invertebrate possible.

The same triclads are present in Windermere as in Esrom Lake, except *Planaria torva* which is much commoner in Sweden than in England (Reynoldson 1958a, b) and no doubt in Denmark too. None were found in Ennerdale (Reynoldson, personal communication). In contrast there are ten species of Hirudinea in Esrom Lake as against six in Windermere, if *Trocheta bykowskii* which has been found only on deltas (Mann 1959) be disregarded. The six in Windermere are found also in Esrom and the four other species are *Hemiclepsis marginata*, *Glossiphonia heteroclita*, *Theromyzon tessulatum*, and *Erpobdella testacea*. *Haemopis sanguisuga* and *Batrachobdella paludosa* occur in both lakes but have been omitted from table 3 because they are scarce. A further indication of the status of Windermere is that *Erpobdella octoculata* is numerous, whereas Mann (1955a) failed to find *Helobdella stagnalis*, the characteristic species of hard productive waters, and includes it only on the strength of an earlier record. No collections have been made in any of the Lake District lakes less productive than Windermere.

According to Boycott (1936, p. 172) a 'good' place for molluscs is warm, with slight flow, clear, weedy, large and with hard water. Such a place will contain many species and deterioration in any factor will be accompanied by a reduction in the number of species. The calcium concentration is particularly important; a number of 'calciphiles' do not occur where there is less than 20 mg/l and the rest drop out one by one as the concentration falls below that figure. Windermere lacks most calciphiles such as *Planorbis planorbis*, *Neritina fluviatilis*, and both species of *Bithynia*, all of which occur in Esrom Lake. Its main species are *Ancylus fluviatilis*, *Acroloxus lacustris*, *Limnaea pereger*, *L. palustris*. *Physa fontinalis*, *Valvata piscinalis*, *Planorbis albus*, *P. contortus*, and *P. carinatus*. The last is noteworthy, and is quoted by Boycott (1936) as an example of a calciphile occurring in a lake which, though soft, is large. Most other lakes in the English Lake District contain all these species but the last. In the most barren and rocky, however, only *Limnaea pereger* has been recorded.

The corixid fauna of the two lakes is discussed in chapter 12.

Both species of *Asellus* occur in Windermere and Moon (1957a) has made a detailed study of their distribution. *Gammarus pulex* was the only amphipod until quite recently when *Crangonyx gracilis* was found.

More species of Coleoptera were taken in Windermere than in Esrom Lake but most were in reed-beds. Berg found *Deronectes depressus* and *Oreodytes halensis* (which he assigns to the genus *Deronectes*) on exposed shores, but not *Platambus maculatus*, a well-known inhabitant of such places and running water too in the Lake District. Other species of the exposed lake shore of Windermere were *D. depressus*, *O. septentrionalis*, *Haliplus fulvus*, and *H. flavicollis* (Macan 1940). Moon's *Elmis* sp. is possibly *Stenelmis canaliculatus*, which was recently added to the British list by Claridge and Staddon (1961), who found it on an exposed shore.

Moon's list (1936) of mites after reidentification is:

Midea orbiculata (Müll.)
Diplodontus despiciens (Müll.)
Atractides anomalus Koch
Piona coccinea (Koch)
Hygrobates fluviatilis (Ström)
Mideopsis crassipes Soar
Midea orbiculata (Müll.)
Frontipoda carpenteri Halb.
Limnesia koenikei Piers.
Megapus pavesii (Maglio)
Teutonia cometes (Koch)
Limnesia maculata (Müll.)
Oxus oblongus Kram.
Forelia lileacea (Müll.)
Arrenurus knauthei Koen.

In collections made by me on shores of the same type Mr Williamson found twelve species of which only three are in the list above. Of the twenty-four species recorded from each of the two lakes, only nine are common to both.

Ancylus fluviatilis and *Limnaea pereger* are the only species found on bare rock exposed to wave action. Two Trichoptera join them where the rock face is fissured. The number of species increases as shelter increases, rising suddenly when cover is provided beneath slates derived from the disintegration of rock. The same species are found among the boulders and stones derived from the erosion of glacial drift and others are encountered for the first time, and then the list starts to shorten as, in more sheltered places, stones become fewer and sandy areas more extensive. It reaches a nadir in bays sufficiently sheltered to be floored entirely with sand and then increases where there is more protection and *Phragmites* can establish itself. Here, however, many species adapted to a stony substratum are absent and species such as *Leptophlebia vespertina* and *L. marginata* (Ephemeroptera), *Enallagma cyathigerum* (Odonata), *Corixa* spp. (Hemiptera) and *Acroloxus lacustris* (Mollusca) come in.

In conclusion, attention must be drawn to the much smaller numbers in Windermere. The densest population was in sand, mud, and plant detritus, where a restricted fauna of chironomids, oligochaetes and *Pisidium* totalled $315/ft^2$. On the rocky and moraine shores the highest total was $119/ft^2$. If 1 square foot be treated as a tenth of a square metre, the error is insignificant compared with that of the sampling method. There is therefore a maximum of some 3,000 animals/m^2 and at most stations it is under 1,000/m^2, whereas in Esrom Lake the maximum is 10,000m^2 and the number does not drop below 5,000 m^2 till a depth of nearly 20 m is reached.

The animals on a stony lake shore are on a substratum like that in a swift stream and they must be able to cling tightly or seek refuge beneath the stones when an onshore wind is blowing. There are similarities between the faunas of the two but also some marked differences. *Simulium* and most of the net-spinning Trichoptera are not found on the lake shore, presumably because, being dependent on current to bring them food, they would starve in water that may be still for long periods. Ecdyonurids are prominent in both communities, but *Rhithrogena semicolorata* is confined to streams. As will be seen in chapter 9, it needs a flow to bring in contact with its body surface the amount of oxygen which it requires. There are probably many animals limited to running water in the same way. In groups not subject to restrictions of these kinds, the species is not always the same; *Ecdyonurus dispar* inhabits lake shores and *E. torrentis* and *E. venosus* are the species of streams and rivers.

Hodson's Tarn is an example of a small lake or large pond and one could use a lot of ink arguing what exactly it should be called—the most descriptive name for it is probably 'moorland fishpond'. It is artificial and lies in the English Lake District at an altitude of 190 m on rock thinly covered in places with boulder clay. It is just under 0·5 hectares in extent and about 3 m deep. The drainage area is poor fellside now used extensively for plantations of *Picea* and *Larix*. Though the single inflow is small, there is always a fair flow through the pond. Around the inflow and at places along the sides there is emergent vegetation, mainly *Carex rostrata*. Elsewhere the bottom in shallow water is covered with a thick sward of *Littorella uniflora*. In deeper water *Potamogeton natans* grows near the inflow, *Myriophyllum alternaeflorum* and other species elsewhere. The calcium concentration ranged from 3·8 mg/l in a wet spell in winter to 5·7 in a dry spell in summer (Macan 1950a). pH appears never to have been measured, but that of the similar Three Dubs Tarn was 6·7 (Macan 1949) and this is a typical figure for Lake District tarns. The water is not coloured. A recording thermograph has been operating in the tarn since 1957, and table 6 presents the maximum temperatures during five years. Fine spells were unusually long in 1959 though at no time was an exceptionally high temperature reached.

Table 5

HODSON'S TARN

Number of specimens per square metre

The sampler covered an area of 75 cm^2. The catch has been multiplied by a factor that brings it to numbers/m^2. September 1958 was chosen because more samples were taken then than on any other occasion. Species against which no figure is shown were taken on some other occasion or in the net collections only. The names of the Trichoptera are those of the adults captured in an emergence cage.

			Carex Sept. 1958	*Littorella* Sept. 1958
Platyhelminthes	Rhabdocoelida		–	–
Annelida	Oligochaeta		–	–
	Hirudinea	*Erpobdella octoculata* (Linn.)	54	–
		Theromyzon tessulatum (Müller)	–	–
		Glossiphonia complanata (Linn.)	–	–
Mollusca	Gastropoda	*Limnaea pereger* (Müller)	45	–
Crustacea	Amphipoda	*Gammarus pulex* (Linn.)	18	–
Insecta	Plecoptera	*Nemoura cinerea* (Retz.)	–	–
	Ephemeroptera	*Leptophlebia vespertina* (Linn.) } *marginata* (Linn.) }	9954	4158
		Cloeon dipterum (Linn.)	198	–
		C. simile Eaton	–	–
	Odonata	*Enallagma cyathigerum* (Charp.)	54	462
		Pyrrhosoma nymphula (Sulzer)	2808	1056
		Lestes sponsa (Hansem.)	–	–
		Aeshna juncea (Linn.)	–	–
		Cordulia aenea (Linn.)	–	–
		Sympetrum scoticum (Leach)	–	–
		Libellula quadrimaculata Linn.	–	–

			Carex Sept. 1958	*Littorella* Sept. 1958
Insecta	Hemiptera	Corixa nymphs	–	15
		C. castanea (Thoms.)	27	–
		C. scotti (D. and S.)	–	–
		C. dentipes (Thoms.)	–	–
		C. linnei (Fieb.)	–	–
		C. distincta (Fieb.)	–	–
		C. sahlbergi (Fieb.)	–	–
		C. praeusta (Fieb.)	–	–
		Notonecta obliqua Gall.	–	–
	Lepidoptera	*Nymphula nymphaeata* (Linn.)	9	–
	Trichoptera	*Triaenodes bicolor* (Curt.)	18	92
		Phryganea varia Fabr. *striata* Linn.	126	52
		Holocentropus dubius (Rambur)	9	–
		Limnophilus marmoratus Curt.	9	–
	Coleoptera	*Haliplus fulvus* (Fabr.)	–	13
		Deronectes assimilis (Payk.)	–	6
		D. duodecimpustulatus (Fabr.)	–	–
		Gyrinus larvae	–	–
		Haliplus confinis Steph.	–	–
		Rantus exsoletus (Forst.)	–	–
		Acilius sulcatus (Linn.)	–	–
		Dytiscus semisulcatus (O.F.M.)	–	–
Arachnida	Hydrachnellae	*Hydrachna geographica* (Müller)		
		H. conjecta (Koenike)		
		Limnochares aquatica (Linn.)		
		Hydrodroma despiciens (Müll.)		
		Hygrobates longipalpis (Hermann)		

			Carex Sept. 1958	*Littorella* Sept. 1958
Arachnida	Hydrachnellae	*Hydrochoreutes krameri* Piersig *H. ungulatus* (Koch) *Piona carnea* (Koch) *P. nodata* (Müll.) *Arrenurus neumani* (Piersig) *A. leuckarti* (Piersig) *A. ornatus* George *A. globator* (Müll.)	216	79
Vertebrata	Amphibia	*Triturus helveticus* (Raz.)	–	–
		Bufo bufo (Linn.)	–	–
		Rana temporaria (Linn.)	–	–

Table 6

HODSON'S TARN

The number of hours water temperature was at the levels shown. (In fact number of hours at X·5°C is number of hours between X and X + 1°C)

Year	20·5°C	21·5°C	22·5°C	23·5°C	24·5°C	25·5°C
1958	51	13	6	0	0	0
1959	239	84	17	0	0	0
1960	87	68	38	12	17	4
1961	0	0	0	0	0	0
1962	14	8	0	0	0	0

Table 5 presents a list of the species caught during the course of a survey that has extended over several years. The few rhabdocoeles and oligochaetes taken have not been identified. The entomostraca are discussed later. The smaller animals have been ignored too, a regrettable inevitability with the manpower available. The chironomids also have been left unnamed, the reason being that so much time is needed to preserve and label them, even if someone else can be found to identify them. The species in Three Dubs Tarn, which is similar to Hodson's, are recorded by Macan (1949).

The lepidopterous larvae were confined to the *Potamogeton natans* and appear under the heading *Carex* in table 5 because the sample included a piece of *P. natans*. No other species had such a restricted distribution. Most Coleoptera and *Corixa castanea* were rarely or never taken in the *Myriophyllum*; nor was *Pyrrhosoma* whose absence was due to the egg-laying habits of the adults (chapter 5). Larvae of *Holocentropus* were commoner in *Myriophyllum* than anywhere else. *Corixa scotti*, most numerous in the *Potamogeton natans* and deeper *Littorella* stations, and *Cloeon simile*, found mainly in *P. natans* and *Myriophyllum*, occurred in fair numbers and would almost certainly have been taken in a quantitative sample in the right place. Nymphs of *Lestes sponsa* were abundant during the summer, particularly in the *Carex*, where, in May 1961 350/m^2 were found. Tadpoles were also abundant early in the year as long as there were no fish in the tarn.

The predominance of insects and the small number of species and generally of specimens too in all other groups is the first striking thing about the fauna of Hodson's Tarn. *Leptophlebia* is the most numerous species; Odonata are prominent, particularly *Enallagma cyathigerum* and *Pyrrhosoma nymphula*. Groups in which particular species indicate the status of the tarn are: triclads—no species present; Hirudinea—*Erpobdella octoculata*; Mollusca—*Limnaea pereger*; Hemiptera—*Corixa scotti* and *C. castanea*.

Three Dubs Tarn (Macan 1949) is of the same type as Hodson's, though a little larger. Perhaps the most important difference is that six little streams flow into it and bring vegetable debris. As a result the plants are unable to reach boulder clay as in Hodson's Tarn, but must take root, if they can, in this debris. It seems a less favourable soil because extensive areas are without plants. In the deeper water the main vegetation is *Nitella* not *Myriophyllum*. *Apium inundatum* is the commonest plant in shallow water. Another important difference is that, whereas there are many *Salmo trutta* in Three Dubs Tarn, Hodson's has been deliberately kept free of fish during the period of observation.

Although fifteen species of water mite were taken in Three Dubs Tarn and thirteen in Hodson's, only four were found in both. There are two inexplicable vicariants, *Cyrnus trimaculatus* and *Deronectes depressus* apparently occupying in Three Dubs Tarn the places that *Holocentropus dubius* and *Deronectes assimilis* occupy in Hodson's Tarn. *Cloeon dipterum* was not recorded in Three Dubs Tarn. Otherwise the species lists of the two tarns differs little. There are, however, some marked differences in abundance; *Corixa scotti*, for example, is much more numerous in Three

Dubs Tarn. This may be due to the more extensive bare patches. The abundance of the various species of rooted plants is not the same in the two tarns, which must mean a different arrangement of the substratum on which the animals lay their eggs, seek their food and take refuge from their enemies. This no doubt partly causes the differences between the two tarns. Predation must be important too; how important should soon be known as fish have recently been introduced into Hodson's Tarn.

Smyly (1955) collected entomostraca in Hodson's Tarn and also in Scale Tarn, which is close to it and of about the same size but different in that it is not among trees, its bottom vegetation is mainly *Nitella* not *Myriophyllum*, and there is little debris in the *Carex* beds. Scale Tarn also harbours *Asellus*, which few Lake District tarns do, and would appear on this account to be richer than most, though there is nothing in the chemical analysis that reflects this. Wise Een (Smyly 1952) is larger than either of these; most of the land round it is without trees and there is vegetable debris in the *Carex* beds. Table 7 is based on the data in the papers

Table 7

Cladocera, Copepoda and Ostracoda of three moorland fishponds (Smyly 1952, 1955)

	Open Water			*Weeds*		
	Scale	Hodson	Wise Een	Scale	Hodson	Wise Een
Daphnia longispina O.F.M.	+++	+++	+++			
Diaptomus gracilis Sars	+++	++	+++	++		++
Ceriodaphnia pulchella Sars	+++	+	+++	++		++
Chydorus sphaericus (O.F.M.)	+	+	++	+++	+	++
Cyclops leukarti (Claus)	+	+	++	+	+	+
Polyphemus pediculus (Linn.)	+	+	++	+	+	+++
Alona affinis Leydig	+	+		+	+	+
Graptoleberis testudinaria (Fischer)			+	+	+	+
Diaphanosoma brachyurum Liéven		++				
Cyclops albidus (Jurine)			+	+++	++	+++

E

	Open Water			*Weeds*		
	Scale	Hodson	Wise Een	Scale	Hodson	Wise Een
Cyclops viridis (Jurine)			+	+ +	+ +	+ + +
Simocephalus exspinosus (Koch)			+	+ +	+ +	+
Ilyocryptus sordidus (Liéven)			+	+ +	+ +	+
Chydorus globosus Baird				+	+	+
Canthocamptus staphylinus (Jurine)				+ + +	+ +	
Chydorus piger Sars				+ +	+	
Eurycercus lamellatus (O.F.M.)			+	+ + +		+ + +
Sida crystallina (O.F.M.)				+ +		+ + +
Peracantha truncata (O.F.M.)				+ +		+ + +
Cyclops agilis Koch				+ +		+ + +
Acroperus harpae Baird				+		+
Candona candida (O.F.M.)				+ +		
Cyclops macruroides (Lillj.)					+ +	+ + +
Drepanothrix dentata (Eurén)					+ + +	+ + +
Cypria exsculpta (Fischer)					+ +	

quoted and some which Mr Smyly has kindly added for me. It does not include all the species recorded (Smyly 1958b) as it is difficult to know what importance, if any, to attach to rarities. *Ceriodaphnia pulchella* in the open water of Scale and Wise Een is largely replaced by *Diaphanosoma brachyurum* in Hodson's. Smyly tentatively correlates this difference and food supply but produces no evidence except that Scale has more entomostraca in both open water and weeds. He suggests that the difference in the fauna of the weed-beds may be due to low oxygen concentration where debris accumulates, but the measurements that might confirm or disprove this idea have not been made. He also points out that debris as well as different species of plants provide different physical conditions which could influence the fauna. A later study (Smyly 1958a) of 144 tarns brings to light one or two clear correlations: *Alanopsis elongata* and *Bosmina coregoni* var. *lillgeborgi* occurred mainly in high mountain tarns, the first generally off rocky shores; *Simocephalus vetulus* was found mainly in the more calcareous tarns, *Acantholebris curvirostris* in *Sphagnum*. In general, however, the work confirms the impression that the habitat of each species is not easy to define.

Store Gribsø (Great Grebe Lake) (Berg and Petersen 1956) lies in the same moraine soils as Esrom Lake in the middle of a large forest. It is some 600 × 200 m in extent and 11 m deep. That it is not in a region of unproductive soil is indicated by the richness of Esrom lake, but Grib Lake lies in a relatively high depression and in summer has no outflow. It is fed by the drainage from several bogs which have presumably developed in similar situations on a leached soil. This gives it its humic character, indicated by deep-brown water and a pH of about 5. This value is well below that of the Lake District tarns although the calcium concentration, 5-7 mg/l, is higher than in some. There are also 3-4 mg/l Mg and relatively large amounts of sulphate (16-20 mg/l) and chloride (19-25 mg/l). The lake stratifies in summer, and oxygen in the upper layers falls below saturation. The maximum surface temperature was about 22°C in 1941 and a little less in 1942, and, as the lake is sheltered, there may not be so much variation as in the Lake District tarns. On the other hand it is evident from the curves (pp. 99 *et seq.*) that warming up does not start till later. Narrow beds of *Phragmites* occur in the shallow water and thick *Fontinalis dalecarlica* extends down to about 2 m, beyond which there is little vegetation.

There are certain similarities between the faunas of Grib Lake (table 8) and the tarns, the most noticeable being the predominance of *Leptophlebia*. Odonata are also abundant in both and the species lists are fairly similar, though Anisoptera are most numerous in the lake and Zygoptera in the tarns. The absence of the common *Enallagma cyathigerum* could be related to its oviposition habits (chapter 5). *Cyrnus* and *Holocentropus*, of the Trichoptera, show a common abundance. *Phryganea* is represented by two species in both but the species are different. There the resemblance ends. *Glyphotaelius* larvae make cases of dead leaves, *Molanna* of coarse sand grains, neither of which materials are available in the tarns, but how far this limits their range is not known.

Table 8

STORE GRIB LAKE

Numbers of animals per square metre in the reed-swamp and in the *Fontinalis* beds beyond. The table is based on Berg and Peterson's table Xa. Names against which there is no figure are animals not in their table but mentioned by them in the text.

			Reed-swamp	*Fontinalis*
Porifera		*Spongilla lacustris* (Linn.)		
Coelenterata		*Hydra vulgaris* Pallas		
Platyhelminthes	Tricladida	*Planaria lugubris* O. Schmidt.	7	7
		Dendrocoelum lacteum O. F. Müller	17	3
		Polycelis nigra (Müller)		
Annelida	Oligochaeta	*Stylaria lacustris* (Linn.)	75	35
		Others	15	22
	Hirudinea	*Erpobdella testacea* (Savigny)	57	5
Bryozoa		*Plumatella repens* (Linn.)		
		Fredericella sultana (Blumenb.)		
Crustacea	Isopoda	*Asellus aquaticus* (Linn.)	437	198
Insecta	Ephemeroptera	*Leptophlebia vespertina* (Linn.)	456	919
	Odonata Anisoptera	*Cordulia aenea* (Linn.)	79	205
		Aeshna sp. *Sympetrum* sp. *Libellula quadrimaculata* Linn.	17	3
	Odonata Zygoptera	*Lestes sponsa* (Hansem.) *Erythromma najas* (Hansem.) *Pyrrhosoma nymphula* (Sulz.) *Ishnura elegans* v.d. Lind. *Agrion pulchellum* v.d. Lind.	12	25
	Hemiptera	*Nepa cinerea* Linn.		

			Reed-swamp	*Fontinalis*
Insecta	Coleoptera Dytiscidae	*Noterus crassicornis* Fabr. *Hyphydrus ovatus* (Linn.) *Hygrotus inaequalis* (Fabr.) *H. versicolor* (Schall.)	25	17
	Chrysomelidae	*Plateumaris sericea* (Linn.) *Donacia crassipes* Fabr.	12	5
	Megaloptera	*Sialis lutaria* (Linn.)	3	3
	Trichoptera	*Molanna angustata* Curt.	–	3
		Glyphotaelius punctatolineatus Retz. *G. pellucidus* (Retz.)	7	7
		Phryganea grandis Linn. *P. obsoleta* Hag.	–	5
		Holocentropus dubius (Rambur) *Cyrnus* sp.	91	247
		Anabolia laevis Zett.		
		Limnophilus nigriceps (Zett.)	30	–
	Diptera	*Phalacrocera replicata* (Linn.)	10	3
		Corethra flavicans Meigen	3	–
		chironomids	47	176
Arachnida	Hydrachnellae	*Limnochares aquatica* (Linn.)	10	86
		Diplodontus despiciens (O.F.M.)	5	–
		Unionicola crassipes (O.F.M.)	10	–
		Limnesia maculata (O.F.M.)	–	5
		Neumania vernalis (O.F.M.)	15	17

			Reed-swamp	*Fontinalis*
Arachnida	Hydrachnellae	*Arrenurus adnatus* Koenike *forcipatus* Neuman *bicuspidator* Berlese *tricuspidator* (O.F.M.) *robustus* Koenike *albator* O.F.M. *pustulator* O.F.M.	62	116
	Araneae	*Argyroneta aquatica* (Linn.)	12	3
Vertebrata	Pisces	*Esox lucius* Linn. *Perca fluviatilis* Linn.		
	Amphibia	*Triturus vulgaris* (Linn.) *Bufo vulgaris* (Laur.)		

Certain differences suggest that Grib Lake is more productive than the tarns, notably the occurrence of *Asellus* and of triclads. Incidentally it is surprising to find that the two most abundant species are those found by Reynoldson to be the most exigent.

There are certain other differences for which an explanation can at least be offered in the light of existing knowledge even though it must remain to some extent speculative. The only leech is *Erpobdella testacea*, shown by Mann (1955a, 1961a) to be typical of emergent vegetation, which its greater tolerance of poor oxygen conditions enables it to inhabit in the face of competition from other species. The presence of *Gammarus* in the tarns may be due to continual recruitment from flourishing colonies in the inflowing streams. It is never abundant in them except under stones. There are stones in neither Grib Lake nor its inflowing ditches, which could be why *Gammarus* has not been found in them. Commenting on the absence of *Limnaea pereger*, Berg quotes an experiment whereby Lang showed that eggs of this species would not develop at high temperature in waters with much humic matter, although the adults were not affected. I have found that *L. pereger* tends to be absent from places in woods (table 35, chapter 12). In woodland ponds, however, *Planorbis albus* has always been abundant and, therefore, the humic content seems to be the most likely explanation of the absence of snails in Grib Lake. The larva of *Phalacrocera* bears an astonishing similarity to a piece of moss. I have found specimens only in *Sphagnum* pools. It is clear from Berg's account

(p. 184) that they are equally well concealed in *Fontinalis*, and this record arouses in me a strong curiosity about the egg-laying habits of the adults. In Lake District lakes, where the common emergent is *Phragmites*, I find *Notonecta glauca*; in tarns, where the emergent is *Carex*, I find *N. obliqua*. Berg records *N. glauca* which, therefore, is perhaps associated with the nature of the emergent rather than with the general nature of the water body.

For the remaining differences not even the most tentative explanation is offered. No *Cloeon* has been found in Grib Lake, and Berg records only a few corixids under ice (p. 178). I have been to Grib Lake to seek these ubiquitous animals, which are such useful indicators, and did not find any at all. Only one species is common to the long list of beetles recorded from the lake and from the tarns. Berg expresses gratitude to Professor Lundblad for his assistance with the hydracarina of Grib Lake, and Mr T. Gledhill, who has named the mites of Hodson's Tarn, has worked under the same authority. The identification of this difficult group is therefore not in question, yet out of thirty-three species in Grib Lake and thirteen in Hodson's Tarn only seven are common. This, however, has been true of all the comparisons made in this chapter. Mites travel easily, but only as passengers, on chironomids, dragonflies, water-bugs and other insects to which the larvae attach themselves. They have, therefore, no say in the selection of a destination and cannot leave it once they have quitted the host. The species of mite may, therefore, depend more on haphazard distribution than the species of any other group.

Of twenty-six species of entomostraca in Grib Lake only twelve occur in the three English tarns mentioned.

Little work has been done on ponds of other type—horse ponds, duck ponds, village ponds and similar small pieces of water characteristic of different kinds of countryside. Elton records that he was attracted to ponds once because he thought that the communities in them must be determined by a limited set of factors, but repelled when he had found as many distinct communities as he had visited ponds. Doubtless between some of the ponds there were basic differences but it is possible too that small rich ponds are liable to catastrophe rather often. If they are large enough not to dry up, many must be rich enough to suffer complete deoxygenation in hot still weather. This would kill or drive out most of the animals and, for a period thereafter, what a collector found in the pond would depend largely on what had invaded it. The problem of measuring oxygen continuously, particularly throughout the night, is an important one.

GENERAL CONCLUSIONS

These, with the omissions noted at the beginning, are the main surveys of modern times; older ones are not comparable on account of the recent completion of certain taxonomic studies. The lists are perhaps longer than need be but it is better to err on the cautious side at this stage; sometimes a species that is always scarce but consistently present may be a useful indicator. One conclusion from the lists that needs no stressing is that correct naming is of fundamental importance. As more is discovered and works of reference improve, taxonomy becomes a less formidable obstacle, but there are still some major gaps to fill. It is worth emphasizing that, notwithstanding the lists of Trichoptera quoted, the larvae of a number of species are still undescribed.

The idea of arranging lakes in a series according to various features that probably give an indication of productivity was originated by W. H. Pearsall. Esrom Lake can be added to the series and placed above the Lake District lakes, with which Pearsall worked. There is no diminution in the number of triclads till somewhere in the series below Windermere but a number of leeches and snails drop out between Esrom and that lake. The corixids, a group different in that each species has a well-defined habitat, change, over this range, at points which do not correspond with any of thosc indicated by other groups (table 34, chapter 12). This seems to suggest that there were good grounds for the fears expressed at the beginning of this chapter that each species has its own range along each gradient with the result that communities merge into each other in an infinite variety of ways. Berg's study of a river led to a similar conclusion (p. 35). In consequence any attempt to discover a set of 'typical' biocoenoses will be fruitless. On the other hand the definition of a few convenient biocoenoses by means of scrupulous taxonomy and quantitative collection, after the manner of Berg, is likely to prove a useful starting point for other work. If deviations from the type can be correlated with differences in the environment, and these correlations can be shown to be real by experiments, useful knowledge about the structure of communities should have been gained. The pioneer can do little more than describe the environment, list his species and record how many of each he took. Each successor can make more comparisons and can begin to put forward explanations of the differences which become apparent. Strong circumstantial evidence can sometimes be brought forward to explain the range of a given animal, but generally confirmation from experimental work of some kind is desirable. It is with this circumstantial evidence and experimental proof that the rest of this book is concerned.

SYNONYMS

Species	Group	Other generic names	Other species names
Acroloxus lacustris (Linnaeus)	Moll. Gastropoda	*Ancylus*, *Velletia*	
Acroperus harpae Baird	Crust. Cladocera	*Lynceus*	*leucocephalus* Koch
Alona affinis Leydig	,, ,,	*Lynceus*	
Amphinemura sulcicollis (Stephens)	Ins. Plecopt.		*cinerea* (Olivier)
Ancylus fluviatilis Müller	Moll. Gast. Pulmonata	*Ancylastrum*	
Aphelocheirus aestivalis Westwood	Ins. Hemipt.		*montandoni* Horváth
Bithynia leachii (Sheppard)	Moll. Gastropoda	*Bulimus*	
B. tentaculata (Linnaeus)	,, ,,	*Bulimus*	
Chydoros globosus Baird	Crust. Cladocera	*Lynceus*	
C. piger Sars	,, ,,	*Lynceus*	*barbatus* Brady
C. sphaericus (Müller)	,, ,,	*Lynceus*	
Cyclops agilis Koch	Crust. Copepoda	*Leptocyclops*, *Eucyclops*	*serrulatus* auct
C. albidus (Jurine)	,, ,,	*Pachycyclops*	*annulicornis*
C. leukarti (Claus)	,, ,,	*Mesocyclops*	*obsoletus*
C. viridis (Jurine)	,, ,,	*Megacyclops*	*vulgaris*
Diaphanosoma brachyurum Lieven	Crust. Cladocera	*Daphnella*	*wingii* Baird *brandtianum* Fischer
Drepanothrix dentata (Eurén)	,, ,,		*homata* Sars
Erpobdella octoculata (Linnaeus)	Hirudinea	*Herpobdella*	*atomaria* (Carena)
Glossiphonia complanata (Linnaeus)	,,	*Glossossiphonia*	
G. heteroclita (Linnaeus)	,,	*Glossossiphonia*	
Graptoleberis testudinaria (Fisher)	Crust. Cladocera	*Lynceus*	
Hydroptila femoralis (Eaton)	Ins. Trichopt.		*longispina* McLachlan

Species	Group	Other generic names	Other species names
Laccophilus hyalinus (Degeer)	Ins. Coleopt.		*interruptus* Panzer
Leuctra fusca (Linnaeus)	Ins. Plecopt.		*fusciventris* Stephens
Limnaea auricularia (Linnaeus)	Moll. Gast. Pulmonata	*Radix*	
L. palustris (Müller)	Moll. Gast. Pulmonata	*Stagnicola*	
L. pereger (Müller)	Moll. Gast. Pulmonata	*Lymnaea, Radix*	*peregra, ovata* Drap.
Micronecta poweri (D. and S.)	Ins. Hemipt.		*borealis* Lundblad
Micropsectra brunnipes (Zetterstedt)	Inst. Dipt. Chironomidae	*Tanytarsus*	
M. subviridis Goetghebuer	Inst. Dipt. Chironomidae	*Tanytarsus*	
Nemoura cinerea (Retzius)	Inst. Plecopt.		*variegata* Olivier
Nepa cinerea Linnaeus	Ins. Hemipt.		*rubra* Linnaeus
Neritina fluviatilis (Linnaeus)	Moll. Gastropoda	*Theodoxus*	
Noterus crassicornis (Fabricius)	Ins. Coleopt.		*clavicornis* (Degeer)
Notonecta obliqua Gall.	Ins. Hemipt.		*furcata* Fabricius
Peracantha truncata (Müller)	Crust. Cladocera	*Perata-cantha Lynceus*	*truncatus* N and B.
Perla bipunctata Pictet	Inst. Plecopt.		*carlukiana* Klapálek *marginata* Panzer
Perlodes microcephala (Pictet)	,, ,,		*mortoni* Klapálek
Phalacrocera replicata (Linnaeus)	Inst. Dipt.		*diversa* (Walker)
Planaria alpina Dana	Platyhelminthes	*Crenobia*	
P. lugubris Schmidt	,,	*Dugesia*	
Planorbis albus Müller	Moll. Gast. Pulmonata	*Gyraulus*	*glaber* Jeff.
P. carinatus Müller	Moll. Gast. Pulmonata	*Tropidiscus*	

Species	Group	Other generic names	Other species names
P. contortus (Linnaeus)	Moll. Gast. Pulmonata	*Bathyomphalus*	
P. corneus (Linnaeus)	Moll. Gast. Pulmonata	*Planorbarius, Coretus*	
P. crista (Linnaeus)	Moll. Gast. Pulmonata	*Armiger*	*nautileus* (Linnaeus)
P. planorbis (Linnaeus)	Moll. Gast. Pulmonata	*Tropidiscus*	*umbilicatus* Müller *complanatus* Stud. *marginatus* Drap.
Polycelis felina (Dalyell)	Platyhelminthes		*cornuta* (Johnson)
Salmo trutta Linnaeus	Pisces	*Trutta*	*lacustris* Linnaeus *fario* Linnaeus
Sialis lutaria (Linnaeus)	Ins. Megalopt.		*flavilatera* (Linnaeus)
Simocephalus exspinsosus (Koch)	Crust. Cladocera	*Simosa*	
Simulium venustum Say	Ins. Dipt.		*austeni* Edwards
Stenophylax latipennis (Curtis)	Ins. Trichopt.	*Potamophylax*	
S. stellatus (Curtis)	,, ,,	*Potamophylax*	
Sympetrum scoticum (Leach)	Ins. Odonata		*danae* Sulzer
Theromyzon tessulatum (Müller)	Hirudinea	*Protoclepsis*	*tessellatum* (Müller)
Velia caprai Tamanini	Ins. Heteropt.		*currens* (Fabricius)

CHAPTER 4

Transport

The heading has been changed more than once during the course of the composition of this chapter. Several of the words apparently suitable at first sight (migration and dispersal, for example) have had special meanings attached to them by other writers, and should no longer be used in a general sense. Travel, a word I have used elsewhere in this context, carries perhaps too much suggestion of volition, conjuring up a picture of man setting out with a precise idea of where he wants to go and how to get there. The word chosen must be applicable alike to the water-beetle which, impelled by urges that we do not understand, crawls out of the water, sits in the sun to get dry and warm, and then flies off under the guidance of a further set of impulses that we know nothing about, and to the snail which, crawling on solid surfaces under water, passes on to the foot of a bird and is carried to another piece of water before it can pass off again. I hope that the word transport will be found acceptable. In general it evokes an idea of passive conveyance from one place to another but there is nothing incongruous about the idea of an insect opening its wings and transporting itself, and moreover its derivation from words meaning 'carry across' is particularly apposite to the study of an environment consisting of small isolated pieces with barriers between them.

As stated in the first chapter, it is imperative to discover the extent to which these barriers are crossed, for, unless it can be established that the community in any one place is related to the environmental conditions and does not depend mainly on what has happened to reach it, the line of thought which it is proposed to follow in these pages is not worth following. The problem can be approached directly and indirectly: animals can be observed moving from place to place, though unfortunately most of our species except the Odonata are too small to be identified on the wing and must be captured in traps or on arrival in a natural water; or indirectly, deductions about their ability to cross land barriers being made from their distribution in a given area.

Many species are cosmopolitan, but nearly all these are small and have an egg or some other stage which, tolerant of desiccation, can be blown about by the wind (Gislén 1948). Larger animals are usually confined to one of the great zoogeographical regions of the world, and sometimes do not range over the whole of that part of it which is apparently suitable for them. There may be historical reasons for this. Species characteristic of the north and middle parts of the palaearctic region were driven from high latitudes as the ice advanced during the last Ice Age, but were able to spread into southern regions which had probably been too hot for them previously. When the ice started to retreat, ranges started to extend northward again, but some species did not gain as much ground as others because they failed to pass a certain point before some critical event. For example those that reached the Rhine while the Thames was still one of its tributaries had no difficulty in crossing over into England, but those that came later found the way barred by the sea and some never succeeded in making their way across. The British fauna is today an attenuated version of that of neighbouring Europe and the Irish fauna is somewhat poorer still. The history of the freshwater fauna of Europe has been written by Thienemann (1950). It provides an essential background which merges with the present theme, particularly when the limits of a range are due, not to historical events, but to conditions obtaining to-day.

It is preferable to consider separately the transport of animals that are aquatic all their lives and the transport of those that possess wings when adult. The former, the wingless ones, are the less easy to observe, and the most spectacular data about ability to travel from place to place are provided by new arrivals that apparently find an unoccupied niche. A small freshwater shrimp, *Crangonyx pseudogracilis*, was brought from America, probably in a consignment to an aquarist, and escaped in this country in about 1930. The exact date of introduction of such foreign species is rarely known, and the initial population undoubtedly takes a long time, probably a decade or more, to build up numbers before any spectacular advance is made. Now *Crangonyx* occurs over most of central England. It can thrive in canals, and the pattern of its distribution suggests that it is largely along them that it has spread (Hynes 1955a), but it has certainly crossed land barriers also, for in 1960 it was found in Windermere, which is not connected with the canal system.

Potamopyrgus jenkinsi was first recorded in the Thames estuary in 1883, and ten years later it was taken in fresh water. To-day it occurs in almost every vice-county in England and Wales and in many in Scotland and Ireland (Ellis 1951). Hunter and Warwick (1957) and Bondesen and Kaiser (1949) discuss the unsettled question, which is not relevant to our present argument, of the origin of this species, and the latter also give an account of its spread in other European countries. This little snail and *Crangonyx* show how rapidly a creature that cannot transport itself across land barriers can spread through a country. There have been many other importations of foreign species into Europe, some deliberate, some not, and a full account of them is given by Thienemann (1950). Some have spread and some have done little more than maintain themselves at the point of establishment, but it is impossible to know whether this was due to poor powers of transport or to inability to compete with the native fauna.

The movement of indigenous species is less easy to investigate, since a new arrival in a place is not detectable if members of the same species are there already, and freshly created environments are not very common, though Boycott obtained some valuable data by frequent visits to a number of ponds, many isolated, over a long period of years. On the average he recorded in each pond a species he had not found before once every nine years, which indicates a surprising amount of overland transport when allowance is made for inability to recognize new arrivals when snails of the same species were already present (Boycott 1936).

How animals of this kind traverse land barriers is not known. Mites are parasitic on adult insects at one stage and are transported by their hosts, with whom they should be classified. There are occasional records of ostracods or *Pisidia* gripping part of some larger animal that could easily carry them to another pond (Lansbury 1955). Bondesen and Kaiser (1949) believe that birds may eat *Potamopyrgus jenkinsi* and later vomit them up. Moon (1957a and b) stresses the importance of man, particularly the fisherman who is continually introducing plants and animals that he thinks will benefit his fish, and also the naturalist who does not wash his net carefully after each collection. For the rest we are left with the old familiar bird's foot. That smaller animals do sometimes get carried about on birds is indubitable, but the records are scattered (Gislén 1948) and no systematic study has been made such as would warrant even a guess at the amount carried and how the amount varies from species to species.

Direct observation of the flight of insects should provide answers to the following questions of relevance to the theme of this chapter: how far do they fly; if they are aquatic as adults, how often do they fly; and to what extent are they picked up by the wind and carried considerable distances? Because the insects in the different orders are strikingly unalike in their flying habits, it is better to take them one by one rather than to try to divide them under headings such as the questions set out above suggest.

Thienemann (1950) ignored all the insects when reconstructing the history of the European freshwater fauna on the grounds that they could fly over obstacles that barred the way to other animals, but Illies (1955b) exempts the Plecoptera from this ban on the grounds that they use their wings only reluctantly. Even when provoked to flight by a collector, their reaction is to return soon to the bank and run into the shelter of the vegetation or the stones. As a result of this tendency, populations, isolated in various places by various events, have remained isolated and given rise to subspecies or species that are clearly distinct today (Illies 1953).

The Ephemeroptera are obviously not built for strong flight but may nevertheless cover several kilometres according to Verrier (1949, 1954), who believes that some of those that like to dance at sunset may travel to try to remain in the last rays of the setting sun and also to find surfaces that are radiating heat. She records following *Potamanthus luteus* for 500 metres at a speed of about 6 km/hour. Some places seemed to be particularly attractive to dancing swarms and *P. luteus* travelled five or six km to reach one of these. A fall of *Polymitarcys virgo*, described as 'une chute de manne', was seen eleven km from the nearest known breeding place and the wind was not in the right direction to have carried the specimens.

The Odonata, particularly the larger ones of the sub-order Anisoptera, are powerful fliers and mass flights of a number of species are on record (Longfield 1949, Williams 1958, Corbet, Longfield and Moore 1960). Hundreds of miles may be covered and the colonies of some species are believed to be maintained in Britain only by immigration.

The Hemiptera spend all their lives in the water and therefore it is important to know how often they leave this medium to take a flight. E. S. Brown (1951) has made an important contribution to this problem by endeavouring to obtain an idea of the total population of each species of corixid in a neighbourhood by sampling every piece of water, and comparing this with the number caught in spring and early summer in a temporary pond that dried up later. From the two figures he could calculate a migration rate, and this proved to be high for species of pools and ponds and low for species of larger places such as lakes and rivers. He surmised, however, that the result might be different in the autumn when overcrowding was tending to stimulate emigration from all sorts of places and this was confirmed by light-trap catches, in which *Corixa dorsalis* and *C. falleni*, two lake and river species, were numerous (E. S. Brown 1954).

Light-traps have been used extensively since then, but Richard (1961), comparing light-trap catches and collections made with a pond-net in the neighbourhood, concludes that the light-trap is not a good instrument with which to investigate what Brown called migration rate. *Corixa praeusta* formed 14 per cent of the light-trap catches but was not taken with a pond-net. It was, however, captured by means of an underwater light-trap, and Richard believes that its apparent absence was merely due to its habit of living beyond the reach of a pond-net wielded from the bank. *Corixa fossarum* was not found in the light-trap, though it was numerous in near-by waters. Richard points out that this could indicate that it is not a ready flier but equally well that it is not attracted to ultra-violet light.

Another objection to the light trap is that it functions only at night. Corixidae certainly fly by day, for I have seen them doing it myself, and Fernando (1959) and Weber (1955) are among those who have put observations on record. Still, sunny but, according to Fernando, not necessarily warm days are those on which corixids take to the air. It is evident that some continue flight after dark, but not for long because specimens taken in light-traps all come in soon after sunset. There is, however, no reason to suppose that all do, and the explanation of why *Corixa punctata*, for example, though known to be a ready flier, is hardly ever taken in light-traps, may well be that it ceases to fly well before sunset. The percentage of other species taken may depend less on the number flying than on the proportion which continues to fly after sunset. If this surmise be correct, light-traps will never yield information of the type with which this chapter is concerned.

Probably Brown's technique of catching the invaders of a temporary pond is the best one for the study of migration rate, provided a suitable pond can be found. Fernando tried traps but, catching only species typical of small pieces of water, suggests that corixids from lakes and rivers may be attracted only by large sheets of water. Possibly traps submerged in such places might provide useful information, though it might be difficult to sample a sufficiently large area.

Species of Trichoptera whose larvae inhabit running water tend to fly upstream and may cover 5 km, though observations on distance have so far only been casual (Roos 1957).

Coleoptera, like Hemiptera, are aquatic as adults and their occasional flights have been studied in the same way. Fernando (1958) caught nineteen species in tanks four feet (1·3m) square and in traps whose attraction was the shining surface of a sloping piece of glass. By far the commonest was *Helophorus brevipalpis*, which seems a regular vagrant, continually on the move from one piece of temporary water to another. Balfour-Browne (1940) records captures in tubs and in a canvas tank, and it is possible to list certain species that are frequently on the wing, but observations have not been sufficiently systematic to make the calculation of a migration rate possible. Jackson (1952, 1956a and b) has found that all the specimens of some species that she has dissected have atrophied flight muscles and are incapable of leaving the water. In other species some can fly and some cannot. Accordingly she disagrees with Balfour-Browne's (1953) view that the beetle fauna of the Scottish Western Isles is maintained by flight across the sea from the mainland, and holds that some species could only have reached them via a land bridge. Her argument, however, depends on the assumption that the finding of a hundred or so flightless specimens indicates that the species is 100 per cent flightless and has been so for a long time. This assumption is difficult to accept; any student of the Hemiptera knows that *Velia caprai*, for example, is generally wingless, a condition that is immediately obvious because both wings and wing covers are absent, but that occasionally fully developed specimens are encountered (Brinkhurst 1960).

Culicidae, Order Diptera, are unlike any insect so far mentioned in that they make regular flights between water where they lay their eggs and the haunts of man or beast where they feed. They are also the only group for which there is any significant data about the distance that can be covered, this being a matter of extreme importance to those charged with the control of malaria. Eyles (1944), in a review of the subject, recounts that, in the earliest days of mosquito control, it was reckoned sufficient to take measures within a radius of half a mile of the place to be protected, and that for most species this has proved to be correct. Some, however, especially those of barren lands, will fly much further than this. Eyles quotes a figure of 4·5 km for *Anopheles sacharovi*, a serious vector in Palestine and other countries in what, up to the war, was known as the Near East, and Buxton and Leeson (1949) record that the military authorities were mistaken in assuming that *Anopheles gambiae* would not fly across three miles of open water in Freetown harbour to feed on the occupants of troopships lying at anchor. These two authors also quote some long-distance records. *Anopheles pharoensis* invaded the Allied lines shortly before the battle of Alamain and must have travelled 30 miles, since there were no breeding places nearer. They had a following wind. *Anopheles pulcherrimus*, related to *A. pharoensis* and, like it, an inhabitant of desert regions, once boarded in strength a vessel lying 15½ miles off the Persian coast. Less is known about culicines but Provost (1952) has obtained some startling results marking *Aedes taeniorhynchus* by feeding the larvae on radioactive phosphorus and catching the adults in light traps. Males were not taken more than 2 miles from the point of origin but some females were recaptured 20 miles away. Nothing is known about the significance of this flight.

In addition to the regular migration between feeding and egg-laying place, some species undertake a longer journey in the autumn in search of a place to hibernate. The best known example is *Anopheles sacharovi*, whose dispersal from the marshes in Palestine has been studied by Kligler and Mer (1930). It covers a distance of 14 km and at the end of the season brings malaria to villages that are free of the disease all through the summer.

It is likely that insects, such as mosquitoes, which make regular flights, or Ephemeroptera, which engage in nuptial dances before mating, are sometimes caught by sudden gusts of wind and carried for considerable distances. Certain chironomids that breed in rock pools where the water is fresh have a mating swarm, as is common in the lower Diptera, but others that breed in brackish rock pools mate and lay eggs before taking to the air at all. Stuart (1941) suggests that the occasional snatching up of some freshwater species is of advantage because the chances of their landing near a breeding place are reasonable, and the incident may lead to an extension of range. For brackish-water species, on the other hand, the chances would be poor, and it has been of advantage to the species to minimize the risk of being swept away by reducing flight before mating. Hardy and Milne (1937, 1938) trapped the insects carried along in the air and did take a fair number of chironomids, which indicates that these flies are liable to be caught up and blown away. The only other aquatic animals were one caddis and one *Helophorus brevipalpis*. That they did not take more is hardly to be wondered at, for their nets were very small in relation to the total volume of air and this postulated sweeping away of insects by sudden gusts of wind presumably happens only infrequently.

Though it is the crossing of land barriers that provides the theme of this chapter, this is an appropriate point at which to interpose some remarks about transport by water currents. Many marine animals depend on the transport by currents of a planktonic juvenile form to keep them well spread over the region in which conditions are suitable for them, but relatives that have colonized fresh water have suppressed the motile stage, because, it is commonly asserted, anything floating in fresh water is sooner or later carried to the sea. The only freshwater animal that still has a floating larva is the Zebra Mussel, *Dreissena polymorpha*, which is thought to be too recent an immigrant to have had time to suppress it. Weerekoon (1956) was puzzled to account for the fauna of a bank in Loch Lomond, which rose to sufficiently close to the surface to resemble a stretch of bottom in shallow water near the edge, but which was separated from the nearest piece of such a bottom by deeper water over mud. Ten species of Trichoptera occurred on the bank and near the lake's edge but not in between. The females did not fly out to lay eggs over the bank but remained near land, and the eggs were too heavy to have been washed across by currents. Larvae were never found crawling across the mud to the bank. The only positive observation was that, when eggs of *Phryganea* were hatched in a beaker, the young larvae were seen swimming like planktonic crustacea, and some evidence was found in the literature that other species do this. Weerekoon suggests that these swimming larvae were carried across to the bank by currents. Support for this suggestion has been brought forward by Mundie (1959), who, towing plankton nets during the night near the surface of a Canadian lake, caught surprisingly large numbers of animals which had previously been believed to spend their whole lives on the bottom—among the Ephemeroptera, for example, a *Caenis* and a typically flat stone-clinging ecdyonurid nymph. *Caenis* in particular is a poor swimmer and, if the nymphs have the habit of leaving the mud at night, they will certainly be transported by any currents that may be flowing. The extent to which species may be dispersed in this way from the place where the eggs were laid awaits thorough investigation. The phenomenon is of general interest as it seems to be a reversal of what has hitherto been regarded as an evolutionary trend.

These direct observations are not yet beyond the casual 'natural history' stage, and must be taken much further before they can give precise and quantitative answers to the questions that ecologists pose. It remains to see how much can be gained from the indirect approach.

Some species have a notably restricted distribution. *Mysis relicta* is believed to have been a brackish-water coastal species that was pushed inland by advancing ice in the Ice Age, and stranded in a number of basins which continued to hold water after glacial conditions had gone. It adapted itself to the gradual reduction in the salinity but did not extend its range, for nowhere is it found outside the area once covered by ice (Thienemann 1950). Within the area too it seems to have gained little ground. In Denmark it has been recorded only in Fure Lake, although there are several others similar and apparently suitable. In the English Lake District it is known only in Ennerdale Lake (Holmquist 1959). Not all of the other lakes have been searched with a thoroughness that warrants a categorical negative, always a difficult thing to obtain in ecology, but there is another survivor from the Ice Age in Ennerdale, *Limnocalanus macrurus*, which, being a planktonic copepod, is easier to catch. Plankton hauls have been made fairly often in all the other lakes but no *Limnocalanus* has been taken, and it may be asserted with fair confidence that it is confined to Ennerdale, which is some confirmation that *Mysis* is also. Presumably some accident in the Ice Age—it is not necessary to review the various theories here—brought *Mysis* to the Ennerdale basin but not to that of any other lake. In the intervening 20,000 years it has not succeeded in crossing the land barriers between it and any of the other lakes. Evidently it is a species with very poor powers of transport.

Near the ponds that Boycott had under observation was a sluggish river and a reservoir. Unknown in any of the ponds, *Planorbis corneus* occurred in both river and reservoir and *Bithynia tentaculata* occurred in the river. Boycott (1927) tried the experiment of transplanting 100 specimens of these species to each of fourteen ponds that appeared to be favourable for snails. In seven of the ponds *Planorbis* flourished, but the best that *Bithynia* could do was to survive in one pond for four years. Boycott concludes that there is something about the pond environment that is unfavourable to it, but that *P. corneus* does not occur in such places because it does not reach them.

The flora of ten rock pools that filled at spring tides and dried out a few days later was very diverse, though conditions in each appeared similar. Bourrelly (1958) believes that each pool was colonized by whichever species had survived in the mud since the last filling or had been blown in immediately after the water came, and that the differences were therefore a matter of chance and not related to any differences between the pools. The oxygen in brackish-water lagoons is used up by decomposition every so often, and Schachter (1958) notes that the composition of the plankton is frequently quite different before and after the catastrophe, though conditions appear to be unchanged. Thienemann (1948) studied the fauna in four consecutive years in a small concrete garden tank, which was drained of water each winter. He recorded 103 species altogether but only ten of these were taken in each of the four years, fourteen species were taken three times and twenty-two twice. Undoubtedly the fauna and flora of a place that has just been created, or whose population has been wiped out, depends considerably on what has chanced to invade it. Unfortunately nobody has watched such a place for a period of years to see how long it takes the fauna and flora to become stable, though work of this kind is being carried out in Russia at present. Also wanting is a series of surveys of places of the same type; if their faunas were found to be similar, it would be a strong argument that environmental conditions, not the accidents of arrival, determined the composition of the population. The reverse would be less conclusive as there would undoubtedly be a school of thought that attributed the faunistic differences to the slight environmental differences which there would certainly be. Godwin (1923) studied the plants of ponds which had been dug out of the flood plain of the River Trent to provide ballast for railways. He found that there was no strong resemblance between the communities of the various ponds, and that the older an excavation was the more species it tended to have. He attributes this to the slow rate at which aquatic plants cross land barriers. A ballast pit starts with straight sides and a flat bottom but, with the passage of time, the sides will slip in at places and the whole pond will become less uniform. A more varied flora is to be expected according to Thienemann's first law of biocoenotics (Thienemann 1950, p. 43). Moreover, as it fills up, fluctuation in water level may become more important because, the flora now consisting largely of marsh plants, conditions never remain constant long enough for one species to become dominant. This, as Hutchinson (1957b) points out, leads to richness in species. Professor Godwin, in a letter, agrees that a lengthening of the list of species could be accounted for partly in these ways but maintains that slow immigration is the most important factor.

Studies of particular groups have contributed most to an understanding of the problem under discussion. K. H. Mann (1955a) on leeches, Reynoldson (1958a, b, c) on triclads, Jewell (1935) on sponges, Hynes (1955a) on Amphipods, Pacaud (1939) on Cladocera, and Macan (1950a) on Mollusca are some of the papers on animals that cannot fly. A similar list could be compiled for insects, but they are of much less interest in the present context. Some account of the findings of these authors is given later in this book, particularly in chapters 11 and 12. The point to be made here is that all these workers are able to group the various pieces of water that they visited according to the species in them, and find that each group can be characterized by certain physical and chemical features. Anomalies, such as would be expected if occurrence was due largely to chance, are few, and the general impression is that most species soon reach places that are suitable for them.

There has been sharp disagreement over *Asellus*. Moon (1957b) collected from a large number of Lake District tarns and found *Asellus* in only a few of them. These seemed to have nothing in common except the likelihood that man had interfered with them, and accordingly Moon concludes that *Asellus* has very poor powers of getting from tarn to tarn and that it occurs where it does because man, wittingly or otherwise, has introduced it. Reynoldson's (1961a) observations were made over a wider area than Moon's, and he concluded that the concentrations of calcium and of dissolved organic matter in the water were critical. Where there was less than about 5 mg/l calcium, *Asellus* was rarely found. When the calcium content was between 7 and 12·5 mg/l, *Asellus* might or might not be present and its occurrence was inexplicably sporadic. Nearly all Moon's tarns were in this intermediate category, and Reynoldson believes that that is why he got the results he did. When there is more than 12·5 mg/l calcium, almost every piece of water, even thought it be quite isolated, contains *Asellus* and this is the experience not only of Reynoldson (fig. 5) in Britain but also of Kreuzer (1940) in Germany. It would indicate that *Asellus* crosses land barriers frequently.

General conclusions about a theme like this must be drawn cautiously,

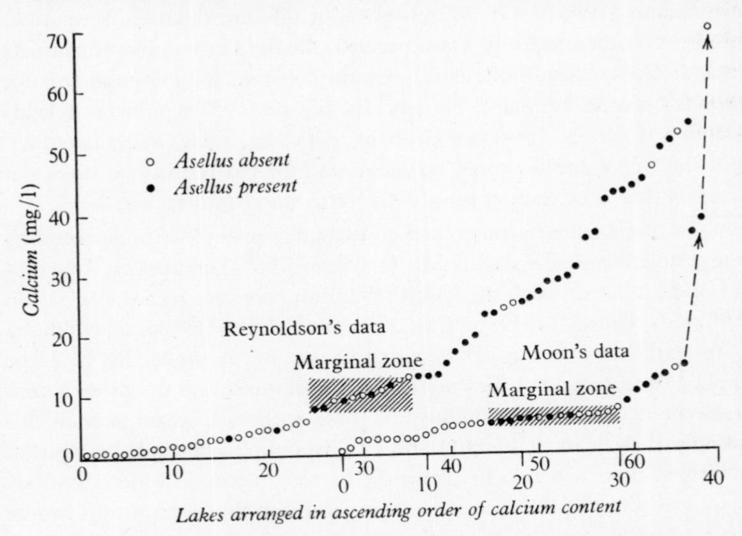

5. Occurrence of *Asellus* spp. in relation to the calcium content of the water (Reynoldson 1961) (*Verh. int. Ver. Limnol.* **14,** p. 989, fig. 1).

for every species may be different and ideally every one should be studied before pronouncements are made. This has certainly not been done, but an examination of such data as exist seems to indicate that, though some species have poor powers of transport and are absent from suitable places for this reason, most travel overland with great, and at present inexplicable, frequency, and are not long in finding places where the conditions are favourable.

SUMMARY

Small animals generally have a resting stage that can be blown about, and they are often cosmopolitan in distribution. Larger ones are generally confined to a single zoogeographical region. The occasional arrival from another region of a species which apparently finds an unoccupied niche and spreads rapidly indicates that snails, crustaceans and similar invertebrates often do surmount land barriers. Little is known of the mechanism.

It is less easy to observe the movement of indigenous species, but collections in a series of ponds over a period of years showed a frequent interchange of snails. Many freshwater animals, being insects, can fly. Facts about the powers of the various groups are presented, but they are still too incomplete to permit generalizations. Some species have poor powers of transport.

The fauna of unstable places is variable. That certain workers who have made special studies of particular groups and the habitats of each species have generally found each species wherever they have expected to, is the strongest evidence that most animals do not take long to find places that are suitable for them.

SYNONYMS

Species	Group	Other generic names	Other species names
Anopheles sacharovi Favr	Ins. Dipt. Culicidae		*elutus* Edwards
Bithynia tentaculata (Linnaeus)	Moll. Gast. Prosobranchia	*Bulimus*	
Corixa dorsalis Leach	Ins. Heteropt. Corixidae	*Sigara*	*striata* (Linnaeus) *lacustris* Macan
C. falleni (Fieber)	Ins. Heteropt. Corixidae	*Sigara, Subsigara*	
C. fossarum (Leach)	Ins. Heteropt. Corixidae	*Sigara, Subsigara*	
C. praeusta (Fieber)	Ins. Heteropt. Corixidae	*Sigara, Callicorixa*	
C. punctata Illiger	Ins. Heteropt. Corixidae	*Macrocorixa*	*geoffroyi* Leach
Crangonyx pseudogracilis Bousfield	Crust. Amphipoda	*Eucrangonyx*	*gracilis* Smith
Dreissena polymorpha (Pallas)	Moll. Bivalvia	*Dreissensia*	
Mysis relicta Lovén	Crust. Mysidacea		*oculata* (Fabricius)
Planorbis corneus (Linnaeus)	Moll. Gast. Pulmonata	*Planorbarius, Coretus*	
Potamopyrgus jenkinsi (Smith)	Moll. Gast. Prosobranchia	*Paludestrina Hydrobia*	*crystallinus carinatus* Marshall
Velia caprai Tamanini	Ins. Heteropt.		*currens* (Fabricius)

CHAPTER 5

Behaviour

The range of every species is tied up to a greater or less extent with its behaviour. Survival must obviously depend on a reaction to the inorganic and plant background that will give an animal good chances of securing food and poor chances of becoming food. For example, notwithstanding the abundance of frog- and toad-spawn in Lake District fishponds, I have seen tadpoles only in one from which all fish have been removed. In this, tadpoles are very conspicuous, swimming slowly in the open water or attached to the upper surface of the plants, a marked contrast to the other inhabitants, which remain out of sight. The tadpole has no reaction to seek safety from predators by taking cover and presumably the entire population is devoured soon after it has started to swim in tarns where fish are present.

A species must also avoid unfavourable physical and chemical conditions. When *Gammarus pulex* is exposed to a current its reactions are to swim upstream and to take refuge underneath a stone or some other shelter as soon as possible. This incidentally keeps it out of the way of its main predator, the trout, but the important effect is that it prevents it being washed away. *G. pulex* is a successful colonist of the swiftest streams. It owes its success to no morphological adaptation, for it is not a good swimmer and has no clinging organs, but to its behaviour.

Investigations of this kind have not been pursued far and only recently has the importance of behaviour been recognized by ecologists. They had forgotten the careful observations of the Victorian naturalists, and the attitude in the intervening period is well illustrated by the work of Beattie (1930) and Howland (1930). They studied six ponds which fell into two groups according to the species of mosquito found in them. One group harboured two species of *Anopheles*, *Theobaldia annulata*, and *Culex pipiens*, and the other two species of the subgenus *Culicella*, two species of *Aëdes*, and a few examples of the species in the first group. Beattie made extensive chemical analyses and Howland a detailed survey of the algal flora. Much valuable information was compiled, but why different species inhabited the two groups of ponds remained undiscovered. The authors mentioned incidentally that the ponds of the first group retained water throughout the year and that those of the second did not. They did not comment on the fact that the permanent ponds harboured species that lay eggs on the water surface and the temporary ones species that lay eggs on damp ground. This is likely to be the explanation they were seeking.

Many species require particular conditions before they will reproduce. This is the most obvious way in which behaviour limits range and therefore the one about which most has been written. The rest of the chapter is devoted to it. Ornithological work, which is frequently ahead of ecological studies of other groups, provides an illustration of a limitation of this kind. Tree pipits and meadow pipits occur together on heaths, but the tree pipit is absent if there are no tall trees. On one occasion Lack (1933) found a tree pipit's nest on a stretch of heath that was bare of trees but where there was a telegraph pole. He could see no explanation of the distribution of this species except that it avoids places where there is no high perch on which the male can alight at the end of his song.

One of the first questions that the freshwater ecologist must seek to answer is whether an observed range is due to random oviposition and mortality in unsuitable places, or to selection by an ovipositing female. This is particularly important in insects, whose adult females are winged, but also of considerable significance in some other groups. Of the insects, the Ephemeroptera have been studied particularly because they are of interest to fishermen, the Odonata because they are large and conspicuous, and the Culicidae on account of their relation to disease.

Certain species of the genus *Baetis* (Percival and Whitehead (1928), Gillies (1950), Harris (1952), and the writer do not all agree about which, but that is irrelevant to the present theme) crawl down stones or other objects projecting out of the water and lay their eggs while submerged. Harris has watched them wandering about for a considerable time, and concludes that some care is devoted to the choice of an oviposition site. The female *Habroleptoides modesta*, which is not a British species, lays her eggs at the bases of stones at the extreme edges of streams. Before any eggs are laid she appears to test the ground with her forelegs and again with the tip of the abdomen, which by now has lost its long cerci, and the whole performance suggests that few eggs are laid in places where they will not have a chance to hatch and the larvae a chance to thrive (Pleskot 1953).

Other species scatter their eggs in a much more haphazard way. Gillies (1950) has seen the bonnet of a car, parked by a river, dotted with little bundles of eggs of *Centroptilum luteolum* and I have noticed similar oviposition by *Ephemera danica*. Verrier (1956) lists a number of records of eggs being laid on wet roads and on cars, and describes her own observations on five species laying eggs on roads, sometimes even dry ones. *Ephemerella ignita* flies upstream and lays its eggs in broken water (Sawyer 1950). There was one place where Sawyer (1952) watched swarms crossing a bridge on their way upstream, but one evening after rain many were apparently confused by the wet road and turned off to fly up or down it, eventually depositing their eggs on the moist surface.

In contrast to those that go into the water or alight at the water's edge, the Ephemeroptera which fly over the water and drop their eggs, or wash them off by immersing the tip of the abdomen, appear to react to little more than a shiny surface. The pattern of distribution of their nymphs must be largely due to survival of those that hatch in suitable conditions.

Oviposition in the Odonata has been studied by Wesenberg-Lund (1913, 1943), Gardner (1950a, b, 1953), Fraser (1952), and N. W. Moore (1952). The large dragonflies (Anisoptera), except those of the family Aeshnidae, fly over the water surface and either wash the eggs off by dipping the tip of the abdomen into the water, or flick them off without touching the water, or, performing a vertical up and down flight, jab their ovipositors into patches of *Sphagnum* or algae. The small dragonflies (Zygoptera) and the Aeshnidae insert their eggs into plant tissues by means of a strong ovipositor, and some of the Zygoptera crawl down into the water to do this. Some variation in behaviour is recorded; for example, according to Wesenberg-Lund (1913), the female *Enallagma cyathigerum* generally immerses no more than the abdomen but sometimes goes right into the water. Fraser and Gardner disagree about whether the eggs of *Sympetrum striolatum* are dropped or washed off. It is, therefore, difficult to make out just how selective oviposition is, and that it may be a somewhat haphazard process is indicated by Gardner's (1950b) frequent observation of eggs of *Aeshna mixta* laid in sedges well away from the water in places where the chances of the nymphs surviving were remote.

Oviposition behaviour is certainly not without influence on the subsequent whereabouts of the nymphs. In a tarn which I have been investigating for some years, *Enallagma cyathigerum* and *Pyrrhosoma nymphula* are both numerous, both have an identical life history except that the emergence period of *Enallagma* starts later and goes on longer, and both have similar feeding habits. The two species occur together in the shallow water except in thick *Carex* where only *Pyrrhosoma* is numerous, but only *Enallagma* is found in the middle of the tarn, where *Myriophyllum* grows thickly. Pairs of *Pyrrhosoma* alight on the floating leaves of *Potamogeton natans*, on the emergent stems of *Carex* or *Eleocharis*, or on the emergent leaves of *Littorella*, and the female feels about with her abdomen under the water until she makes contact with the stem or leaf, into which she inserts her eggs. This done, the pair fly away still in tandem. Pairs of *Enallagma* fly close to the surface of the water and appear to avoid those parts of the tarn where stems of *Carex* emerge sufficiently thickly to impede flight. They alight on stems where these are not close together or on the little inflorescences of *Myriophyllum*, which project only a few millimetres above the surface. The female immediately starts to walk into the water and the male, fluttering wildly, sometimes with sufficient strength to pull the female out again, releases his hold when his wings are in danger of getting wet. The female goes down, perhaps as much as a metre into the water, and then starts to lay eggs. When she has finished she lets go, and the bubble of air between her wings floats her to the surface like a rocket. The nymphs appear to move little from the place where they hatch, and the different distribution of the two species in the tarn is brought about by the different egg-laying habits of the parents.

Nymphs of *Leucorrhinia dubia* are found only in pools on heaths and moors and Steiner (1948) has attempted to show that this is due to the reactions of the ovipositing female. Many specimens of *Leucorrhinia dubia* and *Leucorrhinia rubicunda*, but not one of about twenty other species, were attracted to a white cloth or a piece of white paper placed upon the ground, and Steiner postulates that the water surface of a heath or moorland pond presents a similar contrasting light and dark area that attracts the dragonfly. He found, moreover, that, whereas eighty specimens came to a dish containing water from a *Sphagnum* pool, only forty came to one containing tap-water, and concludes that the taste or smell of the water is attractive. These conclusions are violently attacked by Schiemenz (1954), who maintains that most species are attracted to a white surface, but that only *Leucorrhinia* and *Orthetrum* alight on one that is flat upon the ground; the others either settle on a vertical surface or choose a resting place that is above the level of the ground. He states, furthermore, that this is not an oviposition reaction. He finds it difficult to see why a heath or moorland pool offers a greater contrast of light and dark than some pools elsewhere, and points out that, though more specimens visited the bog-water than the tap-water, a considerable number did go to the latter. Why *Leucorrhinia* is confined to heaths and moors remains unexplained.

Nymphs of *Agrion* (*Calopteryx*) *virgo* and *splendens* are generally found in slowly flowing water, and a considerable contribution to understanding why has been made by Zahner (1959, 1960), who studied both nymphs and adults and took into account physiological requirements as well as behaviour.

He found it difficult to establish a critical oxygen concentration because nymphs varied so much, but he concludes that both species are rather exigent and that in still water they can generally not obtain enough to satisfy their needs at all times of year (chapter 9).

Nymphs showed no special reactions to flow or to poor oxygen conditions; in a fast current they tended to seek something to which they could cling and in low oxygen concentration they showed an increased tendency to wander. Their reactions prevent nymphs being washed away by a steady current but not by one that comes from first one direction and then the other, and they cannot maintain themselves on an exposed lake shore.

Range could therefore be limited by the requirements of the nymphs, but it probably seldom is, because the behaviour of the adults rarely lodges them in intolerable places. Several factors are involved. In the first place the adults rarely fly far, and secondly they are gregarious. They also appear to choose certain types of surroundings. Adults transferred to a strange part of the stream generally settled there, but specimens taken away and released near a pool in the evening left it next morning.

All land up to the water's edge is neutral territory and over it males neither attack rivals nor seize mates. Over the water, however, each male endeavours to establish and defend his own small territory. That of *A. splendens* generally measures roughly $2{\cdot}5 \times 1$ m but that of *A. virgo* is a little smaller, which is perhaps why it inhabits smaller streams. A territory must have a vantage point in the form of an emergent stem or leaf on which the male can rest and from which he sallies to repel intruding males, to seize visiting females, or to feed. There generally seem to be some botanical features which define the limits of the territory and some floating leaves or other substratum to which the male can take his mate to lay her eggs. It is evidently essential that the water should flow because Zahner noticed that a dead loop was treated as neutral territory, even though the pattern of vegetation was not greatly different from that in the stream. On the other hand, when he laid a long curtain in a stream so that all the flow was deflected down one side of the mid-line, territories in the half rendered stagnant were not deserted. It is difficult for the human senses to detect what attracts a dragonfly. Some stretches of stream are obviously more desirable than others because there are more territories in a given area.

The distribution of the species is evidently determined mainly by a complex adult behaviour pattern in which the selection of a territory by the male is particularly important.

The distribution of the larvae of mosquitoes is due largely, if not entirely, to selection by the ovipositing female; at least this is true of the species that have been studied. As long ago as 1922, Lamborn showed that larvae transferred to places in which they were never found would thrive. Later Muirhead-Thomson (1951), able to make more precise observations because he could identify the eggs, failed to find any in places where there were no larvae subsequently. Random oviposition and death of larvae hatching in an unsuitable medium was not taking place.

Buxton and Hopkins (1927), working in the East Indies on two species of *Aëdes*, were among the first to attempt to analyse the factors that attract species to a breeding place. *Aëdes variegatus* lays its eggs in coco-nut shells and husks, in holes in trees and sometimes in crab-holes, but not in ponds or rivers or the axils of certain plants, places where other species are found. *Aëdes aegypti* breeds particularly in artificial containers in the neighbourhood of houses. Buxton and Hopkins placed artificial containers of various sizes containing various infusions in various places where wild females would visit them, and counted the eggs in each in the morning. Few eggs were found after a night of strong wind or heavy rain. There were more in pots in dark and in humid places than in pots in other situations. More eggs were laid in pots 15 cm in diameter than in pots 10 or 30 cm across, and the temperature of the water also influenced the number deposited. More were found on distilled water plus organic matter of various kinds than on distilled water alone, and infusions of grass remained attractive for sixteen weeks. Few eggs were laid in a pot from which ammonia was given off. Evidently the mosquitoes flitted from pot to pot laying some eggs in each, but more where conditions were attractive. A landing place appeared to make a pot more attractive to *A. variegatus* but not to *A. aegypti*. Buxton and Hopkins conclude that a female is guided to a breeding place by tropisms towards darkness, possibly shelter from wind, water vapour, and probably other gases. Physical characteristics such as the size of the breeding place are important too. When water has been found, whether or not eggs are laid is determined by the nature of the surface at the water's edge as well as by the chemical constitution of the fluid and perhaps by the presence or absence of certain micro-organisms. The intensity of these factors also influences the number of eggs laid.

Dunn (1927a), in similar experimental observations on *Aëdes aegypti* in Africa, found that more eggs were laid in sections of bamboo than in tin cans; in an infusion of mango leaves than in tap-water; in the sun than in the shade; and among vegetation than under houses. The total number of eggs counted was not very high, and the exact relation of oviposition to exposure to sun and the nature of the vegetation was far from clear.

The most thorough investigation of any species is that of Muirhead Thomson (1940, 1941, see also his book, Muirhead-Thomson 1951) on *Anopheles minimus* in Assam. This species is hardly ever found in rice-fields or in pools, which, shallow and stagnant, reach during the day a temperature which is lethal to the larvae. Ecological work sometimes stops when such an obvious correlation comes to light, but Muirhead-Thomson pointed out that, since *A. minimus* lays during the night, when temperature of both running and still water is about the same and well below the lethal level, this factor cannot influence the choice of a site.

The larvae occur at the edges of streams and are often eradicated by the planting along their breeding places of a bush that casts a particularly dense shade. That the ovipositing female seeks running water and avoids dense shade was the obvious conclusion to leap to when Muirhead-Thomson started work; he had soon found that in fact she avoids running water and seeks shade. The reason for the disappearance of mosquitoes when the bushes are planted is the shading out of the marginal and over-hanging vegetation which forms, at the edge of the stream, the stagnant places where the female lays her eggs. There still remained, when these preliminary pitfalls had been charted, the problem of why the females went to the streams and not to the ricefields and pools. Muirhead-Thomson investigated the organic matter in the water by measuring the amount of oxygen absorbed from permanganate and the concentration of albuminoid ammonia, and found that, in general, there was three to four times as much in the stagnant as in the running water. He concludes that the distribution of *A. minimus* is related to this gross quantitative difference but brings forward some slight evidence that there is a qualitative relation as well and that particular substances, at present unisolated, may attract or repel female mosquitoes.

Evidently a number of factors combine to make a water attractive to any particular species of mosquito, and some workers have had recourse to laboratory experiments in an attempt to analyse them, confining gravid females in cages at the bottom of which were containers offering different conditions. This has excited controversy. Bates (1949) doubts whether any female mosquito who has discovered that she is confined within a cage reacts in the same way as a free one, and Buxton and Hopkins (1927) take a similarly uncompromising view about laboratory experiments. Kligler and Theodor (1925) found that, when three species of *Anopheles* had four dishes from which to choose, a certain number of each laid eggs in those containing a salt solution of sufficient strength to kill the eggs or any larvae that hatched. On the other hand Muirhead-Thomson (1951) is prepared to accord cautious recognition to some of the results obtained in this way. Bates (1940) found that *Anopheles maculipennis atroparvus* laid more eggs in dishes on a dark background than in dishes set on something light, and Jobling (1935) made a similar observation on species of *Culex*. *Anopheles minimus* tends to avoid water warmer than 35°C (Muirhead-Thomson 1951). *Anopheles maculipennis* never laid eggs in water above about 33°C, rarely in water below 20°C, and deposited most in dishes in which the temperature ranged from 22-29°C. In contrast, no eggs of *A. claviger* were found above 25°C, and most were in a temperature of 10-20°C (Hecht 1931). Later (1934) he showed a slight difference in the temperature at which most eggs of two races of *Anopheles maculipennis* were laid. *A. claviger* is a common and widely distributed species in northern Europe, but is confined to springs, wells, tanks and other cool places along the southern margin of its range in the Near East.

O'Gower (1955) offered *Aëdes scutellaris* cones of filter paper immersed point downwards in beakers in such a way that about half the cone was full of water. The paper was white, grey or black in colour and smooth or rough in texture. The mosquitoes were also offered a moist surface as an alternative to a water surface. In all the experiments few eggs were laid on a white background. More were laid on black paper than on grey when both were smooth, the percentages being respectively 59 and 37 (the missing 4 per cent was on white paper), and more on rough than on smooth when both were black (60 : 40). More eggs were laid on the surface of water in a petri dish than on a piece of filter paper over soaked cotton wool in another petri dish, but again the preference was not outstanding, the percentages being 54 : 46. The colour was white. However, when the alternatives were water surface on a black background and a grey smooth moist surface, 83 per cent of eggs were laid on the water. The proportions were almost exactly reversed (19 : 81) when the moist surface was rough instead of smooth. Evidently the various factors interact and summate. This work brings to light the difficulty of interpreting the results of experiments, those on temperature for example, where not all the components attractive to mosquitoes were taken into account.

O'Gower (1958) tested other species in the same way and obtained the results set out in table 9. *Aëdes aegypti* is similar to *A. scutellaris* but has a more marked preference for a water surface. Nearly every *A. australis* in contrast, goes to a moist surface and most laid eggs on a rough rather than a smooth surface.

Table 9

Percentage of eggs laid on each when a choice of two offered (O'Gower, 1958)

	A. aegypti	*A. australis*	*A. scutellaris*
Rough surface: smooth	59 : 41	80 : 20	60 : 40
Black: grey	73 : 27	68 : 32	70 : 30
Water surface: moist surface	75 : 25	2 : 98	54 : 46
Fresh: saline	74 : 26	67 : 33	Not tested

Experiments designed to study oviposition in dishes differing chemically have given what seem to be less illuminating results than those with dishes differing in their physical conditions. For example, Jobling (1935) found that *Culex fatigans* laid 106 egg rafts on a hay infusion and forty-five on tap-water, which indicates less discrimination than a free female appears to exercise. The same comment may be made of the experiments of Hudson (1956), who offered *Culex molestus* and *Aëdes aegypti* nine dishes each containing 1 g/l NaCl more than the one before it, the first being distilled water. Eggs were found in all, though most were in the five lowest concentrations. To these laboratory observations may be added a few more made in the field but less comprehensive in scope than those described earlier. Occurrence of larvae in shady places does not always mean that the female is attracted by shade. *Anopheles leucosphyrus* is a forest species, but Colless (1956) showed that, when the trees and bushes round a pool were removed, oviposition in it continued undiminished for some time. He concludes that females ultimately ceased to go to the place because it had become less attractive chemically as a result of changes following the clearing, not because the physical conditions had altered.

There may well be subtle physical differences that are of importance to the mosquito but appreciated only dimly if at all by the human observer. Buxton and Hopkins' discovery that the size of the pot made a difference to the number of eggs laid is one instance, and another is Dunn's (1927b) observation that, whereas *Culex nebulosa* frequently breeds inside houses, *Aëdes aegypti* rarely does so. Much easier to comprehend is the reason why larvae of *Anopheles culicifacies* disappear from ricefields when the rice plants reach a height of about 30 cm. The female carries out a hovering flight above the place selected for oviposition, and drops her eggs without touching the water. Vegetation above a certain height interferes with this flight and the females go elsewhere. Egg-laying can be brought to an end by a forest of glass rods (Russell and Rao 1942a). A freshly dug pit is always attractive to *Anopheles culicifacies* but the initial attractiveness diminishes, possibly as plankton develops and organic matter is formed. There is a more definite correlation between absence of attractiveness to mosquitoes and presence of blue-green algae (Russell and Rao 1942b). *Aëdes aegypti* larvae appear at a certain stage in the history of a peptone solution (Margalef 1947). 'Nothing could be more striking than the difference between a wooden and a steel water butt . . . the one will contain *Stegomyia albopicta* and the other *S. sugens*' (Senior-White 1926). This could be due to one of the subtle physical differences mentioned earlier, but it is more likely to be due to a difference in the smell emanating from the two barrels. In Britain larvae of three species of mosquito occur in rain-filled holes in trees and nowhere else, and it is difficult to see how this comes about unless the female is attracted to something, presumably an organic compound, formed where water accumulates in a rotten tree. Probably other species are restricted to a particular habitat because some peculiar odour attracts the female. Such a conclusion has been adumbrated in several of the investigations just described. It is difficult to summarize them because there were so many different species and so many different approaches. Physical factors of various kinds are undoubtedly important, and the influence of any one and of the sum of all may vary greatly according to the species. It is difficult, however, to envisage enough physical variables to account for the restriction of so many species to a particular kind of place, and I believe that our understanding of this restriction will not advance much until the dissolved organic compounds of the various habitats have been isolated.

The adult must feed, mate and rest somewhere in the intervals between laying eggs and, if the requisite facilities are not available near by, larvae will not occur in the most suitable breeding place. The males of each species appear to take up a characteristic position in relation to some object in order to perform the dance which, in some, though not all species, is an essential preliminary to mating, but whether the requirements are sufficiently exacting to limit the range of any species, whether, in other words, there is anything comparable to the pipit and the telegraph pole, is not known.

Oviposition by an animal that is aquatic throughout life is not so easy to watch and less is known about what influence it may have on occurrence.

Notonecta maculata invariably attaches its eggs to something solid, a concrete side, a rock or a stone if one can be found, a branch if it cannot. *N. glauca* inserts nearly all its eggs into plant tissues, but the other two British species both glue their eggs to hard surfaces and insert them into plant tissues. Walton (1936) writes that *N. maculata* is most numerous in tanks or reservoirs built of cement or stonework, and believes that this is correlated with the egg-laying habits.

Salmo trutta congregate off the mouths of streams in the spawning season but do not ascend until the inflowing water is colder than that in the lake. Should this not happen, spawning takes place in the lake, if the females can find beds of gravel; the eggs are resorbed if they cannot (Stuart 1953). Munro and Balmain (1956) did not observe any relation between upstream movement and temperature, but found that the fish entered the streams when the flow increased. Fabricius (1950) observed that *Esox lucius* spawned in a Swedish lake in May, when the melting of the snow in the mountains was bringing in a lot of water. They entered one sheltered bay and spawned in *Equisetum* as soon as it was flooded, the temperature being 8°C, but, in another reed-bed where inflow water kept the temperature at 3°C, they did not spawn till a week had passed. Vegetation in shallow water provides the main stimulus for oviposition, and fish resorbed their eggs when prevented from entering the usual spawning grounds by a lowering of the level of a lake. An increase in temperature provides some stimulus to oviposition, but failure of the temperature to rise only delays and does not prevent it. Chimits (1956), on the other hand, relates spawning to a temperature of 7°C and states that, if it is not reached at the right time of year, females retain their eggs and die. He does not seem to have strong evidence that low temperature had killed the dead gravid females which he found.

Owners of fishponds in America long thought that the buffalo fish would be a desirable addition to their stock, but, although it survived when introduced, it never bred. Eventually it was induced to do so after the sequence of events in the Mississippi, where the species occurs wild, had been observed. A farmer kept the water in a pond low until the buffalo fish were ready to spawn and then raised the level and flooded the marginal land. The fish immediately moved on to the newly inundated areas and made their nests. Another species would not breed in fishponds until it was discovered that it required a tunnel of some sort for spawning, after which good reproduction was obtained in ponds into which lengths of drain-pipe had been thrown (Coker 1954). Another American fish, *Lepomis microchirus*, will not apparently spawn until a high temperature is reached.

though Schultz (1952) and Anderson (1948) do not agree, maintaining respectively that the critical level is 22°C and 26·7°C.

Savage (1939) measured the amount of calcium, magnesium, sodium, potassium, sulphate, and carbonate in a number of ponds in which frogs spawned, and found more frogs than would be expected if chance governed their distribution in ponds with more than the average amount of potassium. There were fewer in ponds with less than 30 p.p.m. carbonate. There is a suggestion, too, that concrete ponds, slightly polluted ponds and ponds with a high pH were more attractive than others. However, the frogs were not highly selective and sometimes their egg-laying seems very haphazard, for, in a wet spring in the Lake District, ruts that will be dry long before development can be completed are regularly used for spawning.

SUMMARY

Some Ephemeroptera appear to select a site to lay their eggs, but others lay them in a haphazard way and many nymphs must die in unsuitable places. The oviposition of Odonata is probably more selective. The occurrence of the nymphs of *Agrion* probably depends less on their own exigent requirements than on the behaviour of the adults, particularly the male who establishes and defends a territory. The oviposition of Culicidae is a highly selective process and the adult female is attracted by a number of factors that differ according to the species. Position in relation to vegetation and buildings, shape, size and temperature influence the choice of a site and it seems likely that particular odours attract certain species. Several species of fish will lay their eggs only under certain conditions. Behaviour at other times must ensure that an animal is well placed to find food and to avoid predators. It also leads some species to forsake certain places when the human observer cannot detect any survival value in the act.

SYNONYMS

Species	Group	Other generic names	Other species names
Aëdes aegypti (Linnaeus)	Ins. Dipt. Culicidae	*Stegomyia*	*fasciata* Meigen *calopus* Meigen *argenteus* Poiret
Agrion splendens (Harris)	Ins. Odonata	*Calopteryx*	
A. virgo (Linnaeus)	,, ,,	*Calopteryx*	
Anopheles claviger (Meigen)	Ins. Dipt. Culicidae		*bifurcatus* Linnaeus

Species	Group	Other generic names	Other species names
Lepomis microchirus	Pisces		*macrochirus* Rafinesque
Salmo trutta Linnaeus	,,	*Trutta*	*lacustris* Linnaeus *fario* Linnaeus
Stegomyia albopicta Skuse	Ins. Dipt. Culicidae	*Aëdes*	
S. sugens Theobald	Ins. Dipt. Culicidae		*vittata* (Bigot)

CHAPTER 6

Interrelationships

An animal may be absent from a piece of water because its behaviour pattern does not keep it out of the way of predators efficiently, as we have just seen. It occasionally happens that an animal becomes so selective that its range is restricted to that of the one or two species on which it feeds. Concealment is not the only defence against predators, and some organisms repel them or poison them. Such methods are used also against what may be a more dangerous enemy, the closely related species that exploits the same resources in the environment. Other methods are employed as well and how a victorious species eliminates its rival is sometimes quite easy to see but sometimes rather mysterious. A good deal has been written recently about the terms to be applied to these relationships, but here I have employed some of them in a rather old-fashioned way. I have used 'predation' when one species reduces the numbers of another by eating it. 'Poisons' seems a reasonably safe non-committal term for the sort of chemical warfare which goes on between some organisms, and 'competition' for the relationship in which one overcomes another without having recourse to either of these measures. Some authors include all three under the head of 'competition'. Few schemes are so perfect that a category entitled 'miscellaneous' can be avoided. The chapter concludes with a discussion of how numbers are controlled.

PREDATION

Kitching, Sloane, and Ebling (1959) describe a good example of predation in the sea. Mussels are common between tide-marks but do not occur below low-water level. They are also absent from sheltered shores except where there are steep rock faces or ledges. When pieces of rock covered with mussels were broken off and taken to places where there were no mussels, predators soon devoured them. The authors suggest that the mussels are confined to exposed shores between tide-marks because their predators are unable to tolerate the pounding of the waves, and to vertical faces and ledges in sheltered places because their predators cannot climb up to them. They admit, however, that, notwithstanding their practical demonstration that adult mussels are eaten when transferred to places where they are not normally found, predation on the young stages could have a greater effect on range than predation on the adults.

Two examples from fresh water have been given in preceding chapters. Tadpoles disappear quickly from Lake District tarns except where there are no trout; and, when a change in a small stream favoured *Polycelis felina*, there was no obvious explanation of the disappearance of some other species except that it had eaten them. Larvae of *Anopheles* are generally absent from tarns but I once found some during a dry summer, when the level of a tarn had fallen sufficiently low to bring the top of a tuft of weed to the surface. It formed what I assumed to be an area of cover without which the larvae would soon have been found by predators. Larvae of *Anopheles* lie at the surface with the breathing tubes in contact with the air, turn their heads through 180°, and then sweep the surface of the water for the minute particles that are their food. With a tuft of weed or a floating mass of alga beneath them, they can pursue this mode of life successfully, but without they are highly vulnerable. In Bengal, during the war, anti-malaria units found that all they need do to protect the troops against malaria, which was carried by mosquitoes breeding in the large ponds numerous in every village, was to remove vegetation breaking the surface; larvae disappeared, and presumably fell a prey to fish and other inhabitants of the pond.

Croghan (1958a), Boone and Baas-Becking (1931) and Hedgpeth (1957) state that sea water is a highly suitable medium for *Artemia*, the brine shrimp, but suggest that it does not occur in the sea because the perpetual swimming in the open water whereby it obtains food makes it too easy for carnivores to see and catch. It is found in small ponds and lagoons, often in highly concentrated water that few other animals can tolerate. Other phyllopods inhabit temporary pools, no doubt for the same reason (Lundblad 1921).

The chironomid *Eucricotopus brevipalpis* feeds only on *Potamogeton natans* (Thienemann 1950), and the range of certain Cladocera is restricted partly by food requirements. This is one of the few groups in which an attempt has been made to evaluate all the limiting factors (Pacaud 1939), and it is preferable to discuss these together at this point rather than to split them up among the various chapters. *Daphnia pulex* and *Moina brachiata* are typical of very polluted ponds, in which abound algae without a hard cuticle, a type of food on which they thrive. They are also able to flourish under these conditions, because, unlike most other species, they can tolerate a low oxygen concentration and their feeding limbs are not readily clogged by the innumerable bacteria in such places. *Moina* is associated particularly with duckponds, and Pacaud thinks that they are favoured by the clay which the activities of the birds keep in suspension, though he did not discover how. *Ceriodaphnia rotunda* is closely associated with ponds covered by *Lemna*. The reason is not wholly clear but that this species is the most tolerant of low oxygen concentration and is one of the few that does well on a diet of vegetable detritus are believed to be important. *Leydigia leydigii* and *Macrothix laticornis* are two other species that can subsist on particulate matter derived from dead leaves. So can *Scapholebris mucronata* but it crawls on the underside of the surface film and requires calm water. *Simocephalus vetulus* can make use of algae with a hard cuticle more efficiently than any other species. Insensitivity to toxic substances produced during decomposition in polluted ponds is also suggested by Pacaud as a reason for the success of species that abound in such places. In a series of experiments carried out later, Pacaud (1949) showed that bacteria of a particular kind were important to certain Cladocera. *Daphnia obtusa* and *Simocephalus exspinosus* were cultured in infusions of lettuce and other leaves. Alone they did not survive long, but in the presence of certain snails—*Planorbis corneus* was found to be the most convenient for the purposes of the experiment—a population could be kept going for several months. A population of *D. obtusa* and *P. corneus* was kept for eight months in a vessel to which nothing but cigarette paper was added, but, if there were no snails, the *Daphnia* were all dead in three to seventeen days. Pacaud suggests that the Cladocera are able to survive on the bacteria that decompose the snail's faeces but not on those that decompose the vegetable tissue.

Specific food requirements appear to be important in the ecology of ciliates. Both Noland (1925) and Fauré-Fremiet (1950), finding that a number of species were tolerant of a wide range of physical and chemical conditions, concluded that the nature and amount of available food had more to do with their distribution than any other single factor. Fauré-Fremiet describes how sulphur bacteria grow on rich black mud in the presence of anaerobic bacteria that form H_2S. The metabolism of the bacteria reduces the pH and this attracts *Colpidium campylum* and *C. colpoda*, which feed on them without reducing their numbers too much. Then predators arrive and reduce the numbers both of *Colpidium* and of each other. Next come species which eat up the bacteria completely, and finally everything left is devoured by *Coleps hirtus*.

A hungry *Stentor* will feed only sparingly on *Gonium pectorale* and certain diatoms, and a full one will not take them at all, whereas ciliates are taken avidly at any time (Hetherington 1932).

There has never been collaboration between somebody making observations like those of Fauré-Fremiet and somebody investigating the occurrence of larger animals, though the result might be very fruitful. Many animals that are quite large when adult but tiny when they first emerge from the egg, are stated to require Protozoa when they start feeding, but there is a dearth of precise observations. Einsele (1961), however, recounts instructive experiences of the rearing of fish in a hatchery. The fry of *Salmo trutta* will take all manner of food including chopped liver, meal, and other fare which certainly never comes their way in the wild. Young *Hucho hucho*, in contrast, are very selective and starve in the midst of a plentiful supply of plankton and fry of *Coregonus*. Their natural food when very small is the fry of *Abramis vimba*. Einsele states definitely that *Leucioperca sandra* is absent from certain places because it cannot find what it requires during the critical period immediately after it has started feeding.

Cladocera are absent from the open water of Lake Tanganyika and representatives of another characteristically freshwater group, the Rotifera, are scarce. Hutchinson (1932) suggests that, on account of the great size, plankton production in Lake Tanganyika is more like that of the sea than of fresh water. It may be deficient in fine organic debris, bacteria or the finest nannoplankton in comparison with smaller lakes, and that is the reason for the absence or scarcity in it of animals so abundant in others.

Nearly all the examples of restriction of range due to restricted feeding habits have been animals that subsist on small particles. Most other animals, whether they are scavengers, herbivores or carnivores will eat anything they can tackle, and their range is unlikely to be related to that of a particular kind of food.

Further work on what is eaten by the smallest specimens is, however, required before a full assessment of the relation between range and food can be made.

POISONS

Lefèvre, Hedwig, and Nisbet (1951) found in a canal great numbers of the alga, *Aphanizomenon gracile*, and little else. They took water from the canal and also water from springs feeding the canal, filtered it and introduced various other algae. Many flourished in the spring water but would not grow in the canal water, owing, the authors presume, to something toxic secreted by the *Aphanizomenon*. Lefèvre, Jakob, and Nisbet (1952) suggest that this is a widespread method whereby species deal with competitors for the same nutrients. Eventually the dominant species inhibits itself with its own products, and there is a rush by other species, one of which gains the lead and overcomes its rivals by means of its toxic secretion. The authors put forward the idea that the outbursts and rapid decline of species of algae in a lake are not due to their being in a suitable medium of nutrient salts which they use up, the commonly accepted explanation, but to the elimination of competitors and eventually themselves by something they release into the water. Lund's (1950) series of observations on Windermere support the first explanation, though it is probable that toxins do play some part in the sequence of events.

Prymnesium parvum is a small alga which, when abundant, kills fish, and it is a pest of some importance in carp-ponds in Israel (Reich and Aschner 1947). Teleologically there seems no reason why it should kill fish, as it is too small to be eaten by them, and the purpose of the toxin, if it has a purpose and is not just a waste product, is more likely to be the neutralizing of algal competitors. There are, however, other species, chiefly among the Cyanophyceae, which are poisonous to animals that might eat them (Shelubsky 1951, Gajevskaja 1958). *Chlorella*, an alga much used by botanists doing culture experiments, produces a bactericide (Pratt *et al.* 1944) and it, or another substance, slows the rate at which *Daphnia* filters water to obtain food (Ryther 1954).

The phenomenon is not unknown in the animal kingdom. There are several races of *Paramecium aurelia* which do not generally interbreed, and Sonneborn (1939) has demonstrated that, in the presence of three of them, others cease feeding and die, apparently killed by some substance secreted into the water. Fish are killed rapidly by a toxin produced by *Limnaea pereger*, but, as the minimum concentration is about twenty snails per litre, a density unlikely to be reached in nature, the significance of this toxin is obscure. It acts on the gills, not on the gut, and fish take no harm from gorging on *L. pereger* (Wundsch 1930).

COMPETITION

Although this book is about freshwater organisms, it would be an inexcusably restrictive practice not to include particularly striking illustrations from other media. No discussion of competition would be complete without a reference to the work of Park (1954) on flour beetles. He used *Tribolium confusum* and *T. castaneum*, both of which spend their entire lives in flour. In one experiment he placed four pairs in each of a number of vials containing 8 gm of flour, counted eggs, larvae and adults at the end of thirty days, and reinstalled them in their vial with a fresh lot of flour. After an initial rapid rate of reproduction, the beetles reached a point of equilibrium when reproduction balanced death and the populations appeared to be viable for ever. In fact the experiment was kept running for 780 days. Batches of vials were kept at three different temperatures, 24°, 29°, and 34°C, referred to as cold, temperate, and hot, and two relative humidities, 30 per cent and 70 per cent, referred to as dry and wet. There were therefore, six climates, hot and wet (HW), hot and dry (HD), temperate and wet (TW) and so on. First the species were kept apart, and the numbers of adults and larvae per gram of flour after equilibrium had been reached were:

	confusum	*castaneum*
HW	41	38
HD	23	9
TW	32	50
TD	29	18
CW	28	45
CD	26	3

T. castaneum eventually died out in the cold dry climate, and always produced a smaller population than *T. confusum* when conditions were dry. It produced more when they were wet, except when they were hot as well, in which case numbers were about even.

In the second series both species were introduced into each vial. There were only five climates because *T. castaneum* died out in the cold dry one in any case. The important result of the experiment was that one species always died out. In a hot wet climate the survivor was always *T. castaneum*. It was the more numerous at the end of thirty days, probably on account of a more rapid rate of growth at the highest temperature, but *T. confusum* was never as numerous as in the control cultures and there was evidently some other factor in operation as well. It might have been expected to be the survivor in the other two wet climates, since its equilibrium population in them was markedly the larger when it was alone, but in fact it was in only 86 per cent of the vials in the temperate climate and in 29 per cent in the cold climate. Similarly *T. confusum* was the survivor in only 90 per cent and 87 per cent respectively of the vials in the hot and the temperate dry climate. In the hot dry conditions, whichever species was destined to die out was the less numerous at the end of thirty days, but in the temperate wet climate *T. confusum* did not establish a lead, on the few occasions when it was on its way to victory, till eight months had elapsed.

In a further series of experiments (Park 1957) different proportions of the two species were introduced at the start, and the vials were kept in the hot wet climate. Three hundred and fifty eggs were placed in each vial and the percentage of one of the species was 10, 30, 50, 70, or 90. In addition there were two experiments when proportions were 50 : 50, but one species was given a six-day start. *T. castaneum* always won, except when it was but 10 per cent of the starting population and even then it came out victor nine times in fourteen tests. Larvae and adults eat eggs and pupae, apparently without regard to their species; therefore the slightly faster growth of *T. castaneum* should give it an advantage if other things are equal. It is, however, still far from clear how it manages to win when heavily outnumbered at the start.

Competition is probably of widespread importance in ecology. Sometimes its mode of operation cannot be explained as in the experiments just described, but sometimes there is an explanation that appears plausible. It is not an easy phenomenon to detect, for how can it be proved, if two species are neighbours, that either would extend its range if the other were not there? The easiest way is to seek similar conditions in a place which only one of the species has reached, for example on an island. The British Isles serve very well. They were repopulated to a large extent after the Ice Age from the mainland of Europe, but they became detached from it before every species had reached them. Some species had got no further than England before Ireland and other islands to the west were isolated by the sea. In England, Wales, and Scotland the Plecopteran, *Diura bicaudata*, occurs on stony lake shores and higher up in streams, but not in the lower reaches of streams, where the large stonefly nymph is *Perlodes microcephala* (Hynes 1941). *Perlodes* is not found in the Isle of Man, and in Manx streams *Diura bicaudata* extends from source to mouth. Its absence from the lower reaches elsewhere would seem to be due to competition (Hynes 1952).

The amphipod, *Gammarus duebeni* occurs in brackish water all round the coasts of northern Europe. It occurs extensively in fresh water all over Ireland, in some streams in the Isle of Man, and in a few streams near the Lizard. *G. pulex* is the common freshwater species in most of the British Isles but it has been introduced into Ireland only during the last few decades. It probably entered England comparatively recently, speaking in terms of zoogeography, not the span of human life. As it spread across the country it drove *G. duebeni* out, and this species now has on the mainland only one small foothold near the Lizard. This area is almost a peninsula owing to the deep incision of the Helford River, and *G. pulex* has not yet reached it. In the Isle of Man the process of replacement is still in progress. One stream with two branches contains both species. Below the fork the two occur together, but, above, each branch is occupied by one species only. Near the mouth of the one in which *G. duebeni* occurs there is a fall, erected in the twelfth century, and this, by preventing *G. pulex* from travelling any further, has protected *G. duebeni* (Hynes 1954). Observations on the two species in years to come should show whether this hypothesis about their relations is correct.

The success of one in competition with the other is easy to understand, for *G. pulex* produces more young during the course of a year (Hynes 1955b), and it can be shown mathematically (Hardin 1960) that, if the populations of two species with a similar mode of life are kept down by indiscriminate predation, whichever reproduces less rapidly will eventually disappear.

Most suitable streams in the Balkans harbour two species of flatworm, *Planaria montenegrina* and *P. gonocephala*, but Beauchamp and Ullyott (1932) had the good fortune to find some in which either one or the other occurred alone. When *P. montenegrina* was the only species, it extended from the spring to a point where the temperature was 16-17°C. *P. gonocephala*, in the same circumstances, extended from the spring to a point where the temperature was 21-23°C. When the two occurred together, *P. montenegrina* occupied the upper reaches, *P. gonocephala* was found lower down, and the zone of overlap was narrow and at the point where the temperature was 13-14°C. Thus one species was kept out of warm, the other out of cold, water in which it could live, apparently by competition with the other. This is probably a widespread phenomenon, and few species extend to the point where some physical or chemical factor would limit them, because as they pass towards the edge of the zone of tolerance some other species with a different optimum prevents further spread. The species therefore never gets a chance to adapt to a higher or lower temperature, or whatever it may be, a point that is enlarged on in the chapter devoted to temperature.

In the rest of Europe there are three species occupying stretches of a stream: *Planaria* (*Crenobia*) *alpina* at the top, *P. gonocephala* at the bottom and *Polycelis felina* in between. Temperature again seems to be the determining factor and *P. felina* extends from about 13° to 17°C. Beauchamp and Ullyott remark that the various workers do not agree on the exact temperature at the boundaries, which does not surprise me at all, because in a small stream there may be a big range during a day, the pattern of change is different in different parts of the stream and the various workers have relied on temperature measurements taken whenever they happen to be collecting. *P. gonocephala* has not reached Britain and neither it nor *P. felina* Scandinavia, from which deductions about stream temperatures at the time when those two areas were cut off from the mainland have been made. In Britain Beauchamp and Ullyott found that *P. alpina* occupied the upper reaches alone down to a temperature of about 13°C, *P. felina* those lower down in stony streams. If the current was less swift, both species might occur together up to the spring, or *P. felina* might be the only species. Leloup (1944) was unable to find, in an area in Belgium, a distribution of stream triclads similar to that described by other workers.

Competition in the genus *Salmo* is also easy to understand; the fish show aggressive behaviour from an early age. Kalleberg (1958) has watched this in a large tank 85 cm wide and 950 cm long with a glass front. Fry of both *S. trutta* and *S. salar* establish territories soon after starting to feed, and each consists of a central refuge, from which sallies are made to catch food, and an area in which the occupier will fly at any invader with a zest that diminishes the nearer it is to the periphery. When 250 fry of *S. trutta* were placed in this tank, all obtained territories, but when more were put in some failed to establish a territory and these spent their time either hiding in some crack or crevice or being chivvied from the territories of successful companions. They were soon obviously smaller than the others. The territory of a month-old fish was 2-3 sq. dm in extent but, as the fish grew bigger, it also demanded more ground, and the territory of a fish 23 cm long was about 4 sq m. The size of the territory varies according to how far the fish can see, and there are more fish per unit area of an irregular bottom than of a regular one. The fish feed on animals washed down in the current and do little to seek out the fauna actually in their territory. All the phenomena observed took place in a current of 18-25 cm/sec. When it was below 10 cm/sec, the fish left their territories, became gregarious, and tended to move upstream.

Salmon rely on stylized threat postures to drive interlopers away, trout more on actual attack, and for this reason as well as faster growth, the latter are the more successful when the two are in competition.

Nilsson (1955) found two lakes in both of which *Salvelinus alpinus* and *Salmo trutta* occurred and a third where there were char only. When both occurred together, char were confined to deeper water, where they fed on bottom fauna in winter, on airborne insects in June, and then on entomostraca, whereas trout in the shallow water fed on bottom fauna all the year round and on airborne insects in late summer. Both species appeared to be feeding on whatever was most easily available in their respective habitats. In the lake with no trout, char occupied the shallow as well as the deep water, and fed on much the same food as the trout in the other lakes. They grew large on this diet, and in shallow water these big fish appeared to belong to a race distinct from small ones behaving as all char did when trout were present.

Vivier (1955) and Dymond (1955) record observations on introductions, the first of Canadian fish into Europe, the second the opposite. Numerous attempts to establish *Salvelinus fontinalis* and *Salmo gairdneri* in France have been unsuccessful except in a few places of which the salient features have been isolation and the absence of *Salmo trutta*. Competition with the indigenous species seems the main bar to successful introduction, but ancillary factors are the great voracity of the American species which makes them easy for the angler to catch, and their lack of resistance to certain diseases. Transported across the Atlantic to play away against *Salvelinus fontinalis*, *S. trutta* is a little more successful, as its slightly higher thermal death point has enabled it to establish itself in some places.

A final example of competition between species, another of those where the mechanism of exclusion has not been discovered, I include on the principle that anything liquid that is not sea water or sufficiently like sea water to be inhabited by at least a few marine species or their near relatives is likely to remain neglected if the freshwater biologists do not claim it. Three species of *Plasmodium*, *falciparum*, *vivax* and *malariae*, transmitted from man to man by *Anopheles*, attack the red corpuscles and cause three clinically distinguishable types of malaria. *P. falciparum* is the most severe but this parasite has the poorest powers of survival and is generally found only where transmission is intense. *P. vivax* can survive longer dormant in the human body and *P. malariae* much longer still, and these are found in populations bitten less often by infected mosquitoes. At the height of an epidemic *P. falciparum* is the preponderant species, even though all three are being transmitted, *P. vivax* is encountered more often as intensity of transmission dies down and *P. malariae* is often not found at all until transmission has ceased and all clinical symptoms are due to relapses (Boyd 1949).

Competition has been one of the main talking-points in ecology since the war, particularly in connexion with what has come to be known in English-speaking countries as Gause's hypothesis. 'How curious it is', writes Hardin (1960), 'that the principle should be named after a man who did not state it clearly, who misapprehended its relation to theory, and who acknowledged the priority of others.' The others were Lotka and Volterra, two mathematicians who have theorized about interrelationships. Who was the first actually to formulate the principle and father it on to Gause is not clear. An anonymous reporter of the British Ecological Society's Easter meeting in 1944 wrote in the *Journal of Animal Ecology* the same year (vol. 13, 176): 'The symposium centred about Gause's contention (1934) that two species with similar ecology cannot live together in the same place. . . .' Hardin thinks that Lack was mainly responsible, and he certainly wrote similar words, but not until later (see Gilbert, Reynoldson, and Hobart 1952 for discussion). However, the main point about a name is that it should convey a certain meaning to everybody and, if at the same time it confers honour where honour is not due, that does not really matter; as Hardin writes, the Fallopian tubes were named after a man who was not the first to see them and who misconstrued their significance. Amerigo is another name that comes to mind. There seems on these grounds some justification for retaining the term 'Gause's principle' but this proves to be one of those cosy compromises that break down at once as soon as the literature not written in English is consulted. The same arguments by different people may be found in French and German but they make no reference to Gause and centre round Monard. This author (1920) has the advantage, not only of fourteen years priority, but also of a clear statement of the principle: 'Principe de tendance à l'unité spécifique. Dans un milieu uniforme, restreint dans le temps et l'espace, ne tend à subsister qu'une espèce par genre.' (There tends to be but one species per genus in a uniform environment in a given region at a given season.) He qualifies it thus: 'Enfin, lorsque le milieu éprouve des variations saisonnières intéressant la température ou la composition chimique de l'eau, il tend a dominer pour chacun de ces états particuliers, une espèce particulière.' (If there are seasonal variations in temperature or in the chemical composition of the medium, a different species tends to be dominant in different conditions.) On competition he writes: 'Toutes les fois que deux espèces voisines sont en présence, l'une d'elles, ordinairement la plus cosmopolite et la plus eurytherme, tend à éliminer la seconde.' (Whenever two related species come into contact, one, generally the more tolerant of a wide range of conditions, tends to eliminate the other.) Monard was studying the fauna of the depths of Lake Neuchâtel. He compared the number of species in various genera in the whole of Switzerland, in the whole lake, and in the profundal region and it was the discovery of progressive diminution that led him to the principle.

According to Illies (1952a), Thienemann regarded Monard's idea, not as a principle, but as an interesting corollary of his own first principle of biocoenotics (Erstes Grundprinzip der Biozönotik) which runs: 'Je variabler die Lebensbedingungen einer Lebensstätte, um so grosser die Artenzahl der zugehorigen Lebensgemeinschaft.' (The more variable the conditions of a place, the more species there are in the community inhabiting it.) We shall return later to this principle. Illies does not agree, and puts forward his own version of 'Das Monard'sche Prinzip': 'In den Biotopen des

Mittelgebirgsbaches findet sich von jeder Gattung gewöhnlich nur eine Art. Wenn weitere Arten auftreten so sind sie entweder erratisch, oder sie besiedeln verscheidene Habitate des betreffenden Biotops, oder aber sie haben deutlich unterschiedene Flugzeiten.' (In the biotopes of mountain streams at moderate heights there is generally only one species in each genus. If there are more, the others are erratics, or each has a different habitat within the biotope, or each is on the wing at a different time.) Erratics are species that have been washed down from a biotope higher up, or have wandered in from an adjoining one. Illies takes various genera to illustrate the second point, among them *Baetis* (Ephemeroptera), and states that *B. pumilus* is a species of the upper reaches, whereas *B. rhodani* is a ubiquitous species that attains its greatest abundance in the lower reaches. In fact the difference between the two is more subtle than that, and illustrates the point more clearly. Pleskot (1953) has shown that *B. pumilus* is what she calls a 'Schlängler', in other words a species whose nymph dwells in the small spaces between pieces of gravel, and Macan (1958a), contrasting the relative numbers of nymphs caught in a net and in a shovel sampler, has concluded that *B. rhodani* is not. The habitat of the two is quite distinct. Illies (1952b) goes more fully into the question of different flight-times. He found, for example, six species of *Leuctra* (Plecoptera) in the Fulda River, but the first emerged in March and April the last in September, October, and November, and the rest at different times in between. Incidentally he claims that there is complete temporal isolation, but Gledhill (1960) did not find this to hold in an English stream.

Monard clearly has precedence over Gause, but Hardin (1960) brings forward yet another claimant, Grinnell, who expressed the idea in 1904, and he shows that it was in Darwin's mind though he never formulated it exactly. Clearly the best solution is to drop names altogether and adopt Hardin's proposal to refer to 'the competitive exclusion principle.'

Before passing on to a discussion of the validity of the principle, we may note two restatements of it by Hardin: 'Complete competitors cannot co-exist', and 'Ecological differentiation is the necessary condition for co-existence'.

Elton supported the principle at the Ecological Society's meeting and published his argument in 1946. In effect he did what Monard had done twenty-six years before, and showed that, in various communities that had been surveyed, the number of species per genus was lower than in the country as a whole. This evoked a reply from Williams (1947), who pointed out that Elton's smaller number per genus was the mathematical result of taking a smaller sample and that what required doing was to work out the relative proportions of species and genera. The statistics demanded rather large numbers, and Williams took units such as Wicken Fen, Windsor Forest, and the county of Hertfordshire. In them he found relatively more species and fewer genera than in the country as a whole. Bagenal (1951) points out that certain important facts had been overlooked in both these mathematical operations. The surveys used by Elton had not covered uniform biotopes; one, for example, comprised fifty miles of the River Wharfe, another eight pools on Bardsey Island that ranged from brackish to fresh. This would tend to raise the number of species per genus. Not all the collections were made in the same way. Thirdly the lists used generally covered a year or more's collections and therefore any difference in time of appearance, a point included by Illies in his formulation, would be concealed. This would also raise the number of species per genus. I would make a fourth point from my own experience of freshwater surveys, of which several were used by Elton. At that time the taxonomy of the immature stages of most of the main groups had not been worked out. Generally the genera with most species are the difficult ones taxonomically and there can be little doubt that the surveys recorded fewer species in these genera than in fact there were.

Williams's areas were too large, and Bagenal concludes that what he had shown was that related species are more likely to be found in similar places than in dissimilar ones. Wicken Fen, for example, is fairly uniform and a particular genus is likely to be well represented in it, but it is sufficiently diverse to provide a distinct habitat for each species.

It is difficult to obtain a valid opinion on mathematical processes and when they should or should not be used. If a worker who does not possess mathematical ability deprecates them, nobody will attach any weight to his opinion. If a worker does have it, the temptation to make full use of his advantage over those who have not whenever he can must be strong. I do not mean to imply that it is never resisted. Lack (1954, p.3) quotes what seems a very fair comment by a statistician, which, together with Lack's own remarks, strengthen my opinion that this is an instance where the approach of the naturalist, who never likes to see the individuality of a species submerged in a formula, is more profitable than that of the mathematician.

It was two field naturalists, Diver and Spooner, who led the attack on Gause's principle at the famous Ecological Society's meeting. Their main contention was that one may see in the field plenty of examples of closely related species living together and apparently having the same habitat.

Biotic or climatic factors keep their populations at such a level that they never come near eating all the available food and therefore do not compete. Neither speaker published his argument, but a recent expression of support has come from Fryer (1959a and b). Nyasa is a remarkably uniform lake, whose shores are nearly everywhere either steep and rocky or shelving and sandy. For reasons that will be discussed presently, there has been much speciation among fishes and considerable adaptive radiation. A number of lines have, however, led to the same end and, off the rocks, there are twelve species that feed largely or entirely on attached algae. Fryer is able to split them up into groups; for example seven eat 'loose Aufwuchs' only and not *Calothrix* as well, and of these, two, one larger than the other, inhabit shallower water than the rest. Five are left and one, having a narrower face than the others, can feed in narrower cracks than they can. The remaining four seem to exist in exactly the same region and to feed in exactly the same way. They are heavily preyed on, and Fryer suggests that this never allows their numbers to become so great that they can reduce the supply of food and that, therefore, they never compete. However, as we have already seen, according to the mathematical argument which may be found set out in Hardin's (1960) paper, if one species tends to reproduce only very slightly more rapidly than another, it will ultimately eliminate it.

The numbers of land animals are probably more often kept low by catastrophes than are those of water animals; a wet summer may drown them, a hot dry summer may desiccate them, and a cold winter may freeze them to death. In such ways, argues one school of thought, numbers are kept well below what the environment can sustain and there is never competition. Such mortality is independent of the size of the population; I avoid using the familiar term 'density-independent' so that I may allow the conflict raging at the time of writing (1960) about exactly how it may be used to pass over my head. But, as has been pointed out, it is difficult to see how a species could survive unless the proportion of mortality was dependent on numbers, even though they had to be very great before the percentage mortality started to rise and very low before it began to drop. If the destructive effect of a climatic aberration was never mitigated when survivors were few, sooner or later one would annihilate the population. But in a hard winter there will be some specimens that have found a place protected from the cold, in a dry summer some specimens will find asylum in isolated places that are damper than the rest, and the species will continue. Under these conditions competition for the few refuges must be severe, and it seems unlikely that of two species which had identical requirements both would survive. A quite small difference might enable them to do so.

The pros and cons having been presented, it remains to mention the conditions under which the principle would not hold. Hutchinson (1957b) mentions three:

1. Fugitive species cannot compete with their close relatives and never establish themselves permanently, but they are always on the move and take advantage of any period of temporary scarcity of other species. Hutchinson instances *Corixa dentipes*, though my experience of this species is different. A better example is, I believe, *Helophorus brevipalpis* (Coleoptera), which Fernando (1958) took in far greater abundance than any other species in traps. It is found in shallow water, frequently in temporary ponds.

2. If each individual or pair of one species establishes a large territory and drives from it all intruders of its own species but not those of a closely related species, the two can exploit the resources of a region in the same way.

3. If conditions are continually changing, they may not remain favourable to one species sufficiently long for it to have time to oust others. This possibility was considered by Monard and his formulation makes allowance for it. Dr Fryer has pointed out to me that one possible change is the extent to which each species is preyed on. That a predator might tend to take proportionately more of whichever species was the more numerous, though unlikely if the two species are similar, is nevertheless possible.

Now that all the views have been put forward, how can Fryer's fish be explained? There are four possibilities. First, Fryer may be right and these four species may be living together in spite of being identical in everything except a few morphological features. This is the original Diver-Spooner line. The mathematical objection to it was based on the assumption that there is a difference in the reproductive rate. If there is not, the competition between the species would not be different from competition between the individuals of one species (Gilbert, Reynoldson, and Hobart 1952). A long co-existence of two or more species under these circumstances does, however, presuppose a degree of genetical stability that is probably rare in natural populations; in other words two species that started identical except for some small morphological details would not remain identical. The second possibility is that there may be differences in the way of life of each species which Fryer did not detect. Thirdly, the present mixed state of the species may be the result of some big change, a possibility that will be explored in the next paragraph, and competition is now in progress which, if the lake remains stable for long enough, will lead to the survival of one species only. There is no reason why elimination should not take a long time, at least in relation to the life-span of the human investigator. Fourthly, it is possible that each one of the species is favoured by the conditions obtaining at a different time of year or by the conditions during different periods of a cycle longer than an annual one and the co-existence is an example of the third of Hutchinson's exceptions. It will be hard to establish convincingly that one or other of these explanations is the right one. This may turn out to be one of those scientific controversies that drag on till both sides tire, or die off and make way for a new generation that finds nothing in it worth disputing. As Hardin (1960) remarks, the validity of the 'competitive exclusion principle' is very difficult to prove.

Fryer was primarily interested in evolution, but his ideas are relevant here in connexion with Thienemann's 'Erstes Grundprinzip der Biozönotik'. Lake Nyasa, as already stated, is remarkably uniform and harbours few species of plants and invertebrates, but it is very large. Fryer's thesis is that in this big lake any species adapting itself to finding food on rocky shores would split into a series of separate populations. As predators evolved, they would keep the population of each rocky area small, and there would be no overcrowding to stimulate wandering to a neighbouring area. Moreover any specimen leaving the shelter of the rock to which it was adapted, and setting out to cross the sandy area separating it from the next outcrop of rock, would be conspicuous to the predators and liable to be caught by them. The lake is old and Fryer believes that the isolation caused in this way could have led to the evolution of different distinct species off rocky areas. However, no such distribution is found to-day and most of the species are widely distributed in the lake, exactly how widely is not yet known. This mixing could have come about during a dry period, when the water dropped to a much lower level. Whatever the explanation, the fact is that Lake Nyasa, though its shore is not at all diverse, contains an astonishingly large number of species of fish—more than 180 in the family Cichlidae alone—and provides a striking exception to Thienemann's principle.

Another was recorded by Macan (1962), who was struck by the capture of more species of corixid in a reservoir than in any other single piece of water in which he had ever collected. The reservoir had a flat bottom and almost sheer sides and was as uniform a living-place as could well be found. It led me to examine the characteristics of other places in which many species had been found. It was not possible to explain the varied population of one or two that other workers had described, but the rest were either new, or temporary, had been recently drained and refilled, or had certainly or possibly undergone some abrupt change, such as deoxygenation or inundation by an exceptionally high tide. The reservoir had been drained and refilled not long before the collections were made, but it was possible that it offered suitable conditions to no species and was therefore continually being entered and quitted. Evidently many species are frequently on the wing and invading new pieces of water. There is no reason to suppose that stable pieces of water are not invaded as often as unstable ones, and the higher number of species in the latter must therefore be due to an earlier departure from the former. It is suggested that a species leaves a piece of water soon if it finds conditions unsuitable, and one of the most important criteria of unsuitability is the presence of an established population of some other species. The well defined habitat of each species, which is a marked feature of the Corixidae, is, therefore, in some measure due to competition.

MISCELLANEOUS

The fauna of trickling filters in sewage works is simple, an advantage for ecological work of which a group of workers at Leeds University have availed themselves. Theirs is another study which must be taken together, not piecemeal in the relevant chapters.

At the Knostrop sewage works of Leeds Corporation there was a roller distributor and the bed was made of water-worn pebbles. At the surface there was a growth of *Phormidium* and other algae, below it a slime of Fungi, Bacteria and Protozoa. This provided a rich food supply, though at times the animals became numerous enough to reduce it to a level where there was not as much as all the grazers could have eaten. It also offered a substratum where eggs could be laid, though not a wholly reliable one as at times the feeding activities of the animals dislodged it and the flow washed it through the bed. The other main hazard, particularly to an egg or a pupa, was destruction by some larva eating its way through it.

The main animals, found elsewhere in places such as piles of rotting seaweed, were one enchytraeid worm, *Lumbricillus lineatus*, and six species of Diptera: two psychodids, three chironomids and one cordylurid. *Lumbricillus* lays its eggs in *Phormidium* or attaches them to stones and, as no other oligochaete likely to find its way into one does this, it is probably why the species has been able to colonize filter beds. All the fly larvae were carnivorous to a smaller or larger extent, the chironomids *Metriocnemus longitarsus* and *M. hirticollis* more so than the rest.

The rate of reproduction and development of the flies at different temperatures was worked out (Ll. Lloyd 1937) and from it the theoretical abundance through the year was calculated. All the curves were evenly rounded humps, each reaching its highest point at a different time of year according to the optimum temperature of the species concerned. In the sewage beds abundance of each rose to a sharp peak and fell equally steeply, which was presumed to be due to the interaction of the species on each other.

The roller took half an hour to traverse the length of the bed, which meant that, whereas the ends were deluged every thirty minutes, the middle was deluged every fifteen minutes. This tended to drive the swarms of *Metriocnemus*, which mated outside, to the ends of the bed but presumably did not affect the parthenogenetic *Psychoda severini*, whose adults lurked inside the bed. Trapping showed that most adults of *Metriocnemus* emerged near the ends of the bed, but that *Psychoda* became more abundant towards the middle, its numbers rising as those of *Metriocnemus* declined. This was attributed to predation of the one on the other.

There were no *Metriocnemus* in the sewage works at Barnsley, probably correlated with which was a greater abundance of *Psychoda severini* and *P. alternata* and, for the former, a curve of abundance more like the theoretical one (Lloyd, Graham, and Reynoldson 1940). *Metriocnemus* lays its eggs near the surface of the bed, it was the practice at Barnsley to rest the beds for periods long enough to allow the upper parts to become quite dry, and this doubtless accounted for the absence of the chironomids.

In the sewage works at Huddersfield (Reynoldson 1947a and b, 1948) *Lumbricillus lineatus* was scarce, and the common worm was *Enchytraeus albidus*, a species not generally encountered in sewage works. The total population of worms and other animals was not as high as in other beds particularly near the surface, and there were fewer species. The substratum was pitted clinker, not smooth pebbles as at Knostrop, but the important feature was effluent from a chemical factory. This was toxic and, of course, strongest near the surface. Laboratory experiments indicated that *Enchytraeus* could tolerate this effluent in the concentration at which it occurred in the beds, but *Lumbricillus* could not, and evidently such few specimens as were found had been washed in.

Enchytraeus bred during the winter and its numbers decreased suddenly in spring, apparently because increasing numbers of *Psychoda alternata* at that time dislodged much of the slime and many of the worms with it. On the other hand, in one bed where *Psychoda* was absent on account of rapid flow, *Enchytraeus* was absent too, because the lack of scouring fauna led to accumulation of slime and the onset of septic conditions which the worm could not tolerate.

A further paper by Reynoldson (1957) may be mentioned here, as it recapitulates the work described in the three papers quoted, though it would be equally relevant in the chapter in which the influence of substratum is discussed. The rest of it is devoted to *Urceolaria mitra*, a peritrich ciliate found only on triclads, which provide it with a suitable substratum, and, apparently, nothing else. The numbers of *Urceolaria* fluctuate with rainfall, which probably means that they are correlated with the food supply, since rain washes bacteria into the water. They concentrate near the head of the animal, where most food can be obtained and, when the food supply increases, they are able to live further back on the triclads.

A not dissimilar relationship is mentioned by Vollenweider (1948) in an article stressing that the sequence of events in the plankton of a lake cannot be understood without some study of interactions of the components in addition to the conventional examination of physical and chemical changes. The presence of certain rotifers in the open water hangs upon the presence of certain algae to which they attach their eggs.

Another, of a different kind, is known only in the sea, though there may well be examples in fresh water. The young of the barnacle, *Balanus balanoides*, are set free at the time of the spring outburst of phytoplankton, irrespective of the latitude and the temperature. What stimulates the emission is not known, but Barnes (1958b) suggests that it may be some substance liberated by the plants.

The caddis *Ptilocolepus granulatus* uses only the 'leaves' of certain mosses and liverworts to make its case and is not found where they are absent (Thienemann 1950).

THE CONTROL OF NUMBERS

The theme of this book is presence and absence, but presence may be bound up not only with tolerance of certain conditions but ability to maintain in them a population above a certain size. Since, furthermore, the discovery of how numbers are controlled is one of the main objectives in current ecological work—many would claim that it is the key to ecology—and many of the factors already discussed in this chapter play a part in it, the last few pages are devoted to this topic. Solomon (1949) has reviewed it and has pointed out how complex it is; the regulation of numbers is brought about by a combination of factors of which the most important will not be the same for different species, nor for the same species at different times and in different places. It must therefore be stressed that the examples about to be quoted are no more than illustrations of the main ways in which numbers are controlled and are not the only factors operating in each particular instance.

Pennington (1941) sterilized two tubs, filled them with a nutrient solution, and then inoculated them from another tub. A small alga, to which she gave the name *Diogenes rotundus*, increased rapidly in numbers and soon attained a population density which remained fairly constant. After six weeks the rotifer, *Brachionus pala*, was introduced into one tub, the other being left as a control. It fed voraciously on the *Diogenes* and was soon so numerous that it was eating them faster than they were reproducing. A point was reached where there were not enough plants to keep the water oxygenated, and all the animals died, whereupon the algal population increased to a level higher than the one that it had reached initially. If the rotifers were introduced much sooner, they ate all the algae before they were sufficiently numerous to use up all the oxygen and then died of starvation. The same sequence of events was brought about by *Daphnia pulex*, and was also observed in a productive pond.

Mathematical analyses of this type of simple relationship between prey and predator have been made but practical demonstrations of the conclusions they have led to have proved less easy. A number of factors modify the course of events and in the tubs a more stable relationship, with consumer and producer waxing and waning one after the other without their numbers ever falling to zero, might have been reached if:

1. the consumers had been less efficient at finding their food, or the producers able to avoid the consumers;

2. the consumers had reproduced less rapidly;

3. the conditions had not suddenly become good for the producer so that it quickly reached a density at which the consumer could reach a density which proved disastrous.

Violent fluctuations are unusual. Defoliating caterpillars were mentioned in the first chapter, and lemmings and locusts come to mind, but generally herbivores do not go on multiplying till they are numerous enough to eat all the available food. Solomon and other authors point out that the more complex the structure of a community the less likely a big fluctuation in the numbers of any one species, but exact analysis of how the various species interact to keep the numbers of each at a fairly constant level has not been made.

The numbers of *Urceolaria mitra*, studied by Reynoldson, varied with the amount of food available, but did not reach a catastrophic level, as did those of *Brachionus*, probably because the animal, being sessile, was a less efficient hunter. It was also limited by its highly exigent substratum requirements.

Population density in flatworms appears to be controlled in the same simple way. *Planaria alpina* goes downstream when it is hungry and this reaction tends to take it to regions where food is more plentiful. Surplus is devoted to egg-production and a physiological change then causes the animal to move against the current. If it cannot find enough to eat in the upper reaches, as often happens when many are going up together, it starts to resorb its eggs and to become negatively rheotropic once more, and this alternation may go on for a long time if food is scarce (Beauchamp 1933). In winter it may go right down, or it may be washed down, to a lake shore, and Beauchamp (1932) observed permanent colonization of a stony delta. When the level of the lake fell, the delta was exposed, and the numbers of *Polycelis nigra* on the neighbouring muddy areas rose considerably. Beauchamp attributes this to the food supply provided by *P. alpina*, which, having apparently lost its usual reaction to flow in the still water, had to move on to the mud and died on this unsuitable substratum.

Populations of *Polycelis tenuis* and *Dugesia lugubris* in a pond were thought by Reynoldson (1960, 1961b) to have few natural enemies. Mortality due to various causes led to a drop in the total population during the winter, when the temperature was too low for reproduction. When breeding started in the spring, not only was this deficit soon made good, but numbers reached a point where they caused a shortage of food. The result of this was that many specimens not only lost their adult characteristics but became smaller as well.

The relation between supply of food and the rate of reproduction of animals of more complex structure is naturally somewhat less straightforward. The number of progeny for which food can be found is critical for birds and mammals, which feed their young, and this point is stressed by Lack (1954), who postulates that clutch-size in birds has been fixed by natural selection to yield the greatest number of young for which, on the average, the parents can find enough food. It is not, however, so firmly fixed that some species cannot take advantage of unusual plenty to produce more eggs per clutch or more clutches per season than in other years.

Animals which lay many eggs and take no care of their offspring are not subject to this limitation, but a critical factor is often the amount of food available to the young immediately they have hatched. Einsele (1961) finds that the time of hatching of the eggs of *Coregonus* depends on temperature, and that in certain Austrian lakes it takes place before the water is warm enough for an increase in the zooplankton on which the fry feed, with the result that most die of starvation. The result is that there are not as many adults as there is food for in the lakes. Accordingly he has kept eggs in a hatchery at a temperature lower than that in the lakes, and, having thereby delayed hatching, has had fry available to set free at a time when there is plenty of food for them. It is too early yet to decide whether this measure has increased the stock of fish, as Einsele hopes.

Le Cren (1955) recorded that young perch were much more numerous in 1949 than in any other of eleven years in which he had been observing them. The number of eggs was not. The significant point was that the year-class was outstanding in every one of the eight lakes under observation, and a climatic factor seems to be the only one that could have affected separate bodies of water not all in the same drainage area. It is thought to have influenced the food supply.

Macan (1958b) postulated from the finding of tiny nymphs, and Illies (1959) confirmed by direct observation, that the hatching period of the eggs of certain Ephemeroptera was much longer than the oviposition period. The former suggested that this might be an adaptation to prevent large numbers of newly emerged nymphs competing for some special food all at the same time. Hynes (1961b) believes that, in rapidly flowing water, shelter is a critical need and that, as an animal grows, it requires a larger crevice. It is therefore constantly vacating smaller for larger ones. A long hatching period is an adaptation to a limited number of the smallest crevices rather than to a limited amount of food.

No doubt the very young of some species die in such numbers from starvation that the survivors run little risk of eating all the available food. Other species are free of this risk because they are preyed on heavily, as the fishes observed by Fryer in Lake Nyasa are, or because numbers are kept low by parasites or by some abiotic factor in the environment. The fate that befell the rotifers in Pennington's tubs is also avoided in other ways. Some fish, for example, establish territories. The strongest trout instal themselves in a favourable place and the weaker are driven away to places where they either starve or, owing to inappropriate behaviour, fall a prey to predators. The relation, if any, between size of territory and food supply has not been investigated.

Macan and Mackereth (1957) found that *Gammarus pulex* was fairly evenly distributed along the length of a small stony stream and that its numbers did not alter much from month to month and from year to year. It lives under stones and eats dead leaves and other debris. It is one of the first animals to reappear in a stretch of stream that has been dry, and it tries to swim against the current when exposed by the lifting of a stone. Macan and Mackereth suggest that it is tending to work its way upstream all the time but that, when population becomes dense, there is so much mutual disturbance that many are pushed out from under the shelter of a stone and washed downstream again. They suggest further, though without producing any experimental proof, that, the less food there is, the more pushing out and the smaller the population.

When tadpoles are starved or kept together in water that is not changed, cells of obscure origin appear in their faeces and, when eaten, bring growth to a halt. It starts again in fresh water well supplied with food (Richards 1958). Rose (1959) records that large tadpoles inhibit the growth of smaller ones in the same aquarium. The latter eventually die unless they are transferred to another container, where they may flourish.

Cannibalism is another self-regulating mechanism. Some Swedish tarns were found by Alm (1952) to be dominated by one year-class of *Perca fluviatilis*, which was evident year after year growing slowly and apparently preventing any younger year-class attaining prominence by eating it. Eventually, old age having reduced the numbers of this year-class, a favourable year would throw up a new one, which, too numerous to be devoured completely by the old one, would gradually take its place.

Many insects are parasitized by other insects that lay in each host only one or a few eggs, from which hatch larvae that have eaten all their victims' tissue by the time they are ready to pupate. Studies on their effect on numbers have been made but have not yet extended to any of the several species of Hymenoptera that enter the water to lay their eggs in aquatic organisms. When the parasite is smaller relative to its host, generally it does not cause death, but bacteria may be killers, sometimes bringing about much mortality when the host is numerous and crowded. In those parts of the world where the large carnivores have been exterminated, man can fairly claim to be at the summit of the Eltonian pyramid. Even in the most civilized countries, diseases such as cholera killed many people within living memory, and it is only a few centuries ago that populations were prevented from increasing by periodic epidemics of plague and other pestilences. To-day the spread of knowledge about hygiene to backward lands is leading to an over-population problem that is disturbing many. Populations of *Potamobius* (*Astacus*) *pallipes* increase steadily in Britain and are then reduced severely apparently by a bacterial disease (Duffield 1933, 1936) but otherwise not much is known about infections of this kind in fresh water.

Lastly there are the abiotic factors. Many species of birds are less abundant after a severe winter (Lack 1954), and less mobile organisms may be drowned or desiccated during wet or dry periods. Aquatic animals are less exposed to such hazards, as explained in chapter 2, though sometimes annihilated by lack of oxygen in hot still weather or under ice. Reynoldson

(1957) claims that the population of *Enchytraeus*, which is prevented from becoming excessive by the washing of the humus out of the trickling filters each spring, provides an example of this kind of regulation by an abiotic factor which prevents numbers ever attaining a level where the percentage of mortality starts to rise owing to the density of the population.

SUMMARY

All species are probably limited to places that offer refuges from predators, unless they live in waters which, because they are temporary or offer extremes of some factor such as salinity, harbour no predators. Certain ciliates and cladocerans and the young of some species of fish are limited to places where a particular food is plentiful. Algae produce substances that inhibit the growth of other organisms and animal toxins are known too. *Gammarus pulex* replaces *G. duebeni* in fresh water because it breeds faster; some species actively drive away others; but often it is difficult to explain how one species curtails the range of another. Views on, and the nomenclature of, the principle that 'two species with similar ecology cannot live together in the same region' are discussed, and examples showing how difficult the postulate is to prove are given.

Records of an animal eating all its food and dying of starvation or lack of oxygen exist, but are rare. Population is generally limited before a catastrophic level is reached. Flatworms cease reproduction as soon as food runs short and grow smaller. Lack of food may kill so many of the youngest stages of some animals that it is never a threat to later ones. Fish populations are prevented from becoming too dense by the establishment of territories or by cannibalism. Large tadpoles inhibit the growth of smaller ones. Parasites keep the numbers of some species in check.

SYNONYMS

Species	Group	Other generic names	Other species names
Aphanizomenon gracile (Lemmermann)	Cyanophyta		*flos-aquae f. gracile* Elenkin
Brachionus pala (Ehrenberg)	Rotifera		*calyciflorus* var. *pala*
Diura bicaudata (Linnaeus)	Ins. Plecopt.	*Dictyopterygella*	*recta* Kempny
Dugesia lugubris (Schmidt)	Platyhelminthes	*Planaria*	

Species	Group	Other generic names	Other species names
Hucho hucho (Linnaeus)	Pisces	*Salmo*	
Leucioperca sandra (Linnaeus)	Pisces		*leucioperca* Linnaeus
Limnaea pereger (Müller)	Moll. Gast. Pulmonata	*Lymnaea*, *Radix*	*ovata* Draparnaud *peregra*
Lumbricillus lineatus (Müller)	Olig. Enchytraeidae	*Pachydrilus*	
Metriocnemus longitarsus Goetghebuer	Ins. Dipt. Chironomidae		*hygropetricus* Kieffer
Perlodes microcephala (Pictet)	Ins. Plecopt.		*mortoni* Klapálek
Planaria alpina Dana	Platyhelminthes	*Crenobia*	
Planaria gonocephala Dugès	Platyhelminthes	*Dugesia*, *Euplanaria*	*subtentaculata* (Draparnaud)
P. montenegrina Mrázek	,,		*montenigrina*, *teratophila* Steinmann
Planorbis corneus (Linnaeus)	Moll. Gast. Pulmonata	*Planorbarius* *Coretus*	
Polycelis felina (Dalyell)	Platyhelminthes		*cornuta* (Johnson)
Potamobius pallipes Lereboullet	Crust. Malacostraca	*Astacus*	
Salmo salar Linnaeus	Pisces	*Trutta*	
S. trutta	,,	*Trutta*	*lacustris* Linnaeus *fario* Linnaeus

CHAPTER 7

Physical Factors (1)

WATER MOVEMENT, DESICCATION AND MISCELLANEOUS

EFFECT OF WATER MOVEMENT ON THE SUBSTRATUM

Water movement, whether it be the flow in streams and rivers or the action of waves beating upon a lake shore, is one of the main factors shaping biocoenoses, because of its effect on the substratum. If it be strong, it will carry away finer particles and leave only stones. Where it is weaker sand will settle, and, as it slackens still more, mud particles will drop to the bottom. Moss grows on those stones that are not frequently turned over, in other words on a stable bottom, and higher plants can take root in sand and mud. In a lake the effect of waves diminishes as depth increases and there is a zonation ranging from stones near the water's edge to particles of ever increasing fineness as the bottom slopes downwards. Fine particles are also deposited in bays sheltered from the wind. The sand region is often sparsely inhabited because colonists are confronted with a difficult substratum and little to eat. Beyond it fragments of dead leaves and other fine particles of organic matter fall to the bottom and provide a supply of food.

It is impossible to consider separately the influence on the distribution of organisms of current and the nature of the substratum. Accordingly, in the paragraphs that follow, field observations and then experimental observations on the effect of current are described first, similar observations on the substratum next, and finally such explanation of distribution as is possible in the light of the facts observed is presented.

Nielsen (1950b) gives the current speeds at which objects of a given diameter are moved, and this information is the basis of table 10. This, however, is an engineer's classification and takes no account of various complicating factors which are of significance to the biologist. Much depends on whether the fall is regular. If it is, current diminishes with increasing depth and the substratum is subjected to a current distinctly weaker than that flowing above it. If on the other hand it is irregular, the bottom at the end of a steeper stretch will be subjected to the full force of the current.

Table 10

Size of objects moved by different current speeds (Nielsen 1950b)

Speed of current	Diameter of objects moved	Classification of objects
10 cm/sec	0·2 mm	mud
25	1·3	sand
50	5	gravel
75	11	coarse gravel
100	20	pebbles
150	45	small stones
200	80	stones (fist size)
300	180	small boulders (man's head size)

Another difficulty that confronts the ecologist is that current is very inconstant. Particularly in hilly districts, it may increase suddenly as a result of heavy rain and decrease again quickly when the rain slackens or stops. It is, therefore, one of those factors which should be recorded continuously, but, so far as I know, this has not been done.

The nature of the substratum varies according to the nature of the rock. Fig. 6 is a simplified diagram of a valley in the English Lake District. A glacier has left behind an excavation in the rock, which in the figure is shown filled with material ground off the rocks higher up and carried down by the glacier, though a number are filled with water. Since the ice disappeared there will have been extensive redistribution of the material by streams, which will also have eroded more from the steeper parts of their beds and rolled it down to the valley, a process that continues until the slope of the stream-bed is much reduced. Whether there is a lake or not, the head of the valley will be filled with this mixture of clay and stones rounded by the action of ice and water, and the river will flow over a bottom that shifts every time the current increases. In the middle of the valley velocity is slow, and only mud is moved, but it increases again as the river runs over the lip of the rock basin excavated by the glacier. Moraine material, if it was ever present, has been washed away long ago and the river runs over bare rock. Flat slate-like stones are plucked off this and they cover the bottom further on. They are less easily shifted than the rounded ones at the top of the valley and form a stable bottom generally covered with moss. The streams flowing down the steepest part of the mountainside generally have a stable bottom too, because large stones and boulders become jammed in the channel which the water has carved through the rock. On other geological formations the substratum is probably somewhat different but nobody has yet given careful attention to this.

Percival and Whitehead (1929) describe what they call a 'cemented' bottom consisting of large stones embedded in smaller stones and gravel in

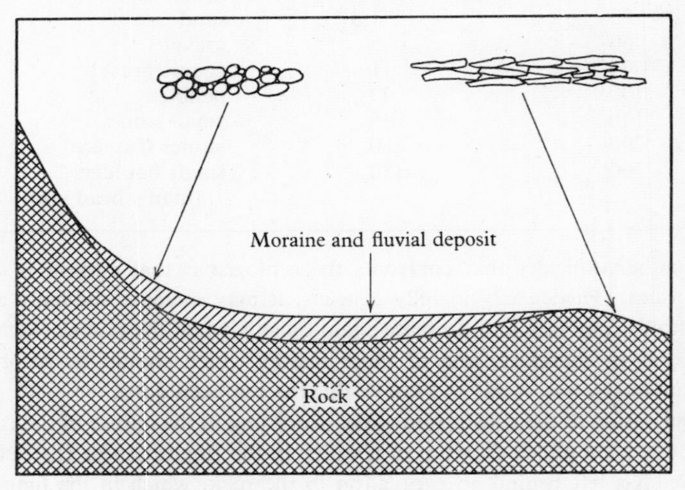

6. Diagrammatic section along a valley (Macan 1961) (*Verh. int. Ver. Limnol.* **14**, p. 590, fig. 2).

such a way that even the severest spate causes little alteration. They do not describe how this is formed but it is probably in morainic deposit where erosion has ceased so that there is no longer a constant supply of small and medium-sized stones being brought down from above.

It is not difficult to measure the rate of flow at the surface of a stream or in midwater, but to find out the strength of the current to which an invertebrate on a stone is exposed is a more formidable problem. Ambühl (1959) has recently made an important advance towards its solution by means of a technique in which acetyl-cellulose powder is photographed under alternating-current illumination. This causes each particle to appear as a series of dots, and, from the distance between them, it is possible to calculate the speed at which a particle is travelling. Ambühl worked with a glass-sided tank on the bottom of which he built stone-like shapes of plaster of Paris. He illustrates the current over one that has an aerofoil section and slopes gently upwards from the front to a highest point near the rear. Current is slower near the bottom just above the upstream end of the obstruction; near the middle the pattern is little modified in comparison with flow in a flat-bottomed channel, but behind it there is a zone of almost stagnant water (fig. 7).

Everywhere there is a considerable reduction of current speed in the few millimetres immediately above a surface, becoming progressively greater

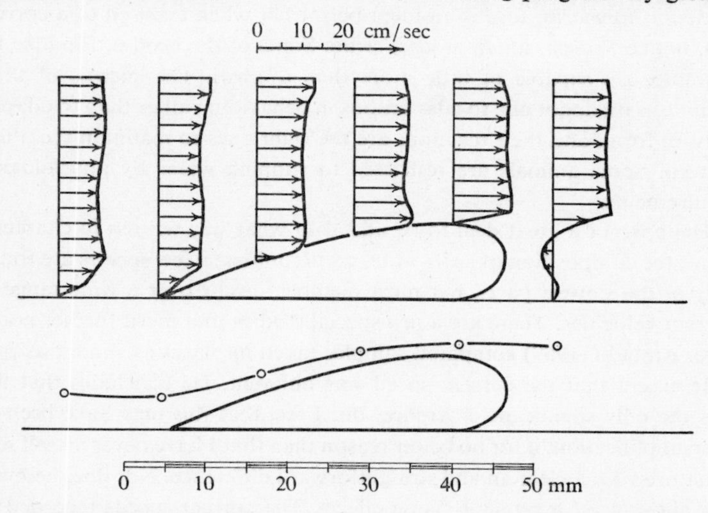

7. The effect of an obstacle on current. The current is flowing from the left. The length of each arrow is proportional to the current speed. The line in the lower diagram indicates where a marked drop in current speed starts (Ambühl 1959) (*Schweiz. Z. Hydrol.* **21**, p. 143, Abb. 4).

as the surface is approached. The thickness of the layer varies with the speed of the current and the thickness of the whole water column, but under natural conditions is as much as 4 mm. This means that it is of considerable importance to invertebrates, many of which are not as high as this, but very difficult, if not impossible, to measure by means of most of the devices at present in use.

DIRECT EFFECT OF CURRENT ON ORGANISMS

(a) *Field observations*

Rapidly flowing water, where the bottom is stony, is the obvious place for the study of the direct effect of current on organisms. To recapitulate earlier remarks on the fauna of stony streams: larvae of Simuliidae are morphologically adapted for the life in that they have means of clinging to a smooth surface and of straining the water to obtain food. The net-spinning Trichoptera also make use of the current to bring them their food. The flattened nymphs of the family Ecdyonuridae, and the freshwater limpet, *Ancylus fluviatilis*, are adapted to cling to hard smooth surfaces, and some species thrive in still as well as in running water. The rest do not show any particular morphological adaption and indeed no design would seem to be less suited to the conditions than that of the fat round larva of *Pedicia*. *Gammarus*, too, is an inept performer when exposed to a current but, unlike *Pedicia*, it can at least swim. Some of the beetles, Elmidae for example, are capable of little more than a crawl. The success of these animals is no doubt due to adaptations in behaviour rather than to adaptations in form, and their reactions are the right ones to maintain them in a current. Some animals are restricted to running water by physiological requirements.

There is not a great deal to be added to what was written in chapter 3 about the composition of swift-water communities; a few species are found only in the slowest parts, but most members occur over a wide range of current velocities. There are a few special studies that merit further notice here. Ambühl (1959) compared samples taken in places as similar as possible except that the current speed was different. He concludes that this was the only significant difference, but I feel that this may have been an oversimplification, if for no better reason than that I have never myself seen in nature such a clearcut and straightforward difference. Nor does he make any allowance for possible biotic effects. The current speeds recorded by Ambühl were those in the water above the bottom, not those to which the animals were actually exposed. They ranged from 0 to 150 cm/sec and most species occurred over a wide range. Most were scarce in the slowest currents, 2-3 cm/sec, rose with increasing speed to a peak which varied considerably with the species, and then decreased in numbers as the current became faster. *Simulium* sp. reached maximum abundance at a speed of about 80 cm/sec; *Hydropsyche angustipennis* at 60 cm/sec, *Baetis* spp., *Ecdyonurus* sp. and *Ephemerella ignita* at progressively decreasing velocities down to about 15 cm/sec where *Gammarus* reached its peak (fig. 8). It must be noted that there could have been more than one species in every genus except *Gammarus*; also that the figure for each species shows its percentage in the whole catch, whose total varies considerably, not how many of it there were in a given area.

Scott (1958) shows by means of histograms the numbers of larvae of various species of Trichoptera along a current gradient divided into units of 10 cm/sec. Most species occur along the whole of the gradient but become progressively less numerous with increasing distance from a mode

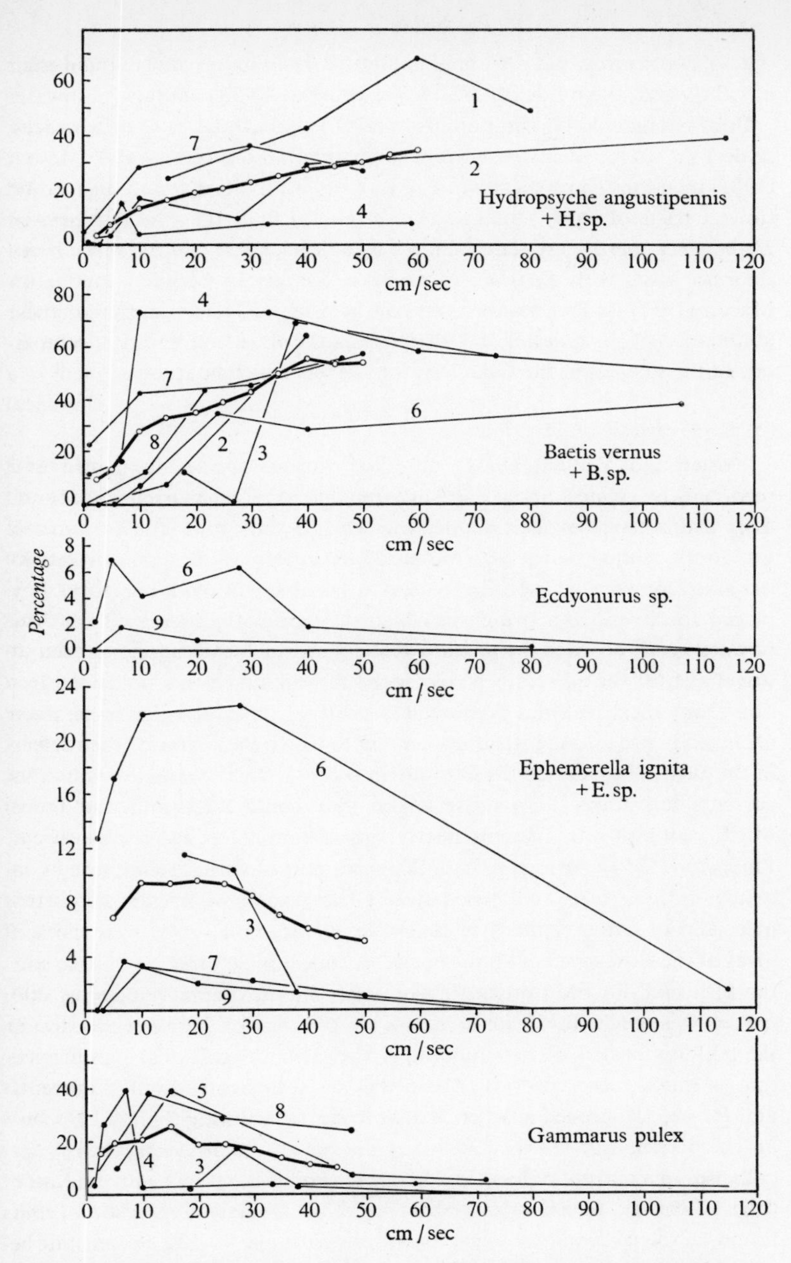

8. Percentage of specimens caught at various current speeds. The thin lines represent the results of individual experiments, the thick line the average of them (Ambühl 1959) (*Schweiz. Z. Hydrol.* **21,** pp. 190 and 191, Abb. 23 and 24 part).

(fig. 9). Scott agrees with Ambühl in finding *Hydropsyche* most abundantly in swift water, the mode for *H. fulvipes* being at 40-50 cm/sec.

Both Grenier (1949) and Phillipson (1956) assert that larvae of *Simulium* settle in a place where the current is running at a certain speed. Macan (1961) mentions some species of Ephemeroptera that are found only in the slowest parts of stony streams. Harker (1953) finds that the numbers of *Heptagenia lateralis* are reduced most by a flood, those of *Rithrogena semicolorata* least, with *Ecdyonurus torrentis* coming in between them, but Macan (1957) doubts whether this can be a major factor determining the abundance of *H. lateralis*, because it was the most numerous species in one of the steepest of the Lake District streams which he studied.

(b) *Experimental observations*

Dorier and Vaillant (1954) observed various species in nature and recorded the greatest and the least current speed at which each was found. They are not very explicit about how near the Pitot tube which they used was to the animals when they measured the current, although, as Ambühl has since shown, this is of the greatest importance. In the laboratory, they placed specimens in a trough in which they gradually increased the flow of water, and recorded the greatest velocity at which each species ceased to travel against the current, and the speed at which it lost its hold. A selection from their readings is shown in table 11, in which the species are arranged in order of the strongest current to which they exposed themselves in the field. They divide the animals into those which ventured only into currents far slower than those which they could withstand and those which ventured into currents nearly strong enough to wash them away. Philipson (1954) performed similar experiments with Trichoptera in a trough with a 'finely roughened floor'. His results were extremely variable, and of five specimens of *Anabolia*, for example, three were carried away at the slow speed of about 10 cm/sec, one hung on till 100 cm/sec, and the fifth had not been washed away when the maximum velocity of 200 cm/sec was reached. His observations on the relation between speed and the building of nets by net-spinning larvae have an ecological significance that is much more obvious. There was no spinning without a current, and *Wormaldia* needed a faster current to start it working than did *Hydropsyche*. Engelhardt (1951) states that the net of *Plectrocnemia conspersa* collapses in a current of less than 13 cm/sec and is burst by a current faster than 17 cm/sec, which must restrict its range severely. Phillipson (1956) found larvae of *Simulium ornatum* within the range 40-120 cm/sec with a concentration between 80 and 90 cm/sec.

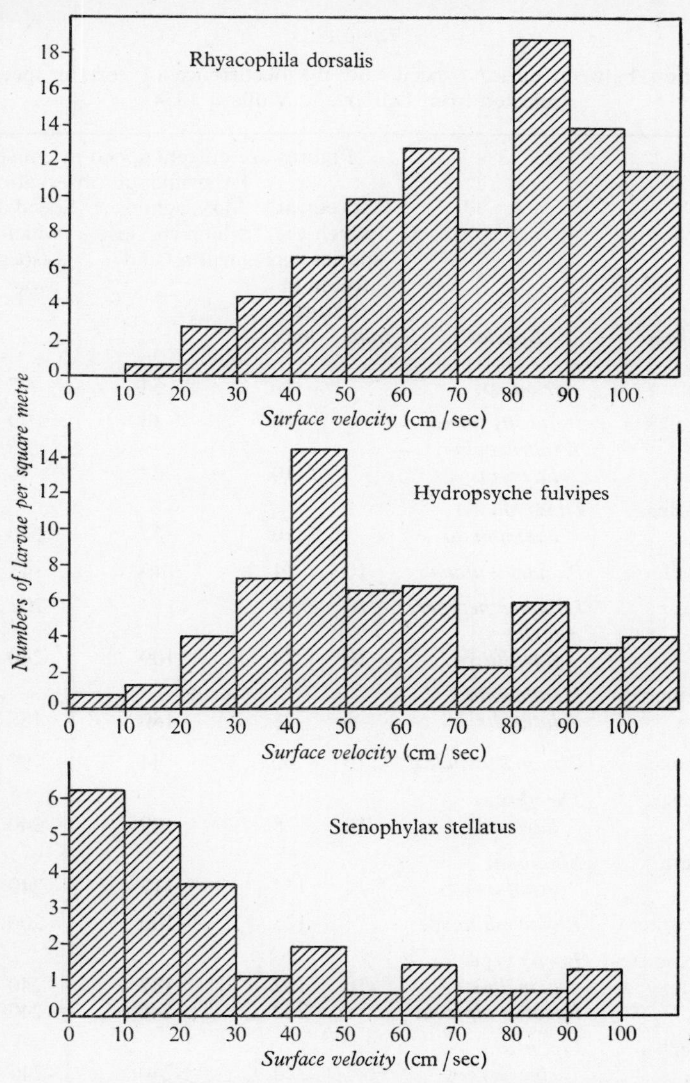

9. Number of specimens per square metre at different current speeds (Scott 1958) (*Arch. Hydrobiol.* **54,** pp. 350–351. Abb. 3, 6, 7).

Table 11

Relation between current speed and the occurrence of certain species (selected from Dorier and Vaillant 1954)

		Figures are current speed in cm/sec			
				Experimental observations	
		Speeds at which the species was found in nature		Max. against which species will ascend	Speed at which washed away
Group	*Species*	*min.*	*max.*		
Odonata	*Agrion* sp.		10	54	77
Turbellaria	*Polycelis felina*		10	44	99
	Dendrocoelum lacteum		10	37	76
Hirudinea	*Glossiphonia complanata*		10	37	240
Turbellaria	*Planaria alpina*	10	14	140	143
Mollusca	*Limnaea pereger*	10	14	117	202
	Ancylus fluviatilis	10	24	109	240
Ephemeroptera	*Heptagenia lateralis*	—	28	140	188
Crustacea	*Gammarus pulex*	10	40	44	99
Mollusca	*Theodoxus fluviatilis*	10	78	109	240
Diptera	*Simulium ornatum*	14	114	117	240
Trichoptera	*Rhyacophila* sp.		125	100	200
Ephemeroptera	*Baetis* type *gemellus*	10	182	187	240
	Epeorus alpicola		222	>240	>240
Diptera	*Liponeura cinerascens*		220	>240	>240

Ambühl (1959) used two tanks in each of which there were four 'stones' of plaster of Paris. These caused a diminution of flow, but conditions were not strictly comparable to those in most stream-beds since the 'stones' were projections from a continuous surface and the animals could not get underneath them (fig 7). For a given input of water the current speed 1 mm above the bottom was measured by means of the photographic technique already described. The gradient between a speed of 0 and the maximum was divided into six categories. Ambühl then prepared a plan of the bottom, dividing it into zones showing how each current-speed category was distributed. In the experiments the maximum speed was either 5 or 18-20 cm/sec.

A number of individuals of one species were introduced into the tank. After some hours in the current the number of specimens in each of the current-speed categories was counted. The results are expressed as histograms showing the percentage in each category of the total number of specimens. Every species tested occurred in every category, that is from places where there was no current to places where the current was maximal, but there was generally a steady rise to a mode in one particular speed category. Ten per cent of *Baetis scambus*, for example, occurred in still water and the proportion rose steadily to reach 26 per cent in the fastest category. The mode of *Baetis vernus* was in the fastest category when the maximum was 5 cm/sec but in the penultimate one when it was 18 cm/sec (fig. 10). *Rhithrogena semicolorata* was similar, but *Ephemerella ignita* was the exact reverse with a big mode in the slowest part of the tank. Most *Ecdyonurus venosus* were found in this category, but numbers were rather evenly distributed among the others. *Hydropsyche angustipennis* showed a preference for the fastest places when the maximum was 5 cm/sec and for the slowest when it was 18 cm/sec. *Rhyacophila nubila* was most numerous in the fastest water in May but was more evenly distributed in June when pupation was imminent.

There were features of the distribution pattern that make it far from certain that current velocity was the only factor to which the animals were reacting. Although there were four ridges crossing the tank, nearly all the *B. vernus* and a large proportion of the *R. semicolorata* congregated at the outflow end of the tank, and several experiments with the latter and with *H. angustipennis* had to be abandoned because so many specimens attached themselves to the glass sides and not to the bottom.

All these data are important and valuable, but, as the authors freely admit, they are of a preliminary nature. To point out where they fail to provide a complete explanation is not, therefore, to criticize. No clear picture of how current affects the distribution of any species emerges from a reading of these works, and the authors do not give the impression that they had one either. General comments that seem pertinent are: first, that all the observations were made on large nymphs and larvae. The problem cannot be finally elucidated until the effect of current on all stages of the life-history has been investigated. The critical stage might easily be the egg

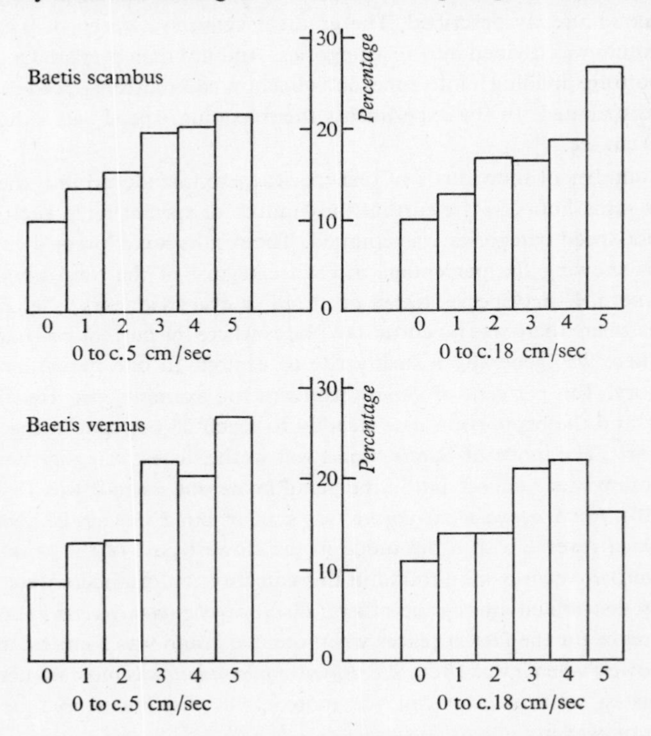

10. Percentage of specimens settling in each of five current-speed categories in an experimental tank. The current ranged from 0 to *c.* 5 cm/sec in the experiments on the left, and from 0 to *c.* 18 cm/sec in those on the right (Ambühl 1959) (*Schweiz. Z. Hydrol.* **21**, pp. 220 and 201, Abb. 27 and 28).

or the very small larva, and selection by the ovipositing female might play a part. Secondly it is puzzling why animals that live among stones should react to the current speed well above the stones. Thirdly, whereas a species that selects a fast current ought to be absent from the slower reaches, a species that selects a slow current ought to occur everywhere because it has only to go deeper into the substratum at high velocities to find the speed of its choice. On this reasoning the population should be more varied the faster the flow and therefore interactions might be a potent factor in determining distribution. Further, if it be postulated that any one species selects a given rate of flow, crowding might be the result and there would be interactions among members of the same species. It is noteworthy that, although in both field observations and laboratory experiments each species was most abundant at a certain current-speed, nearly every one was found over a very wide range of velocities.

There are certain specific points in addition to those already made. It is not wholly clear how Dorier and Vaillant determined the strongest current to which each species exposed itself in nature (*vitesse maximum du courant supporté par les sujets*). It would be unreasonable to expect the animal to oblige just when the observer happened to be looking, and the presence of an observer might affect their behaviour. Hubault (1927) observed ecdyonurids slipping from the surface of a stone to the underside when, at night, he suddenly shone a torch on to them, and I have occasionally seen these nymphs disappearing as I put my head over a stream in the day time. Moreover some of their figures do not agree with the observations of other workers. *Ancylus fluviatilis* comes low on their list, but, in my experience, it occurs in torrential streams and, since it is one of the few species that sit perpetually on the top of stones, it must be able to resist a current much faster than the 24 cm/sec which they give as a maximum. *Gammarus pulex* is another poor performer on their showing and from casual observation I agree that it is, but that does not prevent it colonizing the swiftest streams in the Lake District (Macan and Mackereth 1957), presumably because it never leaves the shelter of moss and stones. Obviously laboratory findings bear no relation to its observed distribution. The figures in their third column presuppose that all animals move against the current, but Beauchamp (1933), as mentioned in the previous chapter, has shown that *Planaria alpina* goes up or down according to its physiological condition.

Ambühl found that *Leuctra geniculata* showed a strong tendency to congregate underneath anything suitable provided for it, and he was unable to obtain satisfactory results when there was nothing. His finding that the reaction of *Rhyacophila nubila* varied according to its physiological state is significant, and there is also the possibility that it might not be the same at all seasons and at all times of day.

THE SUBSTRATUM

Field and experimental observations

The substratum is the place where most animals find their food, and a second essential requisite is that it should provide protection against enemies. Different behaviour patterns mean that different species will find the best conditions on or in different substrata. This is a commonplace idea but data to substantiate it are not forthcoming from freshwater studies. All that can be done here is to describe various apparent correlations between the occurrence of certain species and the nature of the substratum. At the point where the bottom of a river begins to change from stony to weedy, all species of ecdyonurid begin to disappear (Macan 1957). There seems no reason why they should not cling to flat stems or leaves as well as to stones but I have only once found one of them on such a surface. Likewise *Ancylus fluviatilis* is found only where there are bare stones or rock to sit on. It must be tolerant of a wide range of conditions, for I have found it in a quite small quarry pool somewhat polluted by cattle (Macan 1950a), and in a small roadside horse-trough. It eats whatever algae grow on the substratum where it is (Geldiay 1956) and it must be able to survive on a wide variety. The most likely interpretation of its range in the absence of experimental evidence is that it has an innate urge to keep moving on any substratum except a hard one. *Acroloxus lacustris*, placed by some in the same genus, is found on the stems of plants.

Percival and Whitehead (1929, 1930) made a study of the fauna of different types of substratum. Their results are instructive in that they reveal the complexity of a situation which some other authors have simplified to excess, but are difficult to interpret because there are obviously a number of variables. Some of the stones were bare, others covered with vegetation, some of which accumulated detritus and some of which did not. The vegetation no doubt harboured species that could not cling to bare stones, and food supply varied from place to place. Further no specific identifications are given in the tables and those given in the text must be suspect, since the keys available when the work was done were incomplete; also some large groups such as the water-mites and the family Chironomidæ are not split into smaller taxa at all.

The types recognized (1929) are: 1. loose round stones more than 5 cm in diameter forming an unstable bottom and carrying no vegetation visible to the naked eye; 2. similar stones 'cemented' in place by a matrix of finer material, consequently stable, and covered with diatoms; 3. small stones, up to 2·5 cm in diameter, and gravel, unstable and without visible vegetation; 4. stones of variable size and shape set in a matrix, stable, and covered with *Cladophora*; 5. large stones and rock covered with moss not thick enough to accumulate much debris; 6. stones and rock carpeted with moss thick enough to accumulate debris; and 7. stones of various shapes and sizes, stable, and with *Potamogeton perfoliatus* growing from between them.

Numbers per sq. dm varied enormously, being about thirty-three on the two unstable bottoms and forty-six on the cemented bottom, ten times greater in the *Cladophora*, and 100 times greater in the thick moss. Large numbers of Chironomidae, Naididae, Coleoptera and water-mites contributed mainly to the totals in the vegetation. A few species that could not have been misidentified may be taken to illustrate the differences found, the figure in parentheses being the number per sq. dm. *Rhithrogena semicolorata* was most abundant in the *Potamogeton* (11), and relatively numerous on the unstable bottom (3 and 4), where it was one of the commonest animals. It was absent from the cemented bottom and the *Cladophora*, and scarce elsewhere. *Ephemerella ignita* was numerous in the loose moss (198), the *Cladophora* (45), and the thick moss (39), comparatively numerous relative to other species on the unstable bottom of large stones (3), and the cemented bottom (5), but absent from the unstable bottom of small stones. *Rhithrogena semicolorata* feeds by scraping algae off stones (Strenger 1953), whereas *Ephemerella* eats algae and pieces of moss (Percival and Whitehead 1929), and this difference might account mainly for the different distribution of the two; it is surprising that *Ephemerella*, which has not the flattened form of *Rhithrogena* nor the ability to swim rapidly like *Baetis*, should be nearly as numerous as they are on the unstable bottom of large stones. *Ancylus fluviatilis* is most numerous on the cemented bottom with *Potamogeton* (3) and the cemented bottom with diatoms (4); it was found on the uncemented bottom of small stones (1) but not on the uncemented bottom of large stones, the *Cladophora*, and the thick moss. It may be postulated that it cannot find a suitable smooth substratum where vegetation is thick and, being unable to move fast enough to escape, is crushed when the large unstable stones move. *Gammarus pulex* is most abundant in loose moss (28) and thick moss (8), and absent from both types of cemented bottom. The moss, perhaps, provides it with both cover and food, and it avoids the cemented bottom because it cannot get beneath the stones.

Linduska (1942), in a more superficial survey, also reaches the conclusion that the composition of the fauna is largely influenced by the nature of the substratum. Hynes (1941) writes that *Dinocras cephalotes* is always much commoner in places where the substratum is stable and moss-covered, whereas *Perla bipunctata* is commoner on unstable parts of the substratum, where it is found under the loose stones. This observation also must await analysis until more is known of the habits and requirements of the species involved.

In a stony English river, Scott (1958) found greatest numbers of *Glossosoma boltoni* on medium-sized stones and of other Trichoptera on large stones. Other differences in distribution were more easy to explain. *Glossosoma boltoni* and *Ecclisopteryx guttulata* occur on all parts of stones except the underside, which is not surprising as they feed on algae. In contrast, *Stenophylax latipennis*, *S. stellatus*, and *Odontocerum albicorne*, which eat debris, occur under stones and, as might be expected, in water which is not flowing too fast and from which, therefore, debris is likely to fall to the bottom. *Glossosoma* is believed by Scott to avoid slow water because it does not like silt in its food.

Bovbjerg (1952) is one of the few workers who has attempted an experimental analysis of the relation between a species and the substratum. Of two crayfish, one, *Orconectes propinquus*, occurs in stony streams and on stony lake shores, and the other, *Cambarus fodiens*, inhabits ponds and slowly flowing water where the bottom is muddy. One thousand experiments on 100 crayfish were carried out in a tank where there was a choice of a cinder or a mud bottom. Eighty-eight per cent of *Orconectes* settled on the cinders and 60 per cent of *Cambarus* on the mud. *Orconectes* certainly seems to be making selection, but, like the angels in Heaven who rejoice more over the sinner that repents than over ninety and nine just men that need no repentance, I always find the small percentage that do the 'wrong' thing more interesting than the large percentage that do the 'right' one. What happens to them in the wild? Other factors determining the range of these two species are ability to burrow into mud, which, possessed by *Cambarus* only, enables it to inhabit temporary ponds, and tolerance of high temperature and low oxygen concentration, which is greater in *Cambarus*.

Different species of plants bear different sets of sessile rotifers. Those with finely divided leaves tend to have more rotifers on them than plants with entire leaves, but there seems to be a selection, apparently in the earliest stages, of some feature which is more subtle than this and whose nature is unknown. *Utricularia vulgaris americana* generally bears more individual rotifers and more species of rotifer than any other species of plant. *Ptygora brevis* is nearly always found in the forks of finely divided leaves. Several rotifers are generally attached to green algae and two species are associated almost exclusively with *Gloetrichia* (Edmondson 1944). Oviposition by rotifers on algae (Vollenweider 1948) was mentioned in the chapter on interrelationships. Fishermen and river-keepers are convinced that some species of aquatic plants harbour more of the invertebrates on which fish feed than others, and, as Elton has pointed out, the views of such people are to be respected. No biologist, however, has yet made the quantitative comparative study that is needed.

Leeches were never found by Boisen Bennike (1943) in Fure Lake on purely sandy or muddy bottoms, and he concludes that they avoid places where there is nothing to which they can adhere. Dr Hynes, however, tells me that he has not found this to be universally true. Incidentally the Hirudinea show clearly that life in fast water involves more than an ability to cling on. Few animals are better equipped to do this, as the figures in table 11 show, yet they are rarely found in fast moorland streams. In the richer Susaa, in contrast, they are numerous in fast stretches (chapter 3).

Tufts of algae provide a substratum of great importance to Protozoa (Picken 1937). Diatoms are often seen on the filaments of the alga, possibly attached to them by the attraction of two hydrophilic surfaces. Protozoa roam over the diatoms but are noticeably absent on the mud near by, even though the same diatoms lie thickly on it. Some Protozoa may be bound to the algal surface by the same force that binds the diatoms but there are also others, such as *Paramecium*, which do not go beyond the confines of the algal tuft in spite of being free to swim anywhere. Possibly there is a steep gradient caused by photosynthesis or some other chemical process and the Protozoa react away from this whenever they reach it. Some algae may be positively attractive to Protozoa. Whatever the cause, the whole community is obviously based on a substratum or framework of the right kind.

Popham (1941, 1943) has shown that, in aquaria, corixids that do not match the bottom are caught by fish more often than those that do.

EXPLANATION OF DISTRIBUTION IN TERMS OF CURRENT SPEED AND SUBSTRATUM

It is advisable first to give thought to the possibility that some other factor may produce an effect likely to appear to be due to current. A stream is sometimes both swiftest and coldest at the start, but confusion of these two factors seems unlikely. Many swift stony streams, and almost all those that have been studied, rise on hills covered with soil of low fertility. The absence of certain species in their consequently poor water may have led to an underestimate of the importance of food and an overestimate of the importance of current; for instance, Kaj Berg found typical productive-lake species of flatworms in moderately swift parts of the River Susaa (chapter 3, p. 35). Mackereth (1957) found inside *Perla bipunctata* nymphs more larvae of *Simulium* than there would have been had they been feeding at random on the invertebrate fauna of the stream. These sessile exposed larvae would seem to be easy prey for a large carnivore. A clump of them on the most exposed part of a stone could plausibly be attributed to selection of a current of a certain speed, but the truth could be that the large stone-fly nymphs had eaten all the larvae on the more sheltered parts of the stone but had been forced to leave a patch on that part where the current was too swift for them to be able to maintain a foothold. Huet (1942) noted that *Gammarus* is more abundant in small streams than in rivers and suggests that the greater volume of water in the latter sweeps them away. ('Le gammare se maintient parfaitement dans les ruisselets larges de 0·25 à 0·75 m dont la pente s'élève à plusiers per cent, mais dont l'épaisseur d'eau n'est que de quelques centimètres (1-5 cm). . . . Tandis que dans un ruisseau de 2 m de largeur ou plus, profond de 0·15 à 0·30 m, dont la pente n'est que de 1 per cent, mais dont la force d'entraînement sera plus grande que dans le ruisselet, le gammare sera absent.') Macan and Mackereth (1957) found *Gammarus* in much greater numbers in streams than in rivers but were unable to see how, if the bottom was stable, and they had in mind a place where it was, the speed and volume flowing over it could affect organisms living underneath. They expressed the belief, not supported by observations, that it might be predation that kept numbers low in rivers; for example *Cottus gobio*, which feeds on *Gammarus*, is often plentiful in rivers but is absent from most streams.

Allowance having been made for factors such as these, current could bring about a given distribution pattern in two ways: species tend to wander all over the substratum colonizing places from which they are washed away when current increases during a flood, or they exercise a definite selection. Harker's (1953) observation that floods affect the three genera of ecdyonurids to a different degree supports the hypothesis that current acts directly. Furthermore, Pleskot, contributing to a discussion (*Verh. int. Ver. Limnol.* 14, 604), described how *Torleya major* is washed away when the current rises much above normal. But I find it hard to believe that such a direct effect can be of widespread importance. If it is, each flood should grade the animals in the same way that it grades the components of the substratum, but such an effect has never been described. It would be interesting to know more about *Torleya*, and why, when current starts to increase, it does not retire to more sheltered places, as presumably other species do. As mentioned earlier, the evolution of appropriate behaviour has probably been more important in the colonization of rapid water than morphological adaptations. I doubt if the conditions in simple washing-away experiments have any counterpart in nature, but a combination of current and substratum may be important. A successful colonist of an unstable bottom requires a nimbleness that not all species possess. It may happen that an unstable area is colonized during a long fine spell by a number of species that are crushed, or caught up in the current and swept away in their efforts to avoid being crushed, when eventually a flood comes and sets the stones rolling. In the absence of observation only speculations on this possibility can be offered.

If direct action of current cannot entirely explain the distribution of animals in streams, the alternative must operate too and there is selection, not only of the most favourable but also of the least unfavourable conditions. The faster the water is flowing, the more particles pass a given point in unit time. Creatures that exploit this source of food may seek a compromise between the effort of holding on and full use of a food supply. That *Simulium* larvae place themselves where there is a given rate of flow is therefore easy to understand. The net-making Trichoptera are clearly bound to a place where the current is sufficiently strong to belly out their nets but not strong enough to carry them away, two extremes which may be close together as has been seen. Case-building species that feed on debris are most numerous in those parts of a stream where debris settles, and there are probably species in all groups whose distribution is related to their food supply. Some stream dwellers require a flow of water to bring them all the oxygen they need (chapter 9). This was demonstrated by Ambühl but how far the selection which he found in his experimental tank was for this purpose is difficult to say at the present time. There is certainly some active selection of different types of substratum. *Ancylus*, *Rhithrogena*, and *Gammarus* may be the three commonest species in a fairly fast stream, but there is no doubt from the work of Percival and Whitehead that the substratum requirements of each are distinctly different.

DESICCATION

Temporary pieces of water offer to those animals that can colonize them soon after they fill, the advantage that they are free from the attacks of predators at least for a time, and several species have adapted themselves to this type of place. *Cheirocephalus grubii* was found by Kreuzer (1940) to be a typical species of large temporary ponds in woods, and *Aedes* spp. and *Mochlonyx culiciformis* to be typical of small ones. He thought that the chironomid, *Ablabesmyia nemorum*, and possibly some species of water-mites, copepods and ostracods might be characteristic too. Trichoptera, Mollusca, and Coleoptera were also found. The snails and perhaps some of the others can probably lie dormant in the damp soil at the bottom of a dry pond, but for how long is not known.

The typical animals, fairy shrimps and mosquito larvae, are, as stated earlier, particularly vulnerable to predation, the former because they must swim to feed, the latter because they must come to the surface for respiratory purposes. Both survive waterless periods in the egg, which can live for a long time in dry conditions. The problem that confronts them is to strike a compromise between a quick return to the active stage when water comes and the avoidance of annihilation should the water not last long enough for development to be completed; in other words it is desirable that some eggs should hatch as soon as water returns and others not. Gillett (1955a and b) studied *Aedes africanus* and *A. aegypti*, two species in a large genus that is characteristic of temporary collections of water and of which the females lay eggs on damp soil. He found that various stimuli such as removal from water, changes in temperature or immersion in organic solutions stimulated hatching but that in any one batch of eggs there was a wide range in the intensity of stimulation required to bring about hatching. He experimented with various possibilities and eventually carried out a series of tests in which batches of eggs from the same female mated with a series of males were used. These led to the conclusion that a genetic factor determined how strong a stimulus was needed to cause an egg to hatch.

Eggs of *Cheirocephalus diaphanus* hatched about fourteen days after they were laid if they were kept in water, but a few from each sample had not hatched after twenty days. Eggs dried for a week took about fifteen days to hatch when placed in water but the proportion unhatched after twenty days was larger. On the other hand, after a longer period out of water the proportion whose hatching was delayed was less (Hall 1953). In a later paper (1959) Hall showed that eggs kept in water 20 cm deep or more did not develop and that the development was quicker the shallower the water. Evidently there was both an environmental and an inherent factor determining how soon an egg should hatch, but more than this it is not possible to say.

Stream animals also may have to contend with the disappearance of water. Hynes (1958b) found that Platyhelminthes, Oligochaeta, Ostracoda, Copepoda, Hydracarina, Coleoptera and Chironomidae survived in a stretch of stream that dried up for two months in an exceptionally fine summer. Ephemeroptera, Plecoptera, Trichoptera and other Diptera died out, unless they were in the egg stage which, however, could only tide over the dry period as long as the normal time for hatching did not fall during it. *Rhithrogena semicolorata* was unable to tolerate the conditions in any stage. Probably the survivors went into the soil. Ruttner-Kolisko (1961) has recently reported on the neglected fauna of this biotope. She found little in sand which, being below the water table, was always submerged, but many animals higher up where water was attracted by capillary action. Small animals such as ciliates, nematodes, rotifers and oligochaetes were the commonest, and harpacticoids, flatworms, gastrotrichs and tardigrades were found, though not in large numbers. Sometimes recently hatched chironomids, Plecoptera or Ephemeroptera were numerous. There is evidently continuity between the fauna in the stream and in the wet soil adjoining, and the disappearance of visible water is not as disastrous as the human observer might think for many of the animals.

Animals living near the margins of lakes are generally subjected to no greater rise and fall than they can keep up with; Moon (1935) showed that ground flooded by a rise in the level of Windermere was quickly invaded by the animals of the shallow water. When, however, a lake is turned into a reservoir, fluctuations are so much greater and the time for which the water is not near average level so much longer, that considerable changes are brought about. In Scandinavian lakes a high level in summer raises the lower limit of plant zones and a low level in winter leads to the destruction of those plants that are exposed. The zone of rooted vegetation in these lakes is thus narrower after they have been turned into reservoirs (Quennerstedt 1958). Fauna will be affected by the reduction of the vegetation, possibly by the greater size of the zone which waves erode, and by the inability of some species to keep up with the retreating water. Llyn Tegid (Bala Lake) has recently been used for flood control, and the engineers tend to keep the water low in winter so that heavy rain can be allowed to fill it and flooding lower down is avoided, and high in summer so that water is available should drought threaten. This is the same régime as that described by Quennerstedt. Hynes (1961a), making a survey when the level was low, failed to find most of the species typical of shallow-water regions of a lake that had been recorded in an earlier survey. Animals were, however, not less abundant as there had been a great increase in the number of oligochaetes. He doubts whether a position of stability has yet been reached.

MISCELLANEOUS

Gessner (1955) noticed in South America a strong contrast between the rich fauna and flora of clear mountain streams and the paucity of life in the permanently turbid rivers lower down. It was a casual traveller's observation, and he had no time to establish that there was a correlation or, if there was, how the turbidity affected the fauna and flora. That it may affect the flora is indubitable, and an influence on the fauna, apart from the indirect one through the vegetation, seems sufficiently likely to make the observation worth mentioning.

Finally here is one of those 'natural history' observations that sometimes go further towards explaining distribution than prolonged and elaborate experiments. *Gerris najas* on Windermere is restricted almost entirely to boathouses that have dry-stone walls. It is absent from those whose walls are smooth. The bugs hibernate in the crevices between the stones, and Brinkhurst (1959) believes that this flightless species is confined by the requirements of shelter from storms in the summer and suitable crevices in which to lie up for the winter.

SUMMARY

The intensity of water movement determines the composition of zoocoenoses through its effect on the bottom and on the plant communities. Most animals that dwell in running water show adaptations of behaviour not of structure, and live between stones or in vegetation, where there is little current. They are generally most numerous at one current speed and progressively scarcer in slower and faster water, though the limits may be a long way apart. It is unlikely that there is a direct effect of current unless the bottom is unstable, when agility, or some other ability that enables the animal to avoid the dangers of a shifting bottom may influence its distribution. Species that depend on the current to bring food select an optimum speed. Some animals require a current above a certain speed to bring them all the oxygen they require. The selection of a substratum of a particular type is common in both still and running water. It may be related to feeding requirements but sometimes no reason for the choice is apparent.

Inhabitants of temporary pools tide over dry spells in the egg stage and have developed mechanisms that ensure that one filling does not cause all the eggs to hatch. Small stream-animals are found in ground-water beside the stream, which is probably why many are not affected by disappearance of visible water. Excessive rise and fall in the level of a lake that has been dammed to store water leads to a change in the fauna. The occurrence of *Gerris najas* is related to the presence of conditions suitable for hibernation.

SYNONYMS

Species	Group	Other generic names	Other species names
Acroloxus lacustris (Linnaeus)	Moll. Gast.	*Ancylus*	
Aedes aegypti (Linnaeus)	Ins. Dipt. Culicidae	*Stegomyia*	*fasciata* Meigen *calopus* Meigen
Ancylus fluviatilis (Müller)	Moll. Gastropoda	*Ancylastrum*	*fluviatile* (Müller)
Dinocras cephalotes (Curtis)	Ins. Plecopt.	*Perla*	
Gerris najas (Degeer)	Ins. Heteropt. Gerridae	*Aquarius*	
Glossiphonia complanata (Linnaeus)	Hirud.	*Glossosiphonia*	
Limnaea pereger (Müller)	Moll. Gast. Pulmonata	*Lymnaea*, *Radix*	*ovata* Draparnaud *peregra*
Perla bipunctata Pictet	Ins. Plecopt.		*carlukiana* Klapálek *marginata* Panzer
Planaria alpina Dana	Platyhelminthes	*Crenobia*	
Potamophylax latipennis Curtis	Ins. Trichopt.	*Stenophylax*	
P. stellatus (Curtis)	,, ,,	*Stenophylax*	
Theodoxus fluviatilis (Linnaeus)	Moll. Gast. Prosobranchia	*Neritina*	

CHAPTER 8

Physical Factors (2)

TEMPERATURE

TEMPERATURES OBSERVED IN THE FIELD

The sun is the source of heat that warms most waters and its effect depends on the angle at which it strikes the surface. The amount of energy which reaches and is absorbed by the bottom may be important too in very shallow water. Heat is lost by evaporation, which varies with the saturation pressure at the temperature of the surface, the amount of moisture in the air, and the speed of the wind. There may also be a direct exchange of heat between air and water and between substratum and water.

In most temperate lakes the water is uniformly cold from top to bottom at the end of winter. When the sun has crossed the equator it starts to warm the surface layers, which, after a period, are so much warmer than the lower ones that the fiercest gales will not mix the two. The temperature in the depths of a deep lake in temperate latitudes is never far from 4°C.

Surface temperatures of 23·3°C to 29·5°C have been recorded in Lake Tanganyika in the tropics, and Ruttner records surface temperatures of 21·3°C to 31·4°C in Indonesian lakes (Hutchinson 1957a). The maximum temperature recorded in Windermere during six years was 23·1°C but this was in the Wray Castle harbour and therefore probably a little higher than the maximum in the middle of the lake (Jenkin 1942). Berg (1938) records a maximum of 22·0°C in Esrom, another temperate lake a little further north. Some Swedish lakes in the Arctic unfreeze only in unusually warm summers and the temperature never exceeds 4°C (Hutchinson 1957a). On two consecutive days in July 1952, Dussart (1955) recorded surface temperatures of 18·15, 13·9 and 10·35°C in three lakes lying respectively 1,947 m, 2,060 m and 2,172 m above sea-level. He also recorded on two consecutive days in July 1951 surface temperatures of 5·3°C and 14°C in two lakes a little over 2,000 m above sea-level; the colder was enclosed by high mountains and was open only to the north, the warmer was exposed on all sides. Pesta (1929) records maximum summer temperatures ranging from 3·2 to 17·5°C in alpine lakes.

Some lakes are subjected to special influences which keep the temperature higher or lower than might be expected in the prevailing climate. Cold springs (Dussart 1955) or the proximity of a glacier may be the cause of temperature lower than in neighbouring places. Stratification may lead to accumulation of heat. Anderson (1958) describes a shallow lake with no outflow in an arid region in America. Salinity, due mainly to magnesium sulphate, ranges from about 100 g/l at the surface to just over 400 below 3 m. The upper two metres, being less saline than the lower layers do not mix with them. The surface temperature reaches a maximum of 27°C, which is what would be expected in a lake of that size in that latitude, but the sun's rays, passing through to the lower layers, heat them in the same way that they heat the inside of a greenhouse, and the temperature reaches 50°C. Anderson mentions similar conditions in a Hungarian lake where 71°C were recorded.

An important feature determining water temperature is volume. The surface temperature of a shallow fishpond in Indonesia reached about 32°C at 16.00 hours and then fell until about 08.00 hours when it was 26°C. Its fluctuation was always a little behind that of the air temperature but it followed it closely, and the amplitude was only a degree or two less (Vaas and Sachlan 1955). The more water to be heated, the slower is the heating process. Gorham (1958), analysing the temperature readings obtained during the course of the bathymetric survey of the Scottish lochs, found that a maximum temperature was reached in June in small lakes but not until September in large ones.

The smaller the volume of water, therefore, the greater is its diurnal fluctuation and the higher the maximum that it is likely to reach. Pesta (1933) records a maximum of 20°C late in June in a pool at an altitude of 1,900 m. At the hottest time of year in Assam the temperature of the water in ponds and ricefields frequently exceeds 40°C (Muirhead-Thomson 1951). Assam is just outside the tropics and small collections of water nearer the equator probably get warmer.

Water may emerge from the ground at any temperature up to boiling-point, but it is generally colder than the air and it warms up as it runs away from the spring. This process, however, is influenced by several factors already mentioned. Eckel and Reuter (1950) discuss the formulae involved but point out that the biologist who wishes to discover the temperature at any given point will find it easiest to measure it directly. They, and Eckel (1953), give examples of changing temperature along a stream. Schmitz (1954) took a reading every hour at eleven points along a stream 4·47 km long on a sunny day at the beginning of September. The temperature of the spring-fed pond in which the stream originated ranged from 9·5-10·5°C during the course of the day. Lower down there was a greater range, which reached its peak 2·61 km from the source, where it was 10-13°C. Beyond this point there was no further increase in the maximum and a rise of 0·5°C in the minimum reduced the range. The maximum was later at each station downstream, occurring at 14.00 hours at the source and at 18.00 hours at the mouth.

The temperature of a spring often varies little throughout the course of a year, and, well below air temperature in summer, it may be above it in winter. The stream then gradually cools but there may be a considerable stretch in which its temperature is always several degrees above freezing-point.

During the summer of 1959 which, though at no time exceptionally hot, was notable for a series of fine sunny spells that started in May and went on till mid-October, Macan (1960b) concealed maximum and minimum thermometers at eighteen stations in eight small Lake District streams, none very high up, and visited them at convenient intervals between mid-May and late October, once, twice, or occasionally three times a month, generally towards the end of or after a fine spell. The maximum in mid-May at any one station was generally little below the highest temperature recorded, which, until the end of August or September, rarely differed much from the maximum observed at each reading. When the streams are compared, the range found was considerable, at one extreme being a temperature of 28°C, at the other one of no more than 16°C. The highest maximum was in a stream just below its point of origin in what, once a small lake, had been converted into a swamp, with islands and patches of emergent vegetation interspersed among pools of open water, by a lowering of the level at some time. As it was sheltered on three sides by trees and on the fourth by a hill, it was the sort of place where high temperature was to be expected. The stream with the next highest maximum, 24°C, meandered across a flat open stretch just above the measuring station, and the fourth highest, 22°C, was at the lower end of a beck. Macan (1958c) believes that the temperature here follows air temperature fairly closely. The coolest station, with a maximum of 16°C during the summer, was near the source of a stream that came out of a deep bed of peat probably lying on boulder clay and facing north. All the other cool stations were some distance from the spring, and the low temperature was due to the cooling effect of woods. For example, the upper station of Holbeck, already mentioned as being the second highest with a maximum of 24°C, was just above a long wooded gorge and the temperature at the lower station, not much more than 1 km downstream, never exceeded 18°C. Macan (1958c), making observations in weather when such effects were likely to be at their greatest, found that a stream lost 7·6°C flowing through a wood. Thermograph observations showed that the highest temperature was not reached during hot spells when there was not much water but in sunshine just after rain. Part of the rise was doubtless due to heat from the warm surface layers of soil, the water previously having come from cooler lower ones, but more important probably was the reduction of cooling by evaporation owing to the high humidity.

DISTRIBUTION AND TEMPERATURE

There are many animals and plants whose range appears to be limited by temperature. Though sometimes coincidental, the relation is probably generally real but its exact nature has rarely been investigated. In this section are given some examples of naturalists' field observations on range and temperature, and their speculations about its nature; in the next laboratory findings are described; and in the last some conclusions about how temperature limits range are reached.

Merriam's concept of 'life zones', in which species change with increasing altitude and latitude, and whose limits appear to coincide with 'isotherms drawn through places said to have equal sums of effective temperatures' is discussed by Allee, Emerson, Park, Park, and Schmidt (1949). In Europe it is possible to distinguish: species which abound in the warm south and become progressively scarcer further north; species which are widespread in the north, and further south increasingly confined to mountains, except where streams flowing from cold springs provide suitable conditions for them; and species which, most abundant in about the latitude of England, become steadily scarcer both to the north and to the south. It must be stressed that dividing lines between the groups are bound to be somewhat arbitrary. It is not necessary for present purposes to distinguish the alpine and the northern species or to dwell on other differences that are important in a historical study such as Thienemann (1950) has written.

The temperature measurements in streams described above were made by Macan with the object of finding out whether they threw any light on the distribution of *Heptagenia lateralis*. They did. It abounded at every station where the highest temperature was about 18°C or less and was scarce or absent from all the others except one, which it could have invaded from a cool zone lower down where it was particularly abundant. The species cannot grow in the coldest weather, and Macan's suggestion is that its absence from the warmer becks is due to a rise in temperature from a level below which the nymphs cannot grow to one above which they cannot live in a period so short that development cannot be completed. The occurrence of the species in Windermere, in which the temperature may rise above 20°C, demands only the assumption that the eggs can tolerate a higher temperature than the nymphs, because the lake warms up more slowly than the streams and does not reach the lethal temperature until after the nymphs have emerged. Davies and Smith (1958) had previously put forward a similar explanation of the range of *Simulium hirtipes*. *Rhithrogena semicolorata* is probably limited in the same way but it has a wider range than *H. lateralis* because it can grow at lower temperatures (Macan 1960a) and has a higher lethal temperature. This explanation appears plausible but it makes the assumption that temperature has three critical values; that below which the nymphs do not grow, that above which nymphs die, and that above which eggs die. Not one of these is supported by experimental observations. It also assumes that the time of emergence is related to temperature. This is certainly true of many Ephemeroptera because their life histories vary considerably according to the temperature (Pleskot 1951, 1958, 1961), but may not be true of all species. If emergence were related to some factor such as length of day, temperature might reach the lethal level before emergence started, and the species would be barred from places where this happened often.

Pleskot found a much bigger range of maximum temperature round Lunz in Austria. Fig. 11 is a revised version of her published data, which she has been kind enough to put at my disposal. It includes the streams in the Vienna woods as well as those in the neighbourhood of Lunz. The cold waters were high in the mountains or they originated from an underground source, much of the area being limestone, and the two warmest were outflows from lakes. Where there are two maximum readings, the lower represents an average maximum, the upper an occasional maximum that may be reached for a few hours during an unusually hot spell. The figures are based on spot readings. *Ecdyonurus austriacus* occurs in all the coldest streams, *E. venosus* in most of the warmest, and they overlap in three. How far either might extend in the absence of the other is impossible to say. *E. zelleri* occurs in two of the three streams where the two meet. Temperature affects not only range but time of emergence. In the coldest stream, *E. austriacus* was on the wing in August and September. The effect of higher temperature appears to be first to lengthen the flight period, then to reduce the number emerging in the warmest months, and finally to confine emergence to the spring. There is a similar relationship between the three species of *Rhithrogena* but their times of emergence are

Stream		Temperature Max.°C	Temperature Min.°C	Ecdyonurus austriacus	Ecdyonurus zelleri	Ecdyonurus venosus	Heptagenia lateralis	Rhithrogena hybrida	Rhithrogena alpestris	Rhithrogena semicolorata
Rainerquelle	1100 m	5·5	7·5							
Herrnalm	1300 m	14								
Schreyerbach	800 m	7	5							
Rotmoosbach	1100 m	9·5								
Seebach	700 m	11·5	3·8							
Seebachmündung	600 m	13(19)	2							
Bodingbach	600 m	17	1·5							
Weissenbach	600 m	17	1·5							
Mayrgraben	600 m	15(18·5)	0·0							
Schwarzlackn	1000 m	22·5	0·0							
Wienerwald main streams	300 m	20(27)	0·0							
Unterseebach	600 m	23·5	3·3							
Oberseebach	1100 m	24								
				IV V VI VII VIII IX X	IV V VI VII VIII IX X	IV V VI VII VIII IX X	IV V VI VII VIII IX X	IV V VI VII VIII IX X	IV V VI VII VIII IX X	IV V VI VII VIII IX X

11. Flight periods of Ephemeroptera emerging from Austrian streams. The continuous line indicates the main flight period, dots represent periods when small numbers are on the wing (data supplied by Professor Gertrud Pleskot).

less influenced by temperature. *Heptagenia lateralis* is found only in the warmer streams, where its life history varies according to the temperature in the same way as that of *Ecdyonurus* and is generally unlike that observed in the English Lake District.

Apatidea muliebris is found in Denmark in springs, by which Nielsen (1950a) means not only the actual place where the water emerges from the ground but the resulting stream down to a point where the temperature rises above 12°C. It has a short adult season from late April to late May, quick growth in summer, and a pupal stage which survives the winter. It is a life history which, adapted to the short summer and long hard winter of the Arctic, does not enable the species to exploit to the full the resources of a spring that is rather constant in temperature all through the year. Nielsen (1951) suggests that it was able to advance southwards when Europe was colder, and was left isolated, when the ice retreated, in springs where the temperature remained always fairly low. In other springs occur two related species, which are active as larvae all through the winter and whose adult stage is on the wing most of the summer. They are thought to be mutants of the original immigrant, whose new, much less rigid, life history is suited better to the new conditions. *Odontocerum albicorne* is common in Italy and also certain Danish springs (Nielsen 1950a). Adults of *Agapetus fuscipes* and *Wormaldia occipitalis*, two species found with it in the springs, emerge all the year round though it is doubtful if there is any reproduction in winter. Nielsen (1950a) thinks that these are southern species that do not require a high temperature but cannot tolerate a low one. They spread as far north as Denmark during a period of warmer climate. When colder conditions returned they could no longer inhabit the most northern parts of the territory they had occupied except in a few places. These were springs which, during the winter, remained warmer than other water.

Our largest pondskater, *Gerris* (*Limnoporus*) *rufoscutellatus*, is a northern European species, which does not extend as far south as Hungary and any of the Balkan countries, Italy or the Iberian peninsula. It is rare in northern and western France. It has been taken in the British Isles on fifteen occasions during the last eighty-eight years. On only one of these has more than one specimen been captured and all have been fully winged. All the dates were between 11 March and 14 June. There has never been any evidence of breeding, and this combination of facts leads Leston (1956) to the conclusion that all the British specimens are immigrants from the Continent, and that the species has never become established because our Atlantic climate is not suitable for it, which probably means that the winter is not cold enough. Leston suggests that the rate of change of temperature is insufficient to stimulate some process in embryonic or postembryonic development or that a mild winter upsets diapause development. There is no experimental support for these suggestions, but they are plausible. Lansbury (1961) recorded a male in western Ireland in September 1960 and, on the grounds that the finding of two out of sixteen specimens in the extreme west of the British Isles is not consonant with the postulate that all have come from the Continent, 'feels that Leston's hypothesis is not proven'. This is indisputable, but it is nevertheless one worth bearing in mind till opportunity occurs to prove or disprove it.

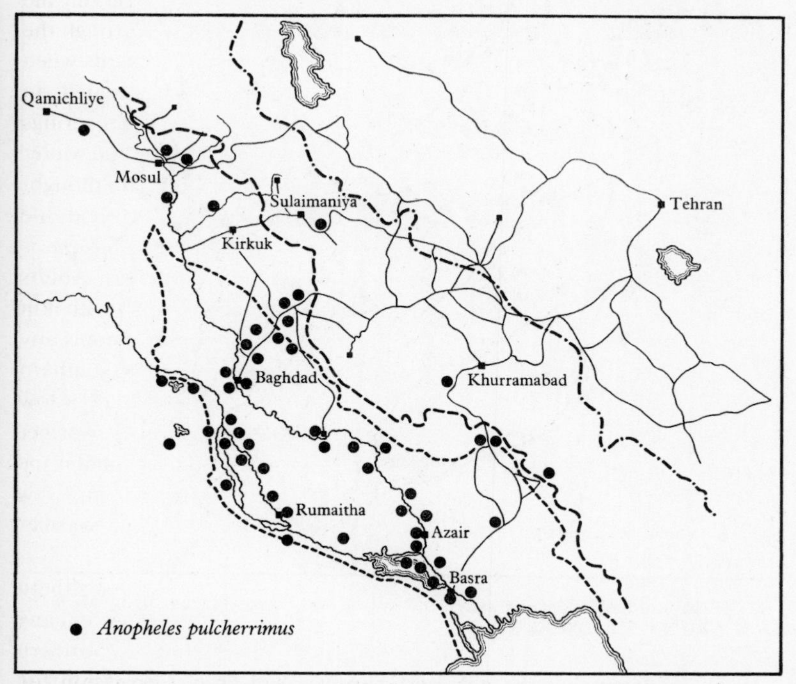

12. Distribution of *Anopheles pulcherrimus* in Iraq and Persia (Macan 1950) (*Memoirs of London School of Hygiene and Tropical Medicine* No. 7, p. 176, fig. 67).

Two oriental species of mosquito, *Anopheles stephensi* and *A. pulcherrimus* (fig. 12), are abundant in the alluvial plain region of Iraq (bounded by the short lines in fig. 12) but scarcely extend beyond a line that runs through the foothills (between the short and long lines on fig 12). On the other side are such palaearctic species as *A. sacharovi* and *A.*

superpictus (fig. 13). There is no physical boundary. Macan (1950b) found that all the species named except *A. stephensi* bred in the ricefields that occurred all over the area. Moreover, the boundary had been crossed, for there were a few scattered colonies of *A. pulcherrimus* in the mountains and of *A. superpictus* in the alluvial plain. It is likely that the boundary was a climatic one. It appeared to coincide with the line along which temperatures below freezing-point occur regularly every winter.

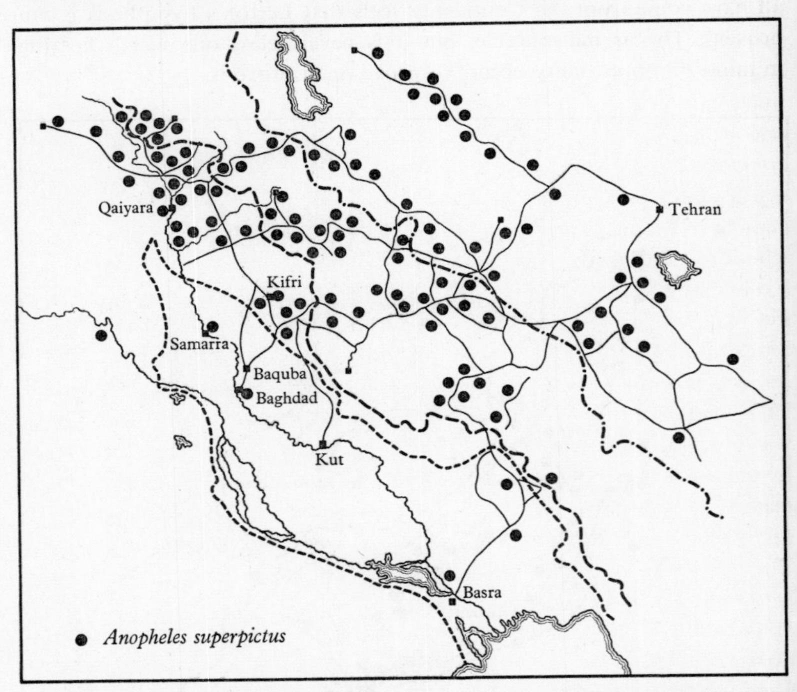

13. Distribution of *Anopheles superpictus* in Iraq and Persia (Macan 1950) (*Memoirs L.S.H.T.M.* No. 7, p. 200, fig. 71).

It is significant that the palaearctic species had special provision for hibernation and the oriental species had not. There are a number of ways in which temperature could have operated. It might, at its lowest, have been lethal either to larvae or to adults; perhaps neither were actually killed but were reduced to such a low level of activity that they died of starvation; another possibility is that activity continued but the life history took so long in winter that destruction by predators exceeded production. Less likely, but not to be ruled out, was an effect of low temperature on larval food more severe than on the larvae themselves. Competition may have prevented species spreading further than they could have done if temperature had been the only factor operating. Some seven possibilities, therefore, must remain until experimental work has been done, and this state of affairs is typical where the limits of the range of a species appear to coincide with some parameter of temperature.

The tortoise, *Emys orbicularis*, occurred in Denmark, southern Sweden, northern Germany and England 5,000 years ago, as subfossil remains show (Degerbøl and Krog 1951); to-day, presumably as a result of a change of climate, it is confined to southern Europe. Smith (1951) states that a certain minimum degree of heat is necessary to hatch the eggs, which one would not wish to doubt, but which has not been established by controlled observation. Rollinat (1934), who studied the animal in the south of France, records that eggs hatch after two months in a hot dry summer but may take four and a half months or more in a cold wet one. The effect of wet seems particularly unfavourable and, when inexperienced, Rollinat used to water the eggs because he thought it would be beneficial. He invariably killed them. Whether water acts exclusively by lowering temperature or in some other way has still to be discovered.

Certain snails found only in warm springs in Europe are believed by Thienemann (1950) to be survivors from a time when the whole continent was much warmer. In an Austrian stream temperature increases suddenly by about 5°C where water from a hot spring comes in. There are marked changes in the fauna, various Plecoptera disappearing, and *Physa acuta* and *Limnaea pereger* becoming suddenly abundant (Starmühlner 1961). The disappearance of the Plecoptera, a group associated with cold water, is to be expected, but there is no explanation of why the two snails become abundant. Mason (1939) has written an account of the fauna of an Algerian hot spring, and Matonickin (1957) gives a list of species that have been recorded in hot springs.

It is not always correct to conclude that the range of a species is limited by temperature just because it is confined to high or low altitudes and latitudes. *Corixa carinata*, *Glaenocorisa propinqua* and *C. wollastoni* all have a boreo-alpine distribution. *C. carinata* occurs in rock pools beside the Baltic though not inland except in the Arctic, and *G. cavifrons* was found by Crisp and Heal (1958) in six ponds not far from the west coast of Ireland. Since neither the Baltic rock-pools nor the Irish ponds can be particularly cold, it is unlikely that either of these species is kept within its range by lethally high temperature. *C. carinata* may be barred by competition with other species from all but cold places and the saline rock-pools along the Baltic coast, and the range of *G. propinqua* may be determined by competition or by choice of particular physiographical conditions.

This, which we may call the speculative section of the chapter, could easily, but not I think profitably, be extended to cover many more pages. Another author would have made the same points by means of a different set of illustrations.

OBSERVED EFFECTS OF TEMPERATURE

Salvelinus fontinalis, known as the speckled trout in America, although it belongs to a genus to which the name char is given in Britain, has the same sort of habitat as the brown trout, *Salmo trutta*, in Europe. Owing to the work of the Canadian School under F. E. J. Fry, more is known about the effect of temperature on this species than on any other.

S. fontinalis is rarely found wild in waters warmer than 19°C though it can withstand a rise up to 27·2°C if of not too long duration (Creaser 1930.) Deaths in a hot spell were observed by Huntsman (1946) in two streams whose temperatures on the day in question ranged from 25·4-31·4°C and from 23·6 to 31·1°C. Elson (1942) observed fish leaving a lake and swimming up a cooler stream as water temperature rose to a value of 21°C.

The method used to determine lethal temperature was described by Fry, Brett, and Clawson (1942), who were working with young goldfish (*Carassius auratus*) on this occasion. Fish were kept in tanks at constant temperature and, if it was desired to 'acclimate' them, to use the American word, temperature was raised 1°C per day. Fish acclimated to various temperatures were then exposed to various temperatures for fourteen hours in order to find out the highest temperature at which all would survive, the lowest at which all would die, and the temperature at which 50 per cent would survive. Table 12 shows a selection of the results taken from their graphs. Two things stand out: first, the difference between the

Table 12

Temperatures at which three levels of mortality were found in *Carassius auratus* (Fry, Brett, and Clawson 1942)

Acclimation temperature	Temperature at which: None dead	50% dead	All dead
1-2°C	26·5°C	28°C	30°C
10°C	29°C	31°C	33°C
17°C	32°C	33·5°C	35°C

temperature at which no fish die and that at which all die is only 3-4°C, and Fry *et al.* feel justified in treating the temperature at which half the experimental animals die in fourteen hours as the lethal temperature; secondly, the lethal temperature rises as the acclimation temperature rises, and in fact an increase of about 3°C in the latter causes a rise of about 1°C in the former. Fig. 14 shows that at 0°C the upper lethal temperature is 27°C, and, with an increase of acclimation temperature to 36·5°C, the

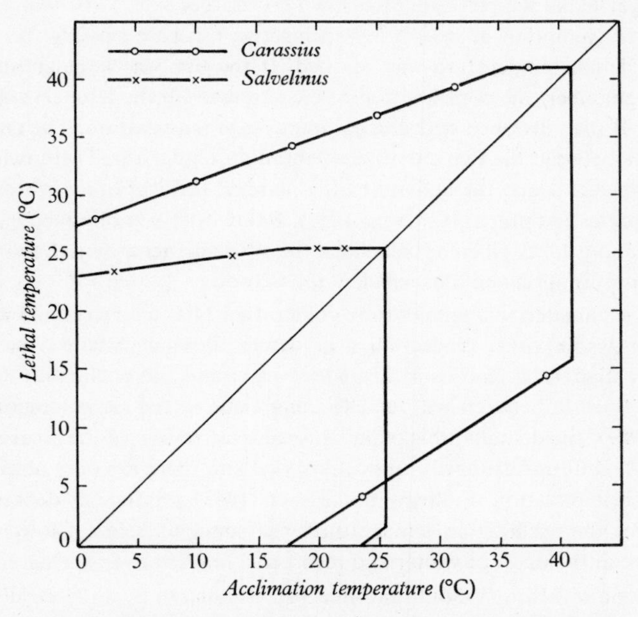

14. Relation between acclimation and lethal temperatures for *Carassius auratus* and *Salvelinus fontinalis* (Fry *et al.* from Macan 1961) (*Verh. int. Ver. Limnol.* **14,** p. 598, fig. 8).

lethal temperature rises to 41°C. Acclimation above 36·5°C does not raise the lethal temperature but prolongs the time of survival in temperatures above it. Fish acclimated to the warmest possible water die at about 17°C, and the lower lethal temperature does not sink to 0°C till the acclimation temperature is down to a level which, by coincidence, is also about 17°C. In fig. 14 a horizontal line joins the point where an increase in acclimation temperature no longer raises the upper lethal temperature and the point where the two are the same. From the latter a perpendicular drops to the point where cold water causes death. The resulting figure encloses 1,220°C², a useful measurement for comparative purposes.

The corresponding figure for young *S. fontinalis* (Fry, Hart, and Walker 1946) is the smaller of the two in fig. 14. The contrast is striking: the upper lethal temperature is only 25·3°C in the trout against 41°C in the goldfish; acclimation is slower, a 7°C increase in the acclimation temperature producing a rise of 1°C in the lethal temperature; the area of the figure, 625°C^2, is but half that of the goldfish; on the other hand it is not till the acclimation temperature reaches 23°C, which is only just short of the upper lethal temperature, that low temperature kills *Salvelinus*.

The consumption of oxygen by an inactive fish rose steadily from 0°C till the lethal temperature was reached. If the fish was kept active in a rotary chamber, the consumption of oxygen rose till the temperature was 19°C and then dropped with further increase in temperature. The greatest distance between the two curves is at about 16°C (fig. 15). There is, therefore, greatest scope for activity at this temperature; in other words it is physiologically optimal (Graham 1949). Baldwin (1957) obtained a lower figure, about 13°C, when he calculated the optimum according to the maximum consumption and most efficient use of food.

Fish acclimated to a temperature colder than 14°C tend to congregate in warmer water when placed in a gradient; those acclimated in water warmer than 19°C tend to seek colder water; and fish acclimated to any temperature in between will tend to congregate in the same temperature (Fry 1951). Incidentally, this paper is a valuable review of the work on *S. fontinalis*, but unfortunately it is cyclostyled and there are only about 100 copies in circulation. Sullivan and Fisher (1953) produce evidence that, whatever the acclimation temperature, fish suddenly select a lower temperature at the onset of winter and revert to a higher one in spring.

Fisher and Elson (1950) found that fish stimulated by an electric shock darted forward, and that the distance covered increased with increasing temperature up to 10°C and then decreased. This is a lower optimum than any of the others, probably because all the fish used had been acclimated to low temperature.

Time of oviposition may depend on temperature, but Fry (1951) writes that the information about *S. fontinalis* is too conflicting to be of value. The same is true of some other fish, but a well-authenticated example is provided by Rawson (1945). The spawning of *Micropterus dolomieu*, the smallmouth black bass, is stimulated by a fairly quick rise in temperature to the neighbourhood of 16°C, but successful development ensues only if the temperature remains constant for about three weeks at a slightly higher level. If it falls, the males desert the nests and the eggs die. Rawson noted that, in two years out of four, there was no production of young for

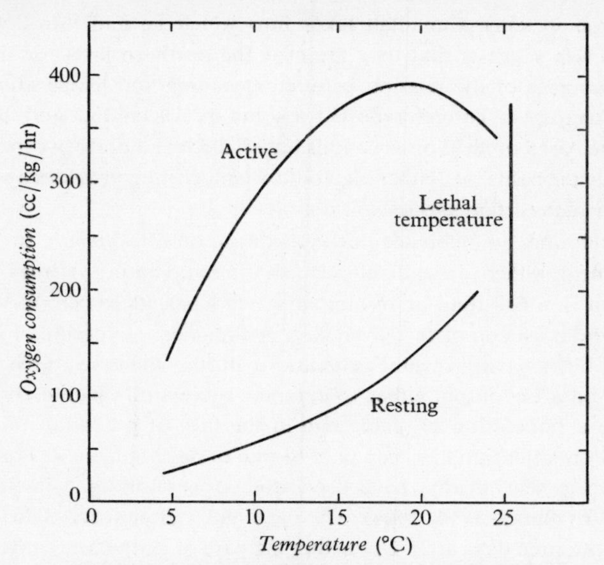

15. Relation between oxygen consumption in active and resting *Salvelinus fontinalis* and temperature (Graham 1949) (*Canadian J. Res. D.* **27**, p. 275, fig. 3).

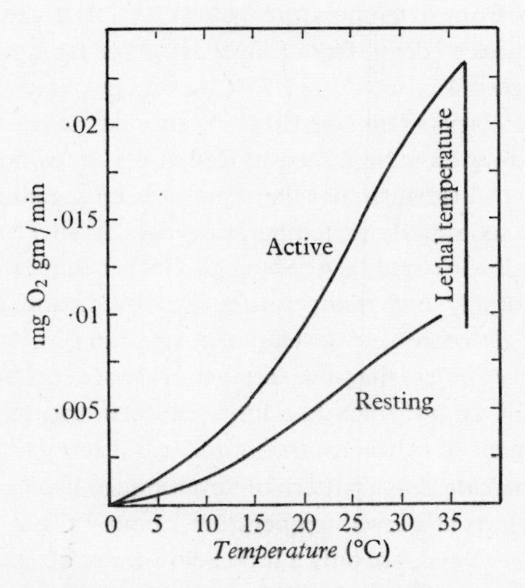

16. Consumption of oxygen by active and resting *Ameiurus nebulosus* at different temperatures (Fry 1947) (*Publ. Ontario Fish. Res. Lab.* **55**, p. 38, fig. 17).

this reason in some Canadian lakes into which he had introduced *M. dolomieu*. He suggests that they are near the northern limit for this fish. Other examples of the relation between spawning and temperature were given in chapter 4. Though reports for some species conflict and those for others are based on field observations unverified by laboratory experiment, it is highly probable that either high or low temperature provides a stimulus without which certain species will not breed.

I have found no reference in the literature on *S. fontinalis* to rate of development, which depends closely on the temperature. It may be less important in a fish than in an animal which has a more complicated life history and must complete a given stage within a certain period of time, or in one which, having several generations during the favourable season, depends on a big output either to compete successfully with a rival or to maintain a population of viable size in the face of predation and other losses. More attention has been paid to rate of development in insects and among them few, if any, have been studied as thoroughly as *Dytiscus marginalis* (Blunck 1914, 1924). The eggs hatch after forty-eight days at 7°C and after ten days at 21°C. Further increase of temperature up to 30°C, which is lethal, reduces the development time by only two days. The optimum temperature lies between 10°C, at which development takes twenty-seven days, and 15°C at which it takes eleven days. The egg itself suffers no harm from a temperature below 10°C but the longer period increases its chances of death from fungal attack or from misadventure to the plant in which it is lying. Above 15°C fewer eggs hatch and more give rise to malformed larvae (Blunck 1914). Young larvae do not grow at all at 10°C, though they have been seen to feed at 6°C, at which temperature, however, they are so sluggish that they cannot keep themselves alive. Rate of growth increases rapidly as temperature rises from 11°C to 15°C and thereafter more slowly (fig. 17). At about 27°C what Blunck (1924) calls the zone of unfavourably high temperature is entered, and the number of specimens that fail to complete development starts to rise. A short exposure to 43°C kills a larva in less than five minutes, a short exposure to 35°C does it no harm, but it cannot tolerate a long period in this temperature. The rate of development of other instars is similar but not identical.

Dytiscus semisulcatus has a pattern of development like that of *D. marginalis* but growth starts at a lower temperature, about 3°C (fig. 17). The upper limits, on the other hand, are only a little below those of *D. marginalis*, and the zone of unfavourably high temperature begins at about 26°C. The amount of food taken increases with increasing temperature up to the zone of unfavourably high temperature when it starts to decrease rapidly.

Dytiscus semisulcatus larvae are found from November to June, that is during the season when larvae of *D. marginalis*, which occur during the summer, could not survive. It is not known whether the different breeding season enables these two species to avoid competing with each other, or eating each other, in the most active stage, nor whether *D. semisulcatus* can extend into colder water. Since *D. marginalis* has been found inside the

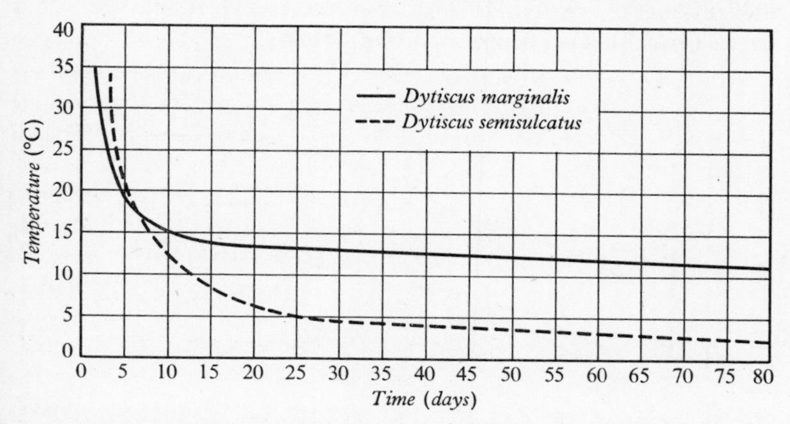

17. Rate of development of first instar larvae of *Dytiscus* spp. at different temperatures (Blunck 1924) (*Z. wiss. Zool.* **121**).

Arctic circle, the adult stage can obviously withstand a long and rigorous winter.

There is less information about other species than about *Salvelinus* and *Dytiscus*, and often it refers to no more than one of the various aspects of temperature so far mentioned. Brett (1944) and Hart (1947) give data about tolerance by other species of fish. They found none with an upper lethal level as high as that of *Carassius* or as low as that of *Salvelinus* (fig. 14). Fry (1947) and Brett (1956) give more general accounts of temperature relations.

The consumption of oxygen by an active *Salvelinus namaycush*, considered in relation to rising temperature, reaches a peak at a temperature below the lethal level and then decreases, as does that of *S. fontinalis* (fig. 15) (Gibson and Fry 1954). The consumption by an active *Ameiurus nebulosus* (Fry 1947) and an active *Salmo trutta* (Brett 1956), in contrast, increases right up to the lethal temperature (fig. 16).

Donaldson and Foster (1941) found that the sockeye salmon, *Oncorhynchus nerka*, grew little at 4-7°C and again at 21°C. At 25·6°C they would not even eat. Rate of growth and efficiency of utilization of food were greatest at about 10°C. Gibson (1954) and Gibson and Hirst (1955) reared

Lebistes reticulatus, the guppy, at different temperatures in fresh water. Growth was faster at 23-25°C than at 20°C and 30°C. Moreover, fish reared at 30°C not only died in larger numbers but neither attained the size nor displayed the activity of those reared at 25°C or 20°C. It was possible with difficulty to bring them to maturity at 32°C but none of the survivors was fertile. Evidently between 25°C and 30°C this fish enters Blunck's zone of unfavourably high temperature and in it rate of growth is reduced, which did not happen in that of *Dytiscus*.

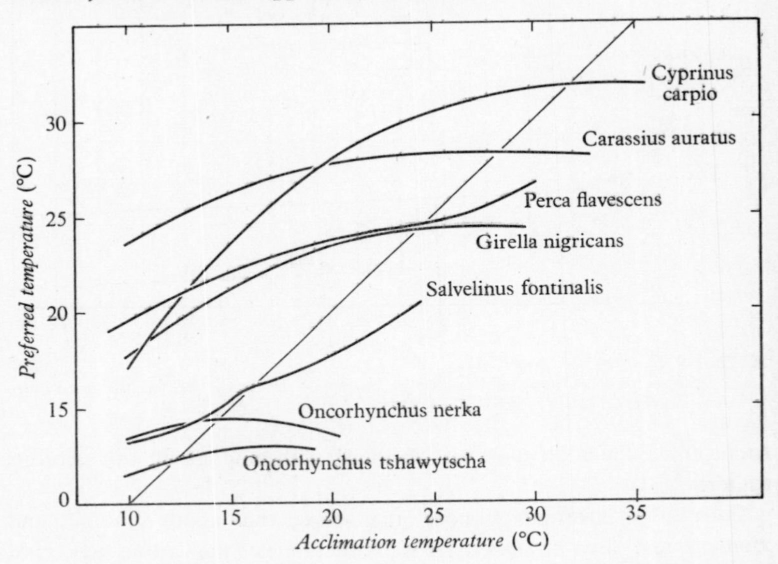

18. Laboratory studies of preferred temperature: the yellow perch (*Perca flavescens*) compared with various other fish. The line at 45° joins all points where preferred and acclimation temperatures coincide (Ferguson 1958) (*J. Fish. Res. Bd. Canada* **15**, p. 611).

The upper lethal temperature of *L. reticulatus* is 33°C and the species appears to be unusual in that this is little affected by the acclimation temperature.

Ferguson (1958) investigated the temperature in a gradient at which various species congregated. He found that the temperature selected rose at first as acclimation temperature increased but gradually became constant. The point where the curve began to flatten was 13°C in *Oncorhynchus tshawytscha* and not till over 30°C in *Cyprinus carpio*, the carp (fig. 18) Pitt, Garside, and Hepburn (1956) found that the carp would choose water warmer than that at which it had been living until this reached 35°C, when it selected water at 32°C. On the other hand Schmeing-Engberding (1953) gives only 21·3°C as the temperature selected by *Cyprinus* in a gradient. Probably his results for this and other species were low because he tested them from one acclimation temperature only. Garside and Tait (1958) acclimated young rainbow trout, *Salmo gairdneri*, to four temperatures, and then transferred them to a gradient. The fish that had been living in cold water went to warm water and vice versa, unique behaviour as far as existing information goes (Table 13).

Table 13

Position in a gradient of specimens of *Salmo gairdneri* acclimated at four different temperatures (Garside and Tait 1958)

Acclimation temperature	5°C	10°C	15°C	20°C
Position in the gradient	16°C	15°C	13°C	11°C

An example of a relationship not yet encountered is given by Fish (1948). The severity of the attack of a myxobacterium on the blueback salmon, whose Latin name is not given, increases with rising temperature. Below 15°C it is innocuous, between 15° and 20°C it invades wounds and sets up secondary infection, and above 20°C it can invade healthy undamaged tissue. A heavy mortality of salmon in an exceptionally warm summer was laid at the door of this myxobacterium. The conclusion does not seem to have been based on post-mortem examination of dead fish.

The only invertebrate which has been studied with an attention to detail comparable to that which is such a feature of the experiments on fish is referred to as the American lobster by McLeese (1956), who quotes no scientific name. Complete acclimation to a new temperature took about twenty-two days and the lethal level was taken as that at which half the animals died in forty-eight hours. Some results are set out in table 14. The

Table 14

Temperatures at which mortality of the American lobster reached the levels shown in forty-eight hours (McLeese 1956)

Acclimation temperature	Temperature at which: None dead	50%	All dead
5°C	24°C	25·7°C	27°C
15°C	27°C	28·4°C	29·5°C
25°C	28°C	30·5°C	32°C

values were lower at the time of moulting and in water of low salinity or oxygen concentration. The paper is particularly valuable because it shows how these three variables interact to produce limiting conditions (fig. 1, p. 5).

Other work has been less elaborate. For example, Muirhead-Thomson (1951) raised the temperature 1°C in five minutes until mosquito larvae died in this period of time. Lethal temperatures established in this way were: *Anopheles minimus*, 41°C; *A. hyrcanus* and *A. barbirostris*, 44°C; and *A. vagus*, 45°C. These values do bear a relation to the maximum likely to be encountered in the usual breeding-places, but, as was stated in chapter 5, the occurrence of the larvae is brought about by the selection of the ovipositing female reacting to a factor unrelated to temperature, and only rarely does *A. minimus* lay eggs in places that become too warm for the larvae; this occasionally happened at the height of the dry season. *A. claviger* dies in five minutes at 37°C and in sixty minutes at 35°C, and larvae have not been found in water warmer than 21°C. It is found in marshes in Britain but not in Palestine, where the temperature of such places ranges from 35 to 39°C. The factors that attract the gravid female of this species have not been investigated.

Whitney (1939) used a longer exposure time, and called the temperature at which 50 per cent of the animals died in twenty-four hours the thermal index. It was 21°C to 21·3°C in *Baetis rhodani* and 22·4 to 24·7°C in *Rhithrogena semicolorata*. The index in this second species varied according to the stream from which the samples were taken and might, it may be deduced, have been more uniform had the material been acclimated before the experiments. In contrast to these two stream-dwelling Ephemeroptera, the pond species, *Cloeon dipterum*, had a thermal index of 28·5 to 30·2°C.

There has been controversy about the temperature relations of *Planaria alpina*. It is rarely found at temperatures above 15°C (Beauchamp 1935, Dahm 1958). According to Beauchamp (1933) sexual development is associated with a temperature of 10°C or less but the extensive researches of Dahm have shown that the relation between sexual reproduction, asexual reproduction and temperature is complex, being slightly different in almost every population. There may be either kind of reproduction or both, in which case sexual reproduction takes place within the range 6-12°C and asexual reproduction at slightly higher temperatures, but amount of food is also a determining factor. Steinböck (1942) denies that this species is a cold-water stenotherm and quotes in support the finding of it in an alpine pool that reached a temperature of 22°C during an afternoon in August. However, he admits that the temperature dropped to a much lower level during the night. Schlieper and Bläsing (1952) made some experimental observations to counter Steinböck's assertion. They showed that temperatures of this order of magnitude could be tolerated as long as the animals were not exposed to them for too long. An exposure to 25°C caused no harm provided it did not exceed twelve hours in winter and forty-eight hours in summer. 20°C could be tolerated for up to a week in winter and 'indefinitely' in summer, but this does not agree with the findings of Beauchamp (1935), who was unable to keep this species alive for a long period in water warmer than 12°C.

Beauchamp (1935) suggests that rate of movement of *P. alpina* depends on a balance between two nervous processes, one excitatory, the other inhibitory. The function of the inhibitor is disturbed by temperature above 12°C. Another effect of exposure to this temperature is to make the animal positively rheotactic (Beauchamp 1937) whether it is fully fed or not.

The rate at which invertebrates feed, and the efficiency with which they convert what they have eaten at different temperatures, have been little investigated, though these factors may be of considerable ecological importance. The number of chironomids which four invertebrates ate in ten days at different temperatures is given in table 15. All are similar in

Table 15

Number of chironomids eaten in ten days at various temperatures (Pacaud 1948)

	5°C	10°C	15°C	20°C	28°C
Calopteryx splendens	1	8	15	18	23
Notonecta glauca	2	8	15	23	28
Erythromma sp.	1	3	7	7	8
Gammarus pulex	6	8	10	12	15

that least is eaten in the coldest water but they differ in the rate at which the rate of feeding is increasing, and *Erythromma* reaches the maximum rate at a lower temperature than the others. There are no data about rate of growth at the different temperatures. Figures of this kind for every species in a community, together with some about rate of growth, particularly of the organisms eaten, might prove highly instructive.

The mouth-brushes of *Anopheles minimus* move twice as fast at 35°C as at 15°C and cease to move at a temperature between 37 and 38°C (Muirhead-Thomson 1951).

An aquatic moth, *Acentropus niveus*, does not feed at a temperature below 11-12°C in Denmark (Berg 1941) though the limit is lower in Finland (Palmén 1953).

Its effect on their rate of development is the aspect of temperature in relation to invertebrates that has received more attention than any other.

Table 16

Number of hours taken by *Anopheles quadrimaculatus* to complete development from the hatching of the egg to the emergence of the adult at the temperatures shown (Huffaker 1944)

Temperature	12·1	15·1	19·2	22·0	24·7	27·2	29·6	32·3	34·6	°C
Hours	1572	768	501	319	245	210	190	178	203	hours
% Survival	1·7	20	84	76	84	88	92	80	44	%

Table 16 shows the data given by Huffaker (1944) for *Anopheles quadrimaculatus*. Hurlbut (1943), making the same kind of observations on the same species, obtained times for development about half as long again, which, together with the high mortality at temperatures above and below the optimum, indicates that *Anopheles* larvae are difficult to rear and perhaps not particularly suitable for this kind of work. Muirhead-Thomson (1951) quotes no figures for the larvae of *A. minimus* for this reason, but gives some information about eggs and pupae. These stages take seven and four days respectively at 16°C, a temperature which is near the lower limit for development. The egg stage is completed in about two days at 30°C and less at 35°C, at which temperature, however, many fail to develop. The pupal stage is passed through in just over a day at 30°C and there is little emergence at 35°C. Fig. 19 shows how long it takes a newly hatched larva of *Aëdes aegypti* to reach successive stages at various temperatures (Bar-Zeev 1958) and it will be noted that the upper and lower limits are nearly the same as those for eggs and pupae of *Anopheles minimus*. At the highest and lowest temperatures there was heavy mortality. Davis (1932) found that 18°C was the lowest temperature at which a colony would maintain itself. It died out in less than two months at 36°C. All these are tropical species. *Culex pipiens*, a species of temperate regions, took twenty-one days to complete a generation at 20°C and eight days at 30°C, times which are not greatly different from those of *Aëdes aegypti* at the same temperatures (Kramer 1915). Other temperate species appear not to have been investigated.

Data more satisfactory, in that the comparison is better, are given for Cladocera by L. A. Brown (1929). *Moina macrocopa* took 297 hours to pass from the first young instar to the first adult instar at 13°C and the time fell steadily with rising temperature till it was thirty-eight hours at 35°C. *Pseudosida bidentata* achieved the same in 517 hours at 13°C and eighty-two hours at 35°C. Times for *Daphnia*, of which three species and two races of *D. pulex* were tested, fell between these two at the lower temperatures, but the time taken at 30°C was only a little shorter than at 25°C in two of them and longer in the other two. None completed development at 35°C. Lethal temperature was defined as that which brought about death in one

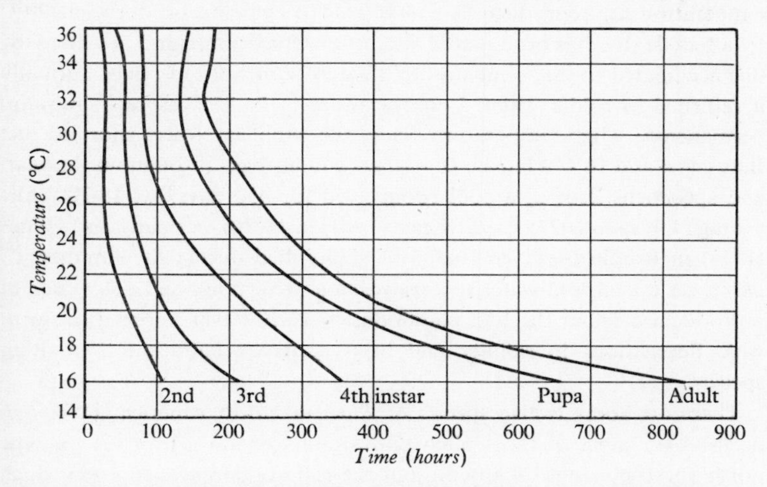

19. Time taken for newly hatched larvae of *Aëdes aegypti* to reach successive stages of development at different temperatures (Bar-Zeev 1958) (*Bull. Ent. Res.* **49,** p. 159).

minute, and was found to be several degrees higher for the first- than for the last-named. The species of *Daphnia* had a more northerly range than the others, which were confined more to the warmest part of the year in colder regions (L.A. Brown 1927).

The eggs of the American freshwater shrimp, *Gammarus fasciatus*, complete development in a week at 24°C and in about three weeks at 15°C; they have been found at 4°C in the wild. The breeding season extends from March to October and Clemens (1950) calculates that seventeen batches of eggs could be produced during this time. The shrimps became mature at the eighth moult and development to this stage took an average of fifty-nine, fifty-three and twenty-eight days at 18°C, 21°C and 25°C respectively. When development was faster, the mature female was smaller and she produced fewer eggs.

The eggs of the north European *G. duebeni* take twelve to thirteen days to complete development at 19°C and forty-five at 6°C. There are fourteen to fifteen moults to maturity and the average time taken over one moult ranges from sixty days or more at 6°C to about ten days at 19°C. Below 6°C there is little development and above 19°C the rate of growth slows down. A steady temperature of 26°C was lethal but a fluctuating one reaching this level occasionally killed few specimens (Kinne 1953). Eggs do not develop at temperatures above 22°C at optimum salinity. The optimum temperature for reproduction is 4-16°C. No eggs are produced at 20°C by a female that has been reared at a lower temperature and not many by one acclimated to that temperature, though both will reproduce normally if returned to colder water. Long exposure to 24-25°C stops oviposition permanently. High temperature below the lethal level also shortens life. Between 4 and 16°C a female *G. duebeni* will lay eggs throughout the year, and specimens kept in a cool room lived for 450 days. At 19-20°C the average life span is 275 days. Kinne (1954) in North Germany and Hynes (1954) in Britain observed that wild adults died during the summer. *G. duebeni* is thus a cold-water species with a narrow range and a slow rate of reproduction under the best conditions. It is, however, more tolerant of wide fluctuations in salinity and in oxygen concentration than other species.

There are some twelve species of *Rana* in North America. *R. pipiens* occurs over most of the United States and Canada except the extreme north and the western seaboard, but the rest are confined to areas which range from just over half to one hundredth of this. Some show a zonation from north to south, and, with decreasing latitude, each has a higher limiting temperature, both high and low, and a higher temperature for development (table 17). Although the temperature range of each species covers a different part of the scale, the range itself is much the same, 22-24°C, and, although at any given temperature, the most northerly species will grow fastest, the time taken to complete development at optimum temperature is not greatly different in the four (Moore 1939). Some of these species at least are constant in their thermal characteristics from the north to the south of their range, but *R. pipiens* is not, and it tends to have a higher threshold for tolerance and development the farther south it occurs. These populations are frequently intersterile, and this appears to be an example of the splitting of one species into several by the selection of genotypes adapted to conditions at different places. This is the aspect which interests Moore (1949). The significance to ecologists is equally great, though for them the picture is not quite complete. It seems

Table 17

Some temperature relations of four American frogs (Moore 1939)

	Rana sylvatica	*R. pipiens*	*R. palustris*	*R. clamitans*
Approx lower lethal temp.	0	2·5	5	10°C
Development starts at	2·5	6	7	11°C
No development above	24	28	30	35°C
Upper lethal temperature	25	30	31	37°C
Time taken to develop from stage 3 to stage 20 at 18·5°C	87	116	126	138 days
Water temperature at time of breeding	10	12	15	25°C
Most northerly record	67° 30″ N	60°N	55°N	50°N
Breeds in	March	April	April	June

to be important that each species breeds at the same time throughout its range, and evidently it cannot extend its range by breeding later further north, as many other animals do. Moore makes no comment on why this should be so and, moreover, gives no information about the requirements of the adults, which might provide the key to the riddle.

TEMPERATURE AS A LIMITING FACTOR

Temperature acts on organisms in diverse ways, and the organisms themselves have diverse life histories and habits. Moreover the consideration of temperature in isolation is an expedient which, adopted here because so much research has been on factors in isolation, is fundamentally deplorable because ecological factors nearly always operate in conjunction with others to produce their effects (figs. 1 and 2). It is to be expected, therefore, that a discussion on temperature as a limiting factor will be involved and often inconclusive.

Every species must have an optimum temperature, but the optimum is difficult to define. *Salvelinus fontinalis* was most active at 16°C, but at a temperature three degrees lower its feeding and utilization of food was most efficient. Rate of growth of one species of insect may not have the same relation to temperature in each instar. Quickest growth in *Gammarus duebeni* takes place at a temperature unfavourable for reproduction, and in many animals is associated with high mortality and the production of monsters. A good illustration of the complexity is provided by D. M. Pratt (1943), who cultured *Daphnia magna*, though the crowded conditions in his containers are probably rarely, if ever, seen in the wild. Growth was more rapid at 25°C than at 18°C, but so much more rapid that a dense population was produced in a short time and the overcrowding caused heavy mortality and brought reproduction to an end. The population generally died out. At 18°C numbers began to depress the rate of reproduction before they had reached a catastrophic level, and a stable population was maintained. Cultures kept at 12°C soon died out. The temperature of 18°C was well below the optimum for growth and reproduction but it was near the optimum for survival. The concept of the temperature optimum is, like many another in biology, useful until attempts are made to define it exactly. Grading into it on either side, rather than bounded by a recognizable and measurable limit, and approaching or retreating as conditions alter, are the zones of unfavourably high or low temperature. Beyond them, with a limit that cannot be transgressed but operating inside it if all other conditions are not at their most favourable, are the lethal temperatures (Blunck 1924, Bates 1949).

Unfavourably low temperature acts by depressing activity, whether it be movement, feeding, growth, or reproduction. Many animals avoid ill effects from it by having a facultative or an obligatory resting phase. *Salmo salar*, for example, lies inactive in pools as long as the temperature is below 7°C (Allen 1940). Vorstman (1951) found *Neomysis integer* commonly in the plankton until temperature sank below 3°C when they went down to the bottom. As soon as the temperature rose above 3°C the shrimp resumed its planktonic way of life. The physiology of this process has not been studied. A regular resting period or hibernation is frequently spent in the egg, which is the most resistant phase in most life histories. Many mosquitoes lay down a fat body in the autumn and tide over the winter as torpid adults. The development of some such provision for the cold season has probably been essential to colonists of the temperate region and may sometimes be the only difference between them and their tropical ancestors.

The zone of unfavourably high temperature is less easy to escape but the quiescent egg stage is sometimes one which is more resistant than any other. *Rhithrogena semicolorata* and *Heptagenia lateralis*, two Ephemeroptera which are found only in the egg stage at the height of summer and which have a longer egg stage in warmer streams, provide an example which, though taken from the speculative section, is reasonably convincing. Eggs of *Artemia salina* can withstand exposure to 103°C (Hinton 1954) which is no doubt an adaptation to life in temporary pools in warm parts of the world.

There are three ways in which a species is prevented from colonizing warmer water.

(1) It is lethal. Many species are found only in cold water; a few, *Salvelinus fontinalis* for example, have been tested and have been shown to be unable to live in water warmer than that in which they occur naturally. The reactions of *Planaria alpina* take it upstream when temperature reaches a dangerous level, and it finds safety, though a warming up of the climate since the Ice Age has no doubt eliminated it from many streams. *S. fontinalis* chooses a temperature well below the lethal level, if it can find one, though occasionally it is caught and destroyed by excessive warmth. Inhabitants of lakes, some of which will be mentioned in the next chapters, seek refuge from the upper layers, if they cannot tolerate the temperatures which these reach in summer, in the cold lower ones. The gradual enrichment of a lake, till it reaches a stage at which all the oxygen in the lower layers disappears, exterminates such species in summer.

(2) There is competition. *Chthamalus stellatus* is a tropical, *Balanus balanoides* an Arctic, barnacle and the two meet along the coasts of Britain. Both eat the same food, of which there may be a shortage where crowding is dense, conditions that are particularly difficult for small new specimens among larger adults. If nauplii arrive when adults are actively feeding, many will be eaten. The cirral activity of *C. stellatus* is the more efficient above 15°C, that of *B. balanoides* below it. The latter sets free one generation in March or April and will, therefore, benefit in competition with *C. stellatus* from a cold summer, particularly one that is cold early on. A warm one, particularly if high temperature persists well into autumn, favours *C. stellatus* by enabling it not only to be more active but also to produce more young (Southward and Crisp 1956). That the boundary between the two, which is fairly sharp, should move to and fro along the coast from year to year in a way that is predictable from the requirements mentioned, leaves little doubt that competition limits the range of both species.

Competition among freshwater organisms has been described in chapter 6. It seems likely that many species do not occupy the whole range which their temperature tolerance would permit because, towards the limits, they encounter some species with the same way of life, but a different temperature optimum. At the temperature in question the second species can move faster, eat faster, utilize food more efficiently or breed faster, and the first cannot compete with it. A relationship of this kind is probably of the widest importance in the ecology of freshwater animals.

(3) The temperature is never low enough to stimulate reproduction. Exposure to low temperature is necessary in the reproductive cycle of *Balanus balanoides* (Barnes 1958a) and the same is probably true of many of the freshwater organisms that breed in winter: for example *Salmo trutta* and *Planaria alpina* already described, *Mysis relicta* (Holmquist 1959), and *Cheirocephalus* sp. (Orton 1919). If such species had a high lethal temperature, there would be large areas of the world barred to them only because the water never got cold enough to stimulate their reproduction. In fact they are cold-water species unable to tolerate high temperatures and, though there undoubtedly are places from which they are kept in this way, they cannot be numerous, and limitation of this kind must be infrequent.

Similarly species are prevented from colonizing colder water because:

(1) It is lethally low at some time of year. Orton and Lewis (1930) attribute the disappearance of several pests on an oyster bed to an unusually severe winter. In fresh water in temperate regions, where winter temperature at or near freezing-point is the rule, such a direct effect must be confined to streams near the source where the temperature is usually higher than it is lower down. Nearer the equator in regions where frost, or even higher low temperatures, is exceptional, it may be important. The range of *Aëdes aegypti*, which is circumterrestrial and which is bounded to both north and south roughly but not exactly by the 10°C winter isotherm, may provide an example, for 10°C is about the lethal temperature for egg, larva, and adult (Christophers 1960).

(2) The threshold for development or activity is not exceeded, or if it is, development or activity is too slow. Between the lethal temperature of 10°C and about 16°C larvae of *Aëdes aegypti* do not grow and adults remain torpid, and Davis (1932) finds that 18°C is the lowest temperature at which a laboratory colony will perpetuate itself. If temperature remains long in this zone, but does not sink below the lethal level of 10°C, distribution may depend on how long the larva can live without growing and how long the adult can survive in a torpor. There is no information on this point.

Another condition critical for survival which, probably of widespread importance but not unequivocally demonstrated for any species, must be fulfilled is that, when the egg does hatch and activity starts, reproduction must be rapid enough to provide sufficient breeding adults after predators and parasites have taken their toll. Presumably, when Davis found a lower limit of 18°C for *A. aegypti* it was sheltered from its natural enemies, and the threshold for a wild population would be higher.

(3) Competition. There is nothing to add here to what has been written in the discussion of how coldwater species are kept out of warm water, because, when two species meet, the ranges of both are curtailed.

(4) The reproduction is not effective. Though modern work has modified the original concept, it is approximately true to say that *Ostrea* breeds as long as the temperature is above 15-16°C (Korringa 1957). No relationship in fresh water quite as simple as this has been described, but there is evidence, already presented and discussed, that certain fish will not breed unless the temperature exceeds a given value. An untimely drop in temperature may also be disastrous to the breeding of warm-water fish, and Rawson's (1945) experience with the *Micropterus* which he introduced into Canadian lakes provides a good example of this.

Lethal temperature is a comparatively simple concept though, since the Canadian workers have shown how much acclimation temperature can influence both the lethal temperature itself and the length of time that an animal can survive exposure to it, not as simple as the death-in-five-minutes school appears to have thought. When temperature acts by slowing growth rate to a point where a population does not survive the onslaught of predators and parasites or the activities of a competitor, its limiting value is less definite and less easy to discover. When its effect is on the reproductive cycle, the problem becomes more complicated still. Many freshwater animals breed at a given time of year, either winter or summer if their entire lives are spent in water, but generally in summer only if they are insects, even though the aquatic stage is confined to cold water. Presumably a given breeding-time has a selective advantage, or had a selective advantage once; of the species of *Apatidea* in springs, one has not adapted itself to changed conditions, but two, thought by Nielsen (1951) to be derived from the same stock, have. Breeding at a given time has been achieved by an increasing dependence of physiological processes on climatic phenomena as a source of the initiatory stimulus. In order that eggs may not hatch or that adults may not emerge in the autumn when temperature may be high but the unfavourable winter season lies ahead, a stage of diapause is entered, and diapause development cannot be completed until low temperature has been experienced. The few freshwater examples described are all quite simple, but in some other animals diapause development is linked to a complex sequence of climatic events (Andrewartha 1952).

The maturation of the gonads of fish and the changes that lead up to courtship and spawning are often governed by temperature and by light. Bullough (1939) caused minnows to spawn in winter by keeping them warmer and exposed to light for longer each day than specimens under natural conditions. High temperature but no extra light caused some development but did not stimulate the completion of maturation. Longer hours of light had no effect on fish kept at normal winter temperatures. Length of day is a more important influence on *Gasterosteus aculeatus*, and under its influence alone the fish will migrate into fresh water and breed, though not as early as when there is a rise in temperature as well as a lengthening of the day. Development of the gonads restarts after breeding but cannot proceed beyond a certain point unless a period of short days and low temperature is experienced (Baggerman 1957).

The more a species comes to depend on a given sequence of climatic events to stimulate physiological events that ultimately result in the production of young at the most favourable time, the more closely must its range coincide with the area in which those climatic events occur. And, of course, the more complex the interrelationship, the harder it becomes to discover the part played by any one.

That temperatures favourable to some animals are lethal to others provokes speculation about why there has not been more adaptation. Living matter is characteristically adaptable and some species have colonized hot springs whose temperature is far higher than that of any water warmed by the sun. The classic experimental study is still that of Dallinger (1887), and his paper is fascinating to-day for reasons other than the interest of the scientific content. Elaborate apparatus constructed on the premises is generally thought of as a twentieth-century development in biological laboratories. Dallinger not only constructed his own water-bath and the thermostat to control it but also had to devise and construct a gas generator to provide the necessary source of heat. This done, he started work and then: 'an accident, which no foresight could have guarded against, happened to the apparatus employed, which, occurring in my absence, brought to an abrupt termination the consecutive observations of nearly seven years'. I do not know in how many modern laboratories a worker could expect a thermostat to work for seven years uninterruptedly without being an optimist beyond all reason; anyhow nobody has repeated Dallinger's experiment. His paper is a delight to read in these days of professionalism when nobody can afford not to think about those on whom he depends for promotion, for it gives the impression that the scientific problem itself was the only thing that held the slightest interest for him.

He started with three species of flagellate, growing at a temperature of 60°F (15·6°C) and quickly raised it to 70°F (21·1°C). At 73°F the organisms showed signs of being adversely affected. The temperature 'was static at this point for two months, and during this time the vital vigour and all the vital activities of the organism were regained. I then ventured to raise the temperature to 74°; at the end of twelve hours there was a visibly adverse influence but not so marked as before, nor so long continued. In four days there was a complete restoration of the vigorous condition. In the course of five months I advanced from 74° to 78°. Beyond this point I could not elevate the heat even half a degree without visible evil influence for 8 months.' After the eight months it was possible to raise the temperature by a degree and in three more months 80° was reached. And so it went on—long pauses when the temperature had to be kept unchanged alternating with periods when it could be raised a little. When the experiment was brought to its unintended conclusion, the temperature of the culture medium was 158°F (70°C). The organisms died at 60°F, their starting temperature.

Failure of an organism to adapt itself at least to the whole range of temperature of waters warmed by the sun in the region where it occurs may be due to inherent inability, or to some other factor, such as competition which prevents it extending its range beyond limits well within those of its thermal tolerance. Which species are affected in which way is unknown.

MODE OF ACTION OF TEMPERATURE

The study of ecology is like climbing a rounded mountain; the horizon retreats as the climber ascends. The students of some groups are still in the foothills engaged on taxonomy, without which no advance is possible. Others, higher up, are collecting in the field and looking for parameters with which range appears to be correlated. Then these field correlations are investigated experimentally and results of the kind that have been described in this chapter are obtained. But such workers are nowhere near the top; they have merely opened up vistas of cell physiology and biochemistry which are almost untrodden at the present moment.

An increase of temperature causes an increase in the rate of output of the contractile vacuole of certain peritrich ciliates without, over a limited period, detectably altering the body volume. Kitching (1948) has shown, by measuring the vacuolar output in sucrose solutions, that the internal osmotic pressure does not increase with rising temperature, and concludes that this must cause the surface to become more permeable. The relation between osmotic pressure and temperature is discussed further in the chapter on brackish water.

Habroleptoides modesta (Ephem.) emerges early in the summer, coming out earlier in warmer water and in warmer seasons. The eggs hatch after two or three weeks and grow steadily throughout the hottest part of the year. The temperature is evidently not lethal to them, but Pleskot (1953) postulates that the early emergence is necessary to prevent the ripe nymph being exposed to too high a temperature. There is no direct effect, but with rising temperature the amount of oxygen in a saturated solution goes down, and the demand for oxygen by tissues increases. Since a nymph about to emerge requires more oxygen than at other times and also may have oxygen transport difficulties as the adult forms inside the nymphal skin, *Habroleptoides* must emerge before a certain temperature is reached or it will die owing to the indirect effect of high temperature on supply and demand of oxygen. Increased oxygen demand with rising temperature affects *Limnaea pereger* which, when full grown, can obtain all the oxygen it needs from solution in the water, as long as the temperature is below 12°C. In warmer water it must come to the surface to fill its lung, a requirement that limits it to shallow water. This may be a disadvantage and cause it to go hungry when it has to move on to bare stones because the level of the lake has risen (Hunter 1953a, b).

Schlieper (1952a and b), Bläsing (1953) and Halsband (1953) have severally and jointly (1952) published the results of work on temperature tolerance. Much of it is described several times in the five papers cited above. They find that tolerance of a big range of temperature is linked with a small effect of temperature on metabolism and the contrary. For example, at 5°C *Planaria alpina* is using less oxygen than *P. gonocephala.* At a slightly higher temperature the consumption is the same and thereafter the rate of consumption increases much more steeply in *P. alpina* than in *P. gonocephala.* It reaches a maximum at 15°C and then drops. That of the latter reaches a maximum of about the same level but not until the temperature is 20°C (fig. 20). The change in the rate of movement is similar (fig. 21). Fig. 22 shows how the rate of oxygen consumption increases much more rapidly in three races of *Salmo trutta*, which have a narrow temperature range, than in *Squalius cephalus* and *Gobio gobio*, which have a wide one. The first three cannot tolerate being transferred suddenly from water at 5°C to water at 15°C in winter, the second two can.

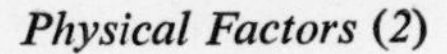

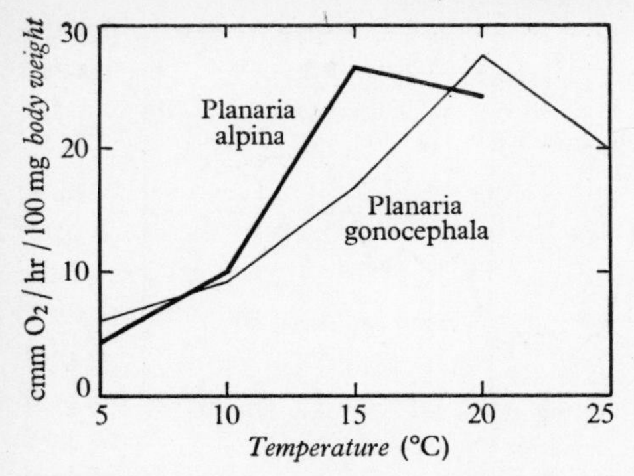

20. Consumption of oxygen at different temperatures by two species of *Planaria* (Schlieper 1952a) (*Biol. Centralbl.* **71**, p. 454, fig. 6).

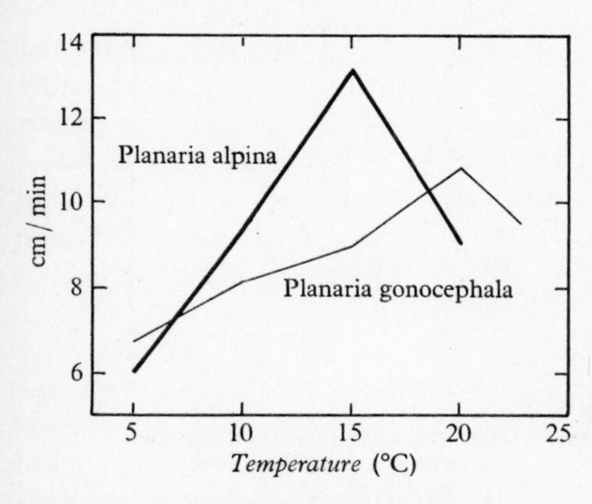

21. Rate of movement with increasing temperature of two species of *Planaria* (Schlieper 1952a) (*Biol. Centralbl.* **71**, p. 454, fig. 5).

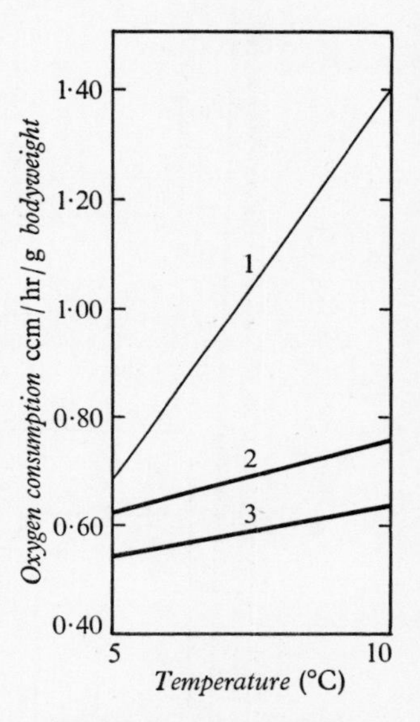

22. Consumption of oxygen with rising temperature by: 1, *Salmo trutta*, 2, *Squalius cephalus*, 3, *Gobio gobio* (Schlieper 1952a) (*Biol. Centralbl.* **71**, p. 452, fig. 1).

The rate of respiration of the stenotherms is increased or depressed more than that of the eurytherms by sudden immersion in dilute sea water (figs. 23 and 24), by a lowering of the pH (figs. 25 and 26) or by a lowering of the ambient concentration of oxygen (figs. 27 and 28). *Salmo* can compensate for a drop in the concentration of oxygen by increasing the rate of its respiratory movements (fig. 27); *Planaria* has no respiratory apparatus and, if oxygen concentration falls, its oxygen consumption falls (fig. 28).

The eurytherms are also less sensitive to poisons and to X-rays.

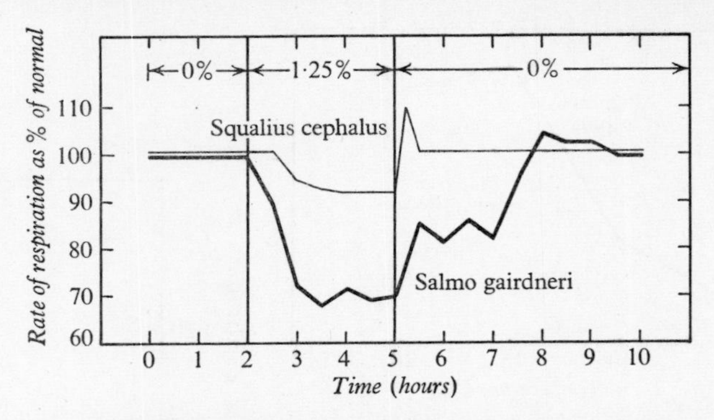

23. Effect of sudden immersion in dilute sea water on the rate of respiration of *Squalius cephalus* and *Salmo gairdneri* at a temperature of 15°C (Schlieper 1952a) (*Biol. Centralbl.* **71**, p. 457, fig. 12).

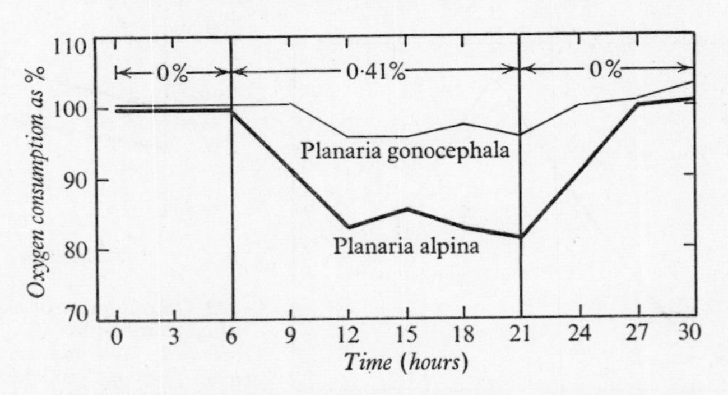

24. Effect of sudden immersion in dilute sea water on the oxygen consumption of two species of *Planaria* at a temperature of 15°C (Schlieper 1952a) (*Biol. Centralbl.* **71**, p. 458, fig. 13).

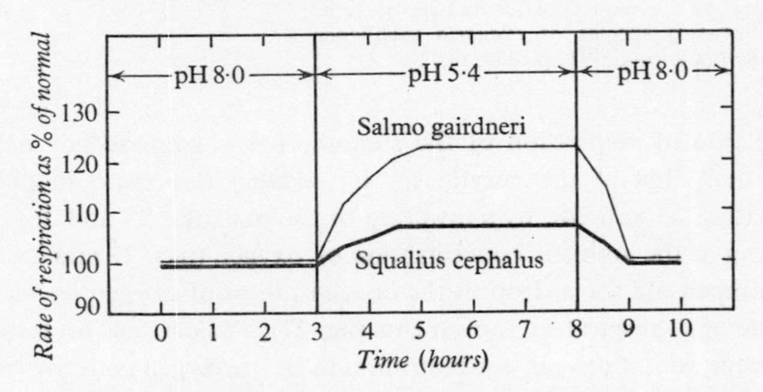

25. Effect of a sudden change of pH on rate of respiration of two species of fish (Schlieper 1952a) (*Biol. Centralbl.* **71**, p. 457, fig. 10).

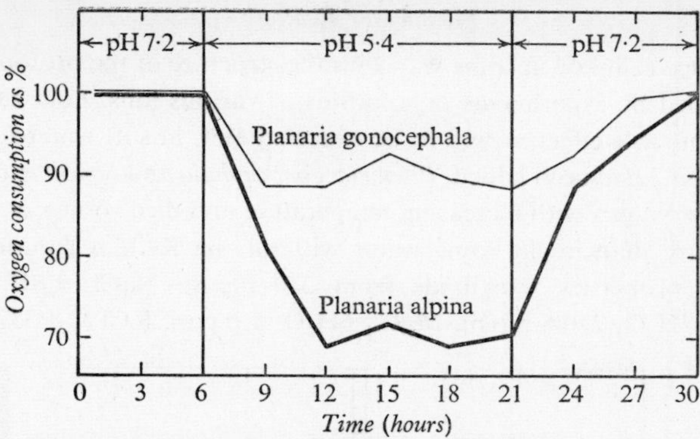

26. Effect of a sudden change in pH on the oxygen consumption of two species of *Planaria* at a temperature of 10°C (Schlieper 1952a) (*Biol. Centralbl.* **71,** p. 457, fig. 11).

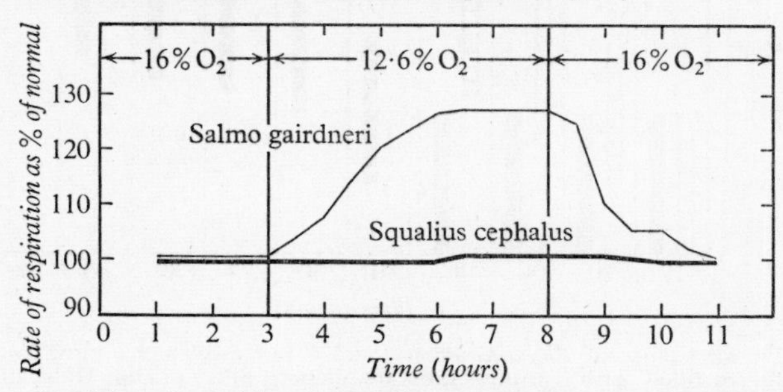

27. Effect of a sudden change of oxygen concentration on rate of respiration of two species of fish (Schlieper 1952a) (*Biol. Centralbl.* **71,** p. 456, fig. 8).

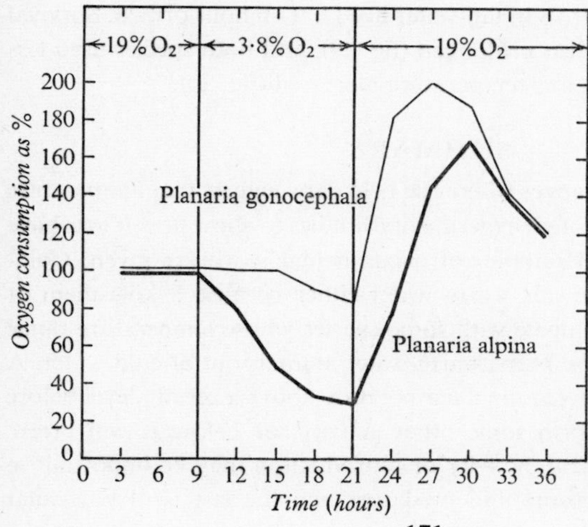

28. Effect of a sudden change of oxygen concentration on rate of respiration of two species of *Planaria* (Schlieper 1952a) (*Biol. Centralbl.* **71,** p. 456, fig. 9).

Tolerance is linked in some way with the structure of the protoplasm as was revealed by experiments in solutions of various ions. Potassium had an unfavourable effect as was to be expected and, in soft water to which 400 mg/l KCl had been added, *Planaria gonocephala* and *Squalius cephalus* used more oxygen with increasing temperature and died sooner at a lethal temperature than in the same water without the KCl. A solution with beneficial properties was made from 230 p.p.m. NaCl, 2,812 p.p.m. $MgSO_4 . 7H_2O$, 2,386 p.p.m. $MgCl_2$ $6H_2O$, 5 p.p.m. KCl and 537 p.p.m.

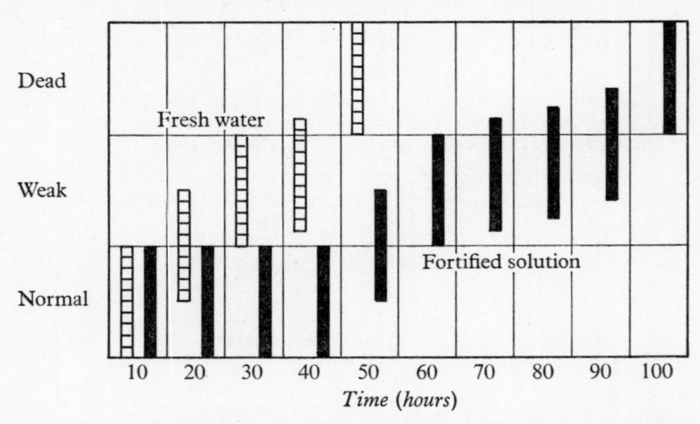

29. Resistance of *Planaria alpina* to 25°C in fresh water and in fresh water fortified with various salts (Schlieper, Bläsing and Halsband 1952) (*Zool. Anzeig.* **149,** p. 165, Abb. 6).

$CaCl_2 . 6H_2O$. In this *Salmo* remained alive for three weeks at 27·5°C, at which temperature controls in tap-water lived but an hour or two. Survival of *Planaria alpina* was also prolonged (fig. 29), and both species used less oxygen than in tap-water as temperature increased (fig. 30).

SUMMARY

Examples are given of ranges where the field data suggest that temperature is the limiting factor, though experimental studies to show how it acts have not been carried out. Examples of experimental work are given. Cold-water species do not invade warm water either because it kills them or because they cannot compete with some species whose temperature range is higher, and the reverse keeps warm-water animals out of cold water. A warm water species may require a temperature above a certain level before it can develop or perform some other activity, or before it will breed. The speed of development, activity or reproduction may be important in the presence of competitors and predators and the last is of particular importance to species that have several generations in a season. Some species which rely on climatic events to stimulate physiological changes that result in the production of young at the most favourable time are restricted to regions where these climatic changes occur.

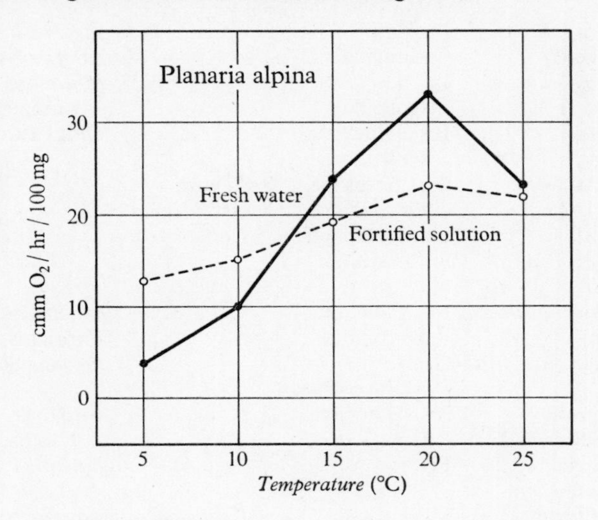

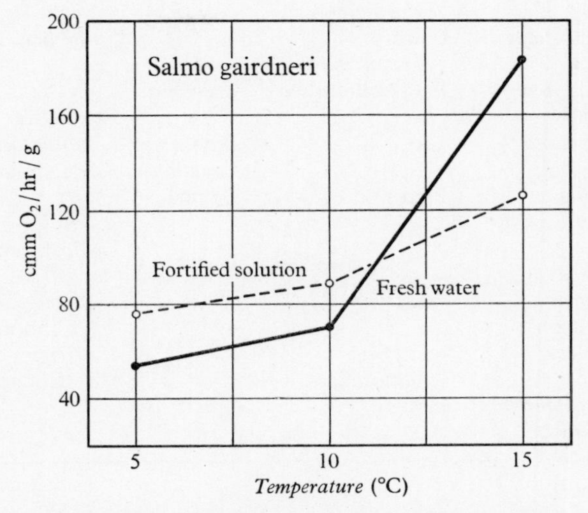

30. Oxygen consumption of *Planaria alpina* and *Salmo gairdneri* in fresh water fortified with various salts (Schlieper, Bläsing und Halsband 1952) (*Zool. Anzeig.* **149,** pp. 166 and 167, Abb. 4 and 5).

SYNONYMS

Species	Group	Other generic names	Other specific names
Aëdes aegypti (Linnaeus)	Ins. Dipt. Culicidae	*Stegomyia*	*fasciatus* Meigen *calopus* Meigen
Anopheles claviger (Meigen)	Ins. Dipt. Culicidae		*bifurcatus* Linnaeus
A. sacharovi Favr	Ins. Dipt. Culicidae		*elutus* Edwards
Corixa carinata (Sahlberg)	Ins. Heteropt. Corixidae	*Sigara*, *Arctocorisa*	
C. wollastoni (Douglas and Scott)	Ins. Heteropt. Corixidae	*Callicorixa*	
Dytiscus semisulcatus Müller	Ins. Coleopt. Dytiscidae		*punctulatus* Fabricius
Gammarus fasciatus Say	Crust. Malacostraca		*tigrinus* Sexton
Glaenocorisa propinqua (Fieber)	Ins. Heteropt. Corixidae		*cavifrons* Thomson
Gobio gobio Linnaeus	Pisces		*fluviatilis* Cuvier
Limnaea pereger Müller	Moll. Gast. Pulmonata	*Lymnaea*, *Radix*	*peregra* Müller *ovata* Draparnaud
Neomysis integer (Leach)	Crust. Malacostraca		*vulgaris* Thomson
Planaria alpina Dana	Platyhelminthes	*Crenobia*	
P. gonocephala Dugès	,,	*Dugesia* *Euplanaria*	*subtentaculata* Draparnaud
Salmo salar Linnaeus	Pisces	*Trutta*	
S. trutta Linnaeus	,,	*Trutta*	{ *fario* Linnaeus *lacustris* Linnaeus
S. gairdneri Richardson	,,	,,	*S. iridaeus* Gibbons
Simulium hirtipes Fries	Ins. Dipt. Simuliidae		*nigripes* (Enderlein)
Squalius cephalus (Linnaeus)	Pisces	*Leuciscus*	

CHAPTER 9

Oxygen

One cubic centimetre of oxygen at normal temperature and pressure weighs 1·429 mg. The partial pressure of oxygen in air at N.T.P. is 0·210 atmospheres, which is equivalent to 159·5 mm Hg. It is useful to have these figures available, since some authors express the amount of oxygen in solution as volume per litre, some as weight per litre, and others as the pressure of mercury which would produce that particular concentration.

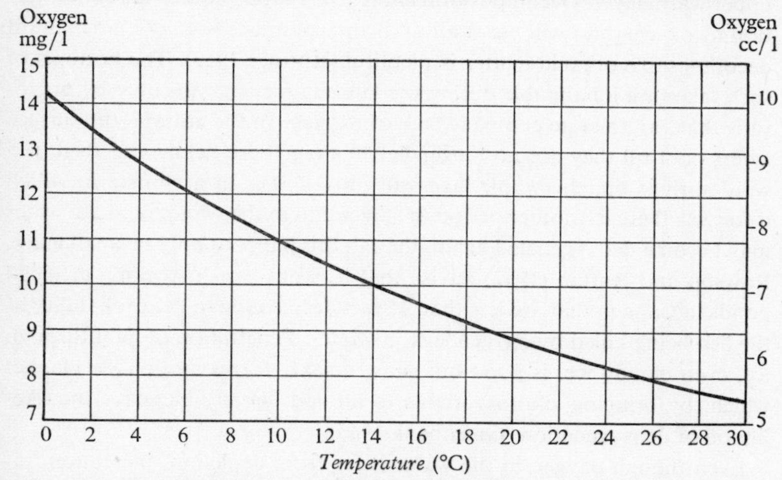

31. The oxygen content of water saturated with air at normal pressure (760 mm Hg) (Mortimer 1956) (*Mitt. int. Verh. Limnol.* No. 6, fig. 1).

The amount of oxygen that a given volume of water will hold in equilibrium decreases as temperature increases, and concentration is often expressed as the percentage of what it would be if the water were saturated at normal pressure and the temperature in question. It is, of course, immaterial to an organism to which 7·48 mg/l O_2 are available for respiratory purposes that that is the saturation concentration at 30°C and only just over half of it at 0°C, but percentage saturation frequently gives to an observer a valuable indication of what has been happening in a sample of water. Fig. 31 is taken from the most recent work on saturation concentration at different temperatures (Mortimer 1956).

CONCENTRATION IN NATURAL WATERS

In extremely unproductive lakes with clear water, values close to saturation may be expected and are in fact generally observed. In the surface waters of productive lakes, photosynthesis may produce supersaturation by day, and respiration may produce a concentration well below saturation by night. Hutchinson (1957a) records that a dense population of *Hydrodictyon* raised the concentration in a lake to 18·7 mg/l, 248 per cent saturation, at 16.00 hours and reduced it to 2·2 mg/l, 27 per cent saturation, at 05.00 hours. Decomposition of big populations of algae may produce a fall in oxygen concentration that is catastrophic to fish.

Most large lakes in temperate regions become stratified in summer, and the lower cold layer of water is sealed off from any source of oxygen by the upper warm layer. Decomposition in the lower layer reduces the concentration of oxygen and will use it all up if the volume is not very large and if decomposable organic matter is plentiful (Ruttner 1953). The bottoms of such lakes are inhabitable during the late summer months only by organisms that can tolerate complete lack of oxygen. In the autumn the surface layers cool till they are no longer lighter than those below and then, if a wind springs up, the whole lake will mix. If it is in a climate in which it freezes, there is another period of stagnation in the winter, and the water may become deoxygenated again, though this happens only in small lakes. Rawson and Ruttan (1952) advise that it is not worth stocking artificial ponds in Canada that are less than 15 feet deep owing to the probability of the fish being killed by oxygen lack in winter. Penetration of light through ice, even cloudy ice, is good but snow quickly forms an opaque blanket which, by bringing photosynthesis to an end, greatly increases the likelihood of deoxygenation (Greenbank 1945).

Even though oxygen in the water is far from depletion, the concentration may fall steeply in the layers immediately above the mud, and Brundin (1951) suggests that chironomid larvae tolerant of low concentrations tend to be large, because respiratory movements of big larvae affect a deeper column of water than those of small ones. The amount of oxygen in solution may also fall within a thick bed of vegetation; Buscemi (1958) records 8·0 p.p.m just above dense *Elodea*, a higher concentration within the top third of the bed, but only 0·4 p.p.m within 5 cm of the bottom.

The smaller the volume of water the more sudden and violent are fluctuations in the oxygen content likely to be, and the cycle of thermal stratification and consumption of oxygen in the lower layer which takes a year in lakes may be completed in twenty-four hours in a pond. Unproductive ponds are stable as are lakes of the same type. Laurie (1942), for example, taking samples from a moorland pond at intervals throughout twenty-four hours, found a lowest value of 66 per cent saturation at 02.00 hours just after the summer solstice. In midwinter at the same time the saturation was 100 per cent. The inflowing stream was nearly always saturated. The highest temperature recorded was 24°C. Whitney (1942), sampling a rich lowland pond found a large difference between readings in spring and in autumn. On 11/12 April there was plenty of vegetation, especially filamentous algae, temperature ranged from 5·8 to 6·8°C, and the oxygen concentration rose from 9 cc/l at noon to 10·75 cc/l at 21.00 hours. At the end of September there was much *Potamogeton natans* and *Lemna* sp., temperature ranged from 11·4 to 12·1°C, and the oxygen concentration was 0·25 cc/l most of the time with a maximum of 0-5 cc/l at 20.00 hours.

Table 18

Amount of oxygen dissolved in the water of a stream at two different points (Hubault 1927)

	Near Spring 18th Sept.		Lower reach		3 July
Time	06.30	15.45	03.45	10.20	17.00
Temperature °C	7·8	13·5	19·6	20	21
cc/l O_2	7·52	6·56	3·96	4·82	5·32
% saturation	89	88	59·8	73·4	82·38

Table 18 shows some measurements made by Hubault (1927) in a stream. Near the source there was least oxygen by day when the water was warmest, lower down there was least by night, owing to decomposition, and the percentage saturation was lower. Butcher, Longwell, and Pentelow (1937) record that the respiration of a thick carpet of *Cladophora glomerata* on the bed of the River Tees reduced the percentage saturation to nearly 50 during the night. As Hynes (1960) points out, the organic matter whose decomposition depletes the oxygen concentration in running water, sometimes reducing it to a critically low level, may be derived from decaying water plants and from dead leaves blown in from the surrounding land. All too often, however, in many European countries, it is the waste from human communities, insufficiently treated or not treated at all. Here also temperature and oxygen concentration are closely related. The warmer the water the quicker the organic matter decomposes and the greater is the likelihood of oxygen being reduced below a critical concentration before tributaries have brought in enough well-oxygenated water.

Recently a continuous recorder has been brought into use by the Water Pollution Research Laboratory (Briggs, Dyke, and Knowles 1958). At the time of writing Mr F. J. H. Mackereth is perfecting a simpler oxygen meter at the laboratory of the Freshwater Biological Association. It is cheaper to construct but requires an elaborate recorder. When apparatus of this kind gets beyond the pioneering and expensive stage, additions to knowledge about fluctuations in oxygen concentration will be considerable. Gameson and Griffith (1959) have reported on six months' records with the Water Pollution Research Laboratory device. Fig. 32 shows the fluctuation of oxygen concentration on two days, one in February, the

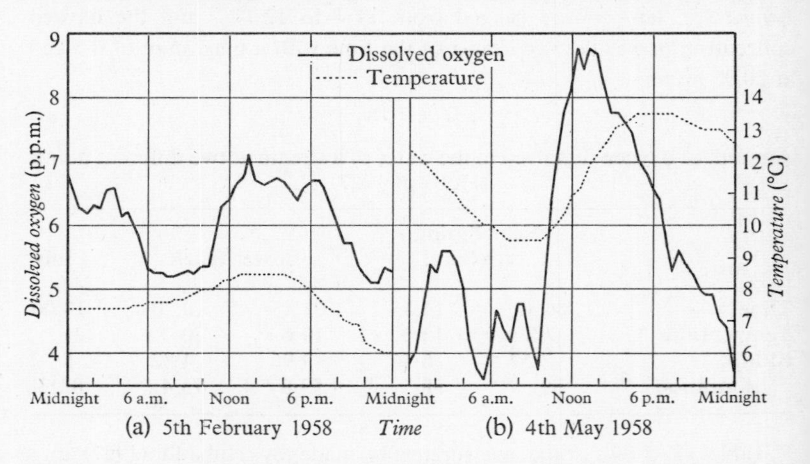

32. Variations in dissolved oxygen and temperature in River Hiz on (*a*) 5 February and (*b*) 4 May 1958 (Gameson and Griffith 1959) (*Water and Waste Treatment Journal* 1959, fig. 1. Water Pollution Research Laboratory, Stevenage, Herts.).

other in May. Fig. 33 shows the daily maximum, the daily mean and the daily minimum for the third week in January, and the third week in May. In January the mean was reasonably steady throughout the week, but in May its fluctuations were large, and the lowest maximum was not much higher than the highest minimum. Fig. 34 shows the number of days on which the maximum (a) and the minimum (b) occurred at any given hour of the day. Both were recorded at every hour but the maximum was generally between about 12 and 6 p.m. and the minimum between the same hours a.m. The stream is polluted by some gas liquor and by several small discharges of sewage and sewage effluent. It may be surmised that the fluctuation in oxygen concentration would be more regular in an unpolluted stream, but that remains to be shown. What these and other sets

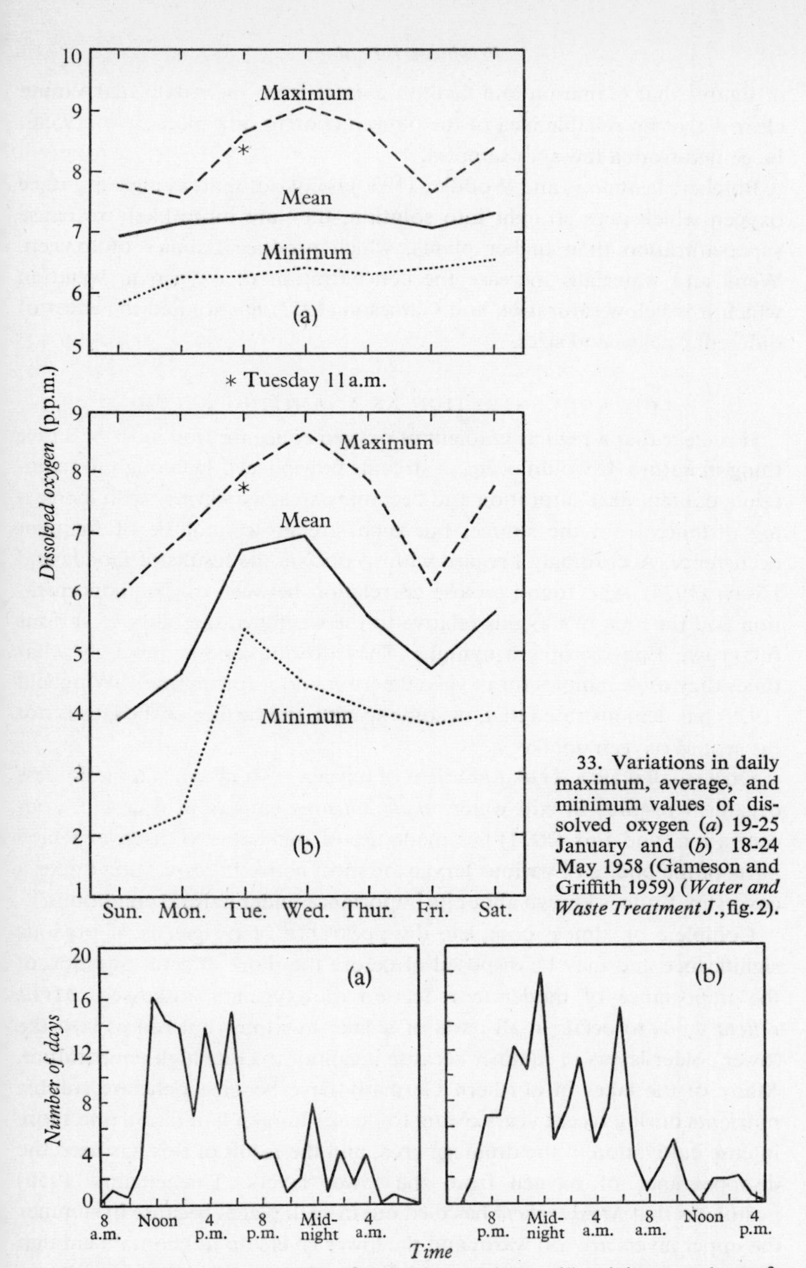

33. Variations in daily maximum, average, and minimum values of dissolved oxygen (*a*) 19-25 January and (*b*) 18-24 May 1958 (Gameson and Griffith 1959) (*Water and Waste Treatment J.*, fig. 2).

34. Frequency of occurrence of daily (*a*) maximum (*b*) minimum values of dissolved oxygen within each range of 1 hour throughout the day. (Gameson and Griffith 1959) (*Water and Waste Treatment J.*, 1959, fig. 3).

of figures that Gameson and Griffith extract from their data make quite clear is that no reliable idea of the oxygen content of a piece of water can be gained from a few spot samples.

Butcher, Pentelow, and Woodley (1930) point out that, as algae produce oxygen which goes straight into solution, they are more likely to cause supersaturation than higher plants, which produce bubbles of oxygen. Weirs and waterfalls increase the concentration of oxygen in water in which it is below saturation and Gameson (1957) has studied the effect of different designs and sizes.

LOW CONCENTRATION AS A LIMITING FACTOR

It is clear that a regular gradient of oxygen concentration must be a rare thing in nature. It would occur in streams flowing swiftly enough to maintain a content near saturation and becoming steadily warmer with increasing distance from the source, but such streams cannot be of frequent occurrence. Accordingly I regard with scepticism the results of Dodds and Hisaw (1924), who found a close correlation between oxygen concentration and the area of the gills relative to the weight of the body of various full-grown Ephemeroptera nymphs. They do not state at precisely what times they took samples for oxygen determination. Furthermore Wingfield (1939) has demonstrated that, in some species, the surface of the gills is not the seat of oxygen uptake.

On a small scale a regular gradient of oxygen is set up when a motionless organism respires in still water. *Bodo sulcatus* gathers in a certain concentration, and Fox (1921) has made use of this habit to discover which parts of the bodies of various larvae are most active in respiration under a coverslip. Little is known about its importance under natural conditions.

Complete or almost complete disappearance of oxygen is of obvious significance and may be disposed of before the more difficult question of the importance of moderate reduction of oxygen is discussed. *Mysis relicta* tends to occur in all parts of a lake in winter but retires into the lower colder layers in summer because it cannot tolerate high temperature. Many of the lakes in northern Germany have become richer in soluble nutrients during recent years owing to denser human habitation and more intense cultivation in the drainage area, and the result of this has been the disappearance of oxygen from the lower layers. Thienemann (1950) postulates that *Mysis relicta* has died out in such places because in summer the upper layers are too warm and the lower layers do not contain enough oxygen; he states that it requires at least 3 cc/l in fresh water though it can survive on 1·6 cc/l in brackish water. Holmquist (1959) quotes various records of *Mysis* being found at the bottom of lakes where the oxygen concentration was well below the level claimed by Thienemann as the minimum, and she also believes that the shrimp is more tolerant of high temperature than he does, chiefly because it occurs in Ireland in two shallow lakes which are stated not to stratify. As, however, continuous records are available from neither lake, it is obvious that the last word in this controversy must await further measurements, preferably in the laboratory as well as in the field.

Limnaea, as already mentioned, can satisfy its oxygen needs from what is in solution provided that the water is cool. It should, therefore, be able to invade deep water and in a number of lakes has done so. Alsterberg (1930) gives a list of the records, and suggests that *Limnaea* is the genus that has most often colonized deep water because its feeding habits are most catholic. Serious depletion of oxygen in the hypolimnion confines snails to the shallow water. When *Coregonus albula* is confined to the epilimnion by oxygen deficiency lower down, it survives but lays down less fat than it does in lakes in which it can avoid the warm upper layers of water (Morawa 1958). Hubault (1955) attributes the failure of a number of introductions of *Salvelinus* spp. to an insufficiency of oxygen in the hypolimnion and an excessively warm epilimnion.

Disappearance of oxygen from rivers is generally caused by pollution, but this is too large a field to start exploring here. Hynes (1960) has recently written a book on the subject.

Oxygen is likely to be the factor of overriding importance to animals when it disappears completely as a result of the decomposition of organic matter. When it is only reduced, the field worker finds it hard to decide how far any changes he observes are due to oxygen and how far to the direct effect of the organic matter.

Before the significance of a lowering of oxygen concentration which does not, however, reach zero or near it can be discussed, something about fluctuations in demand must be known. Berg, Lumbye, and Ockelmann (1958) found that specimens of *Ancylus fluviatilis*, brought into the laboratory and tested at temperatures similar to those of the places where they had been found, used more oxygen in the summer than they should have according to a theoretical calculation from the amount which they used in winter. Accordingly specimens were tested at temperatures which were kept constant at 11°C and 18°C throughout a year. Consumption started to rise above winter level in spring, was about twice as much in May, and then sank again (fig. 35). The increase is thought to be due to greater activity during the reproductive season.

Berg (1953) measured the rate of oxygen consumption of specimens brought from a stream having a summer temperature of about 11°C and from a lake having a summer temperature of about 18°C. At both temperatures the rate of consumption of the lake specimens was significantly higher than that of the stream specimens. Berg is chiefly interested in the physiology, for generally animals living in cold water have a higher rate of metabolism than those from warmer water, and offers no comment on the ecological significance of his results. They show clearly that a given rate of oxygen consumption is not a fixed property of the species.

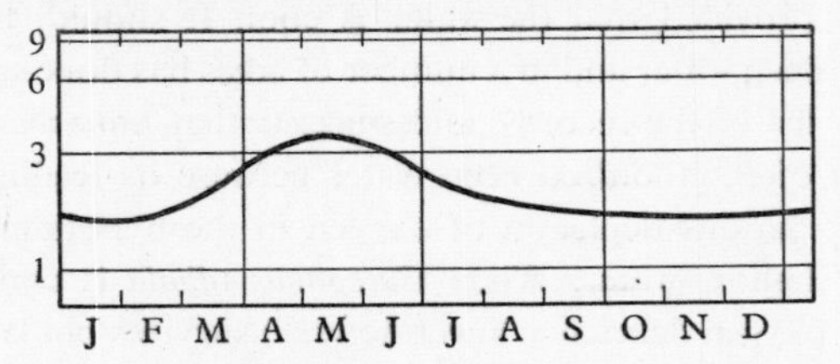

35. Oxygen consumption in l/hr/individual of 20 mg of *Ancylus fluviatilis* kept at a constant temperature of 11°C throughout the year (Berg, Lumbye and Ockelmann from Macan 1961) (*Verh. int. Ver. Limnol.* **14,** p. 595, fig. 3).

Berg (1952) has shown also that starvation reduces the rate of oxygen consumption. He never used narcotics and assumes that the oxygen consumption of fed animals represents an active rate. That of starved animals may be the standard rate, for they were observed to move little.

Oxygen consumption reaches a peak on the day on which the land bug, *Rhodnius prolixus*, undergoes ecdysis. The resting level is similar in each instar but the maximum is higher in each succeeding one (Zwicky and Wigglesworth 1956). The same may be expected in aquatic insects, which are subject to the complication, mentioned in the previous chapter, that, at the end of the last instar, the formation of the flying adult inside the skin of the aquatic nymph not only raises the oxygen demand but makes more difficult the transfer of oxygen to where it is needed. The oxygen concentration necessary at this stage may be a factor determining the range of some species.

The oxygen consumption of *Daphnia obtusa* increases as maturity is reached and then declines. Crowding also causes it to rise, and starvation and the presence of senescent *Chlorella* to fall. A specimen acclimatized to warm water consumes less oxygen than a specimen brought suddenly to the same temperature from a lower one. Conversely in cold water an acclimatized specimen is using more oxygen than one transferred abruptly from warmer water (Vollenweider and Ravera 1958). The oxygen demand of salmon eggs changes greatly as development proceeds (Lindroth 1942).

Larger animals consume less oxygen per unit weight than smaller ones (Berg 1952, Berg, Jónasson and Ockelmann 1962).

Measurements of rate of respiration, particularly if made for the purpose of comparing two species, must therefore take into account the physiological condition of the animal, particularly the stage it has reached in the reproductive cycle if it is adult, its size, and the stage it has reached in development if it is immature. It is scarcely necessary to add that the degree of activity must be controlled. Fig. 36 shows that *Ephemera danica* in a flask floored with sand respires only slightly more rapidly with rising temperature (c). If the substratum is not what it is used to (b), or there is no substratum at all (a), oxygen consumption rises enormously with rising temperature owing to increased restlessness (Wautier and Pattée 1955). Differing rate of change in consumption can also provide a pitfall for the observer who works with one temperature only. It is evident from a glance at fig. 20 that any comment about the relative oxygen demands of *Planaria alpina* and *P. gonocephala* is valueless unless the facts expressed in that figure are known.

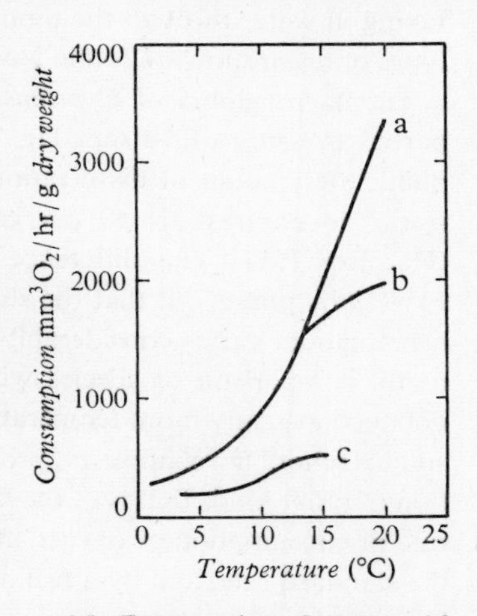

36. Consumption of oxygen with increasing temperature by *Ephemera danica* in a flask with (*a*) no substratum (*b*) some pebbles (*c*) sand (Wautier and Pattée from Macan 1961) (*Verh. int. Ver. Limnol.* **14**, p. 595, fig. 4).

Freshwater animals can be divided into those that must come to the surface to obtain gaseous oxygen and those which can obtain oxygen from solution. Adult Coleoptera and Hemiptera and various larvae, particularly among the Coleoptera and Diptera, belong to the first category. They have the advantage of being able to colonize places where a rich supply of organic matter goes together with a low oxygen concentration as *Tubifera* (*Eristalis*), the rat-tailed maggot, has done. They can also develop a highly impermeable cuticle and live in places with high salt content or even in such improbable places as oil wells (Ephydridae). It is perhaps significant that the air-breathing larvae of the Coleoptera are among the largest and most active aquatic invertebrates. Adult beetles and bugs, lacking any part of the surface specially permeable to gases, do not lose water readily, without which characteristic they would be able to fly less far. Some animals would seem to the human observer to have retained this primitive method of respiration because they have taken to a way of life in which there would be no advantage in evolving any other.

Animals with a truly aquatic mode of respiration may be divided for ecological purposes into those which can maintain a current of water over the respiratory areas and those which cannot, but in the present state of knowledge it is expedient to take fish first, because so much more is known about them, and then invertebrates. All fishes, except lung-fishes, respire by taking in water through the mouth and passing it out over the gills. Both types of respiration are found among invertebrates.

The haemoglobin of *Cyprinus carpio* is still 95 per cent oxidized at a partial pressure of 10 mm Hg, and, as pressure decreases beyond this point, the amount of dissociation increases rapidly. The blood of *Salmo trutta*, in contrast, is 50 per cent dissociated at that partial pressure (Harnisch 1951). This difference is of obvious ecological significance but Fry (1957) points out that the shape of the curve showing dissociation of haemoglobin varies considerably according to the temperature, and that useful comparison of species will not be possible till curves have been obtained at many more temperatures than they have been up to now. He and his school have, however, produced a considerable amount of information of other kinds that is of the highest importance in this connexion. The rate of consumption of oxygen at different temperatures by a resting fish, the standard rate, and by a fish moving at full speed, the active rate, have been discussed in the previous chapter, and here the data for *Salvelinus fontinalis* and *Carassius auratus* are brought together in one diagram (fig. 37). Crucial measurements are the concentrations of oxygen in which the fish can maintain these two levels of activity. It must maintain the standard rate or it will die. Inability to maintain the active rate is less serious, but it is a handicap which, unless effective infrequently and for short periods only, is likely to affect survival. Data for the standard rate are given in the last two lines of table 19. At 5°C *Carassius* can remain alive, by not exerting itself at all, at a partial pressure of 4 mm Hg or 0·3 mg/l O_2, whereas *Salvelinus* requires 30 mm Hg or 2·3 mg/l O_2. Towards the upper lethal temperature the 'levels of no excess activity' are respectively 25 mm Hg or 1·08 mg/l O_2 and 79 mm Hg or 4·01 mg/l O_2. Graham (1949) has investigated the concentrations at which *Salvelinus* fails to survive and finds them to be some 10 mm Hg below those of the standard rate at all temperatures; presumably there is a small margin between possible immobility and the actual immobility of a resting fish in an experiment.

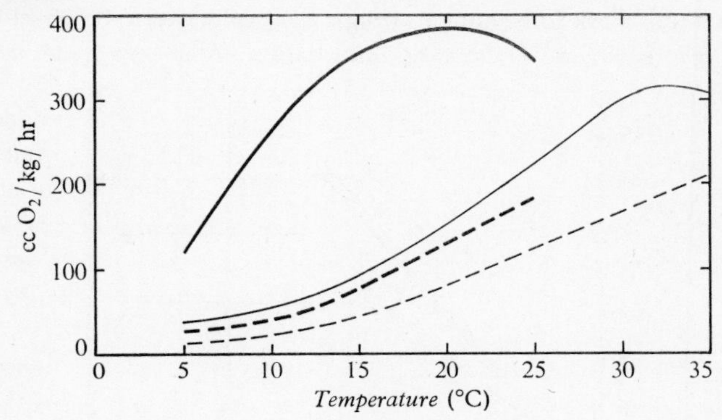

37. Oxygen consumption by active fish (continuous line) and resting fish (broken line) at various temperatures—*Salvelinus fontinalis* thick line (Graham 1949), *Carassius auratus* thin line (Fry and Hart 1948 from Macan 1961) (*Verh. int. Ver. Limnol.* **14,** p. 596, fig. 6).

Table 19

Standard metabolic rate. Consumption of oxygen by a resting fish at different temperatures and the minimum tension of oxygen at which it is possible (level of no excess activity)

	5	9	10	15	16	20	23·5	24·5	25	35°C	
Salvelinus	27	–	59	85	–	140	198	–	–	–	cc O_2/kg/hr
Carassius	8	–	24	50	–	85	–	–	140	225	cc O_2/kg/hr
Salvelinus	30	32	–	–	50	59	–	79	–	–	mm Hg
Carassius	4	–	8	10	–	18	–	–	17	25	mm Hg

Data for *Salvelinus fontinalis* from Graham (1949, tables 1 and 3); for *Carassius auratus* from Fry and Hart (1948, table 2).

As the concentration of oxygen rises, the consumption by an active fish increases to a certain point, 'the incipient limiting point', and then remains constant, because beyond it the amount taken in is limited by the capabilities of the fish and not by the amount available. The incipient limiting point rises with temperature in *Salvelinus* (fig. 38) and in *Carassius* (fig. 39), but again there is a vast difference between the two species, the level near the lethal temperature being about 130 mm Hg and 40 mm Hg respectively. At high temperatures, *Salvelinus* can exert itself fully only in water that is nearly saturated and a comparatively small drop below saturation begins to limit its activity. This is well brought out in table 20 which is based on data obtained by Graham and quoted by Fry (1951).

There is still work to be done to bridge the gap between the physiologist and the field ecologist. There is no information about how much a fish is

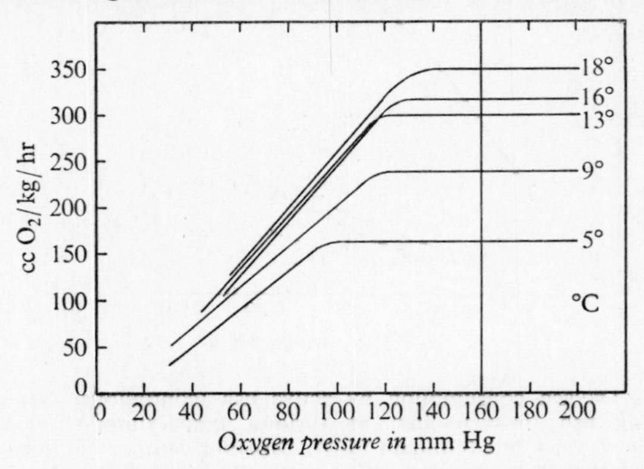

38. Oxygen consumption of an active *Salvelinus fontinalis* at different temperatures and different concentrations of oxygen (Graham from Macan 1961) (*Verh. int. Ver. Limnol.* **14**, p. 596, fig. 7).

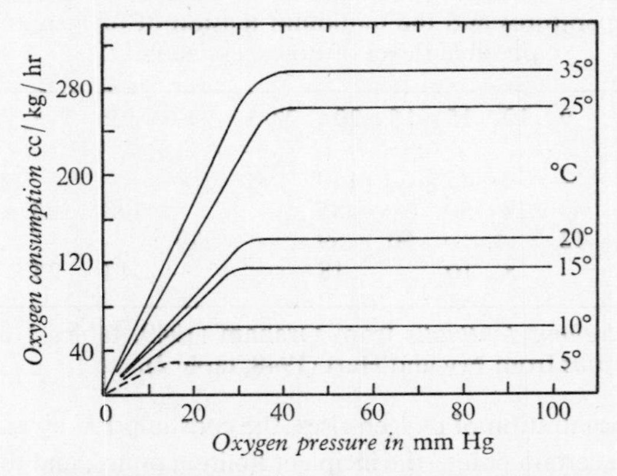

39. Oxygen consumption of an active *Carassius auratus* at different temperatures and different concentrations of oxygen (Fry and Hart 1948) (*Biol. Bull.* **94**, p. 71, fig. 4).

handicapped by inability to move faster owing to shortage of oxygen. Much of the time fish appear to lead leisurely lives, and quick movements are made only to seize prey and to flee from enemies; but how far the efficiency of feeding is reduced or how often a fish that would otherwise escape is caught by a predator because the oxygen concentration is below the incipient limiting point is likely to prove difficult to assess exactly. Nor will it be easy to determine how long a fish can lie inert when oxygen concentration is at the level of no excess activity without being inconvenienced absolutely or in competition with a species that has a lower threshold.

Table 20

Effect of oxygen concentration at different temperatures on the activity of *Salvelinus fontinalis*. The figures show the consumption of oxygen by active fish expressed as a percentage of the consumption at a concentration where oxygen is not limiting activity

Temp.	Concentration of oxygen expressed as % air saturation:			
°C	100	75	63	50
5	100	100	100	64
10	100	100	82	53
15	100	96	65	50
20	100	73	51	25
25	100	57	10	lethal

The table is based on the work of Graham but taken from Fry (1951).

Nevertheless, there can be no doubt about the importance of oxygen as a limiting factor. At 20°C for example, *Carassius* is unaffected as long as the partial pressure is above 40 mm Hg, but *Salvelinus* requires more than that even to remain alive.

There is not a great deal of comparable data for other species of fish. At 20°C *Perca flavescens* consumes more oxygen than *Carassius auratus* and less than *Salvelinus fontinalis*, and its incipient limiting level lies between those of the other two also. In both features it is nearer *Carassius* than *Salvelinus* (Fry 1947). *S. namaycush* has a level of no excess activity like that of *S. fontinalis* at low temperature, but less at higher temperature. Its incipient limiting point is similar but the rate of oxygen consumption beyond it is lower at all temperatures. It also incidentally has a slightly lower lethal temperature (23·5°C) (Gibson and Fry 1954). At 23°C the incipient limiting level of *Leuciscus rutilus* is higher than that of *Carassius*, but at lower temperatures it is nearly the same (Lindroth 1941).

Shepard (1955), experimenting with young hatchery *Salvelinus fontinalis* that had been kept at 9-10°C and 9·1 mg/l O_2 (75 per cent saturation), found that all survived at least five days in 1·9 mg/l O_2 at the same temperature

(17 per cent saturation) and that all died at a uniform and rapid rate at 0·5 mg/l O_2 (5 per cent saturation). The lethal concentration could be lowered by acclimation, and a concentration of 1·05 mg/l O_2 was the lowest at which the fish could be brought to live without significant mortality.

At 13°C *Gasterosteus aculeatus*, the stickleback, will swim into water of low oxygen content apparently without noticing it. After a while there is much random darting about, probably provoked by respiratory distress, and these movements generally took the fish back into well oxygenated water in experiments performed by Jones (1952). At 20°C, however, there was often an avoiding reaction and a fish that entered water with less than 2 mg/l O_2 turned round or backed out of it.

Work on invertebrates has been less elaborate than that on fish, which was true of temperature also. The simplest type of experiment is the determination of the lethal concentration of oxygen, generally only at temperatures likely to be reached in summer. Herbert (1954) imprisoned for four hours various Cladocera in bottles containing known amounts of oxygen at the start and noted how many survived. Generally it was possible to discover in this way a concentration above which there was little, and below which there was much, mortality. The temperature was 22°C for pond species but 18°C for lake species, to which the higher temperature was lethal. *Daphnia obtusa* died at a concentration of about 0·2 cc/l O_2, *D. thomsoni* at about 0·4 cc/l O_2, and *D. pulex*, *D. magna*, *Simocephalus vetulus*, and *S. exspinosus* at values in between. The lake species, *D. hyalina*, had a higher limiting concentration, 0·8 cc/l, and *Leptodora* and *Bythotrephes*, the big carnivorous lake cladocerans, were more exigent still, requiring respectively minima of 1·4 and 1·7 cc/l O_2. The danger to be kept in mind when drawing conclusions from experiments of this kind is confusion of cause and effect. Two Cladocera with equal latent powers of adaptation to low oxygen concentration might have taken, for reasons not connected with oxygen concentration at all, one to life in well-oxygenated lakes the other to life in ponds liable to be poor in oxygen at times. The first would not have developed its latent powers, the second would.

A more comprehensive survey of the same group, discussed already in the chapter on interrelationships, was made by Pacaud (1939). He used the same sort of technique as Herbert but, instead of keeping the animals enclosed for a specified time, he watched them until fair numbers succumbed and then measured the concentration of oxgyen. Many *Sida crystallina*, *Eurycercus lamellatus* and *Daphnia longispina* died at a concentration of 0·6 cc/l O_2. No *Ceriodaphnia rotunda* were dead at 0·08 cc/l. Some *Moina brachiata*, *D. pulex* and *D. magna* died at this lowest concentration. *Simocephalus vetulus*, *Schapholebris mucronata* and *Chydorus sphaericus* survived well at concentrations in the upper part of this range but died at lower concentrations. The tests were carried out at temperatures of 16°C, 19°C, and 22°C and higher temperature was generally found to increase mortality. All are pond species.

Pacaud (1944) also experimented with amphipods. At 11-12°C *Gammarus pulex* survived lowering of the oxygen concentration better than *Echinogammarus berilloni* as long as the water was circulating, but less well if it was stagnant. Probably significant was the observation that at a given temperature the pleopods of *E. berilloni* beat faster. At 17-18°C the survival of *E. berilloni* was better in circulating and in stagnant water. These results tally with the field observation that *G. pulex* tends to occur in the upper cooler reaches and lower down where the water is particularly well oxygenated by, for example, passage over a weir. *E. berilloni* occurs where flow is less, and where flow is fast but temperature high.

Berg (1951, 1952), studying snails, carried out more carefully controlled experiments. He started with the assumption that *Ancylus fluviatilis* is primarily a species of running water, whereas *Acroloxus lacustris* is generally confined to lakes, and set out to discover whether different oxygen requirements would explain this. At 16°C the incipient limiting point for *A. fluviatilis* is 3 ml/l O_2, that for *A. lacustris* 6 ml/l, which is the reverse of what might be expected. Above the incipient limiting point the rate of consumption is similar. Also unexpected was the finding that the consumption of *A. lacustris* begins to fall off at a lower temperature, 31°C, than that of *A. fluviatilis*. On the other hand, in anaerobic conditions *A. lacustris* survives several days, *A. fluviatilis*, which climbs to the surface less, only twenty-four hours. These fresh-water limpets are pulmonates but the lung cavity has largely degenerated and there has been a secondary formation of a gill by means of an extension of the surface of the mantle. Berg concludes that the distribution of the two species cannot be explained in terms of oxygen requirements. This I do not find surprising, as I believe that the ecological premise was faulty, Berg, resident in a moraine-covered country, having no opportunity to become aware of the wide range of *A. fluviatilis* in more rocky districts (chapter 7).

Mann (1956) has applied to the Hirudinea a technique similar to that which Berg used to investigate the Mollusca. The leeches were kept in a stoppered bottle for about one hour, and the oxygen concentration at the beginning and at the end of that period indicated how much had been used. The bottles were wrapped in black paper, which was thought to make the leeches inactive, so that the figures obtained represent standard metabolism. The leeches were of different sizes but a preliminary experiment had yielded information from which the consumption of a leech of known size could be converted to what it would have been had the specimen weighed 30 mg. This was found the most satisfactory method of expressing results. Oxygen consumption is shown in table 21. *Piscicola geometra*, the

Table 21

Oxygen consumption of a leech of 30 mg at 20°C in air-saturated water

Species	μl/hr
Erpobdella octoculata	4·0
Glossiphonia complanata	4·95
Erpobdella testacea	5·98
Helobdella stagnalis	6·1
Piscicola geometra	10·1

one species confined to running water and stony lake shores, has a distinctly higher rate than any of the others, but, as has been stated already, it is difficult to evaluate results of this kind when measurements have been made at one temperature only (cf. fig. 20, p. 169).

In further experiments the oxygen was reduced to a given level before the start of the experiments, and the results shown in fig. 40 obtained. The temperature was, as before, 20°C, which is fairly high, and the highest concentration of oxygen was 6 cc/l, which is below saturation. This is probably why three of the species never reached the incipient limiting point. When the experiments were repeated with leeches acclimatized to a low oxygen concentration, *Erpobdella testacea* showed an incipient limiting level at about 2 cc/l O_2, but *E. octoculata* and *Piscicola geometra* showed none. Without acclimatization, *Glossiphonia complanata* and *Helobdella stagnalis* showed an incipient lethal level at about 2 cc/l (fig. 40), though why the rate of consumption of the latter, having become more or less constant when rising oxygen concentration reaches 2 cc/l, should start to rise again above about 4 cc/l is obscure. *Erpobdella octoculata* is one of the most widespread species but *E. testacea* replaces it in reed-swamps and beds of other emergent plants. Its low incipient limiting point gives *E. testacea* an advantage in such places, and it has, moreover, a life history adapted to the conditions in them. It breeds early in the year, after which it dies, and is in the egg stage or small when the water is likely to be warmest and oxygen most deficient. *E. octoculata* breeds in summer and lives two years (Mann 1961a). *Helobdella stagnalis* is the most abundant leech of rich lakes and ponds and therefore liable to exposure to low concentrations of oxygen. *Glossiphonia complanata* on the other hand, is not characteristic of places likely to be poor in oxygen. *Piscicola geometra*, as noted, is confined to waters likely to be well aerated (Mann 1961b).

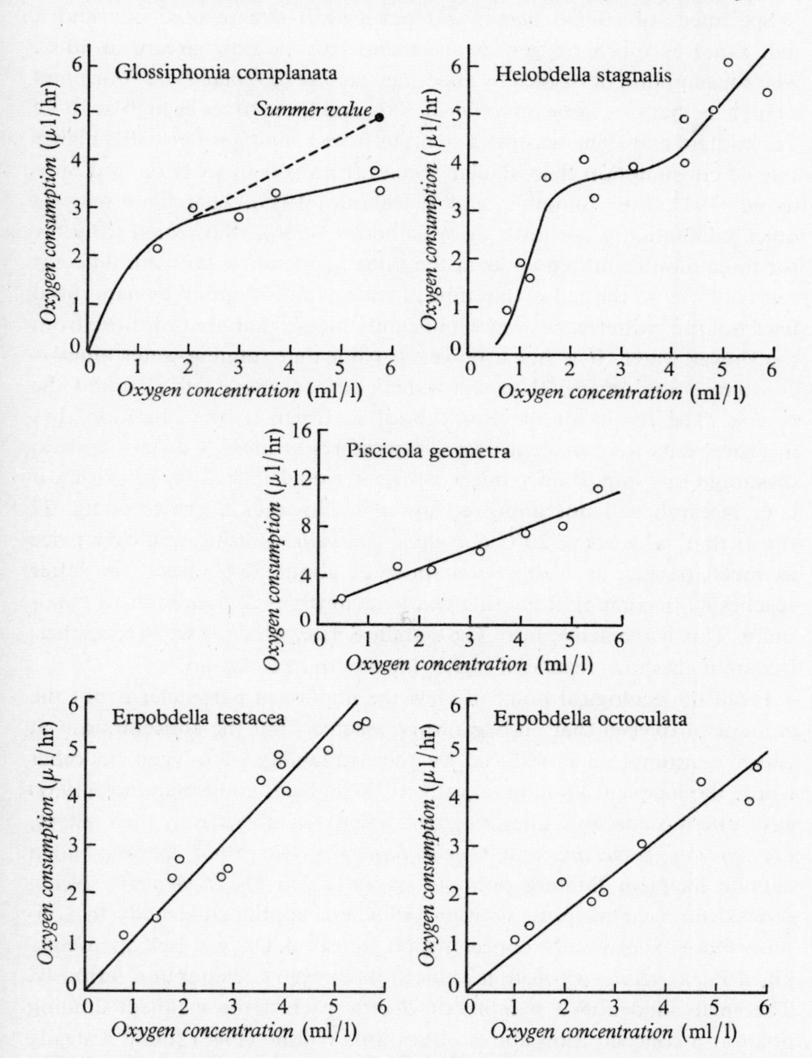

40. The relation between oxygen consumption and oxygen concentration for five species of leech at 20°C with no acclimatization to low oxygen tensions (Mann 1956) (*J. exp. Biol.* **33**, p. 620, fig. 4).

In the discussion that followed this last paper, Mann (1961b) stated that *E. testacea* is capable of living for many days anaerobically. No details are given about the powers of other species to survive absence of oxygen or very low concentrations.

Specimens of *Asellus aquaticus* from a swift stream used one and a half times as much oxygen as specimens from a slow stream at 10°C. The consumption of oxygen by anaesthetized *Baetis rhodani* was four times as high as that of *Cloeon dipterum* at 10°C and three times as high at 16°C. Trichoptera and Ephemeroptera nymphs from running water had a higher rate of consumption than similar species from still water (Fox and Simmonds 1933, Fox, Simonds, and Washbourn 1935). That there may be quick adaptation was shown by Washbourn (1936), who reared trout fry for three months in two tanks at the same temperature but with different rates of flow. At the end of that time the rate of consumption by narcotized fry from the swifter water was significantly higher than that of those from the slower water. It is not a universal truth that running-water animals consume more oxygen than still-water ones, for Berg (1952) found the reverse. The results of the Fox school are open to the objection that measurements were made at two temperatures at most. I do not wish to disparage this important pioneer work; it would indeed be surprising if later research had not improved upon it. However a glance at fig. 37 shows that, whereas at 20°C *Salvelinus fontinalis* is using well over twice as much oxygen as *Carassius auratus*, at about 30°C, when the latter reaches its maximum, it is using nearly as much as *Salvelinus* at its maximum. This is the active rate. The standard rate, which is what anaesthetized animals show, reaches a higher maximum in *Carassius*.

From the ecological point of view the important parameter is not the amount of oxygen that an organism consumes, but the concentration at which consumption is reduced by the availability of oxygen, in other words the incipient limiting point. Fox, Wingfield, and Simmonds (1937) gave attention to this question, and fig. 41 is taken from their paper. *Leptophlebia vespertina* and *Cloeon dipterum*, two pond species, had a definite incipient limiting point at about 2 cc/l O_2. *Ephemera vulgata* showed no clear incipient limiting point, but continued steadily to consume more oxygen as the concentration increased. Oxygen lack apparently killed *Baetis scambus* before it reduced its oxygen consumption seriously. The small unidentified nymphs of *Baetis* reached the incipient limiting point at a concentration above saturation. Walshe (1948) found a steady drop in consumption by two chironomids from a stream with fall in concentration from saturation. Two species from a ditch showed an incipient limiting point at 3 cc/l. That the stream animals were found by neither set of workers to have an incipient limiting point could have been due to the conditions in the experiments. Fox *et al.* placed their test material in a bottle that was turned upside down sharply every five minutes. Walshe's larvae were in a cylinder that rotated slowly throughout the experiment. In neither container could the conditions have been at all like those in a stream and for further understanding of oxygen as an

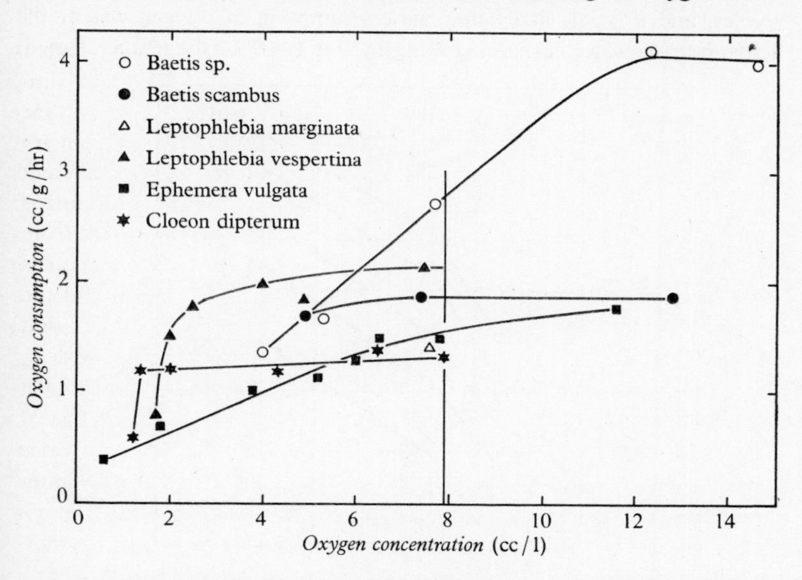

41. Oxygen consumptions at 10°C of Ephemeroptera nymphs at various oxygen concentrations. The vertical line at 7·9 cc/l marks the oxygen content of water in equilibrium with the atmosphere (Fox, Wingfield, and Simmonds 1937) (*J. exp. Biol.* **14,** p. 214, fig. 1).

ecological factor it is necessary to consider work in which flow has been taken into account.

Fox and Simmonds (1933) enclosed nymphs in a jar and noted that *Baetis rhodani* died at a higher concentration than *Cloeon dipterum* did. One reason for the greater tolerance of the latter is that, by beating its gills, it can keep a current of water flowing over its body, whereas *B. rhodani* cannot. Wingfield (1939) showed that a whole nymph of *Cloeon* could survive a lower concentration of oxygen than a specimen without its gills, but that this difference disappeared if the water was kept in motion. Evidently then, for species that cannot themselves keep a current of water flowing over their bodies, not only is the concentration of oxygen important but also the rate at which the water is flowing. This has been investigated by Ambühl (1959), whose work has already been noticed in the section on the effect of current. He placed animals in a tube through which water could be kept circulating at a desired speed. Their movement along the tube was circumscribed by gauze discs and they were provided with fused powdered glass as a substratum. Conditions cannot have been in the least like those of the natural habitat of any of the species, and it seems likely that the consumption of oxygen was at the active rate. In water containing 8 mg/l O_2 at 18-19°C, the temperature at

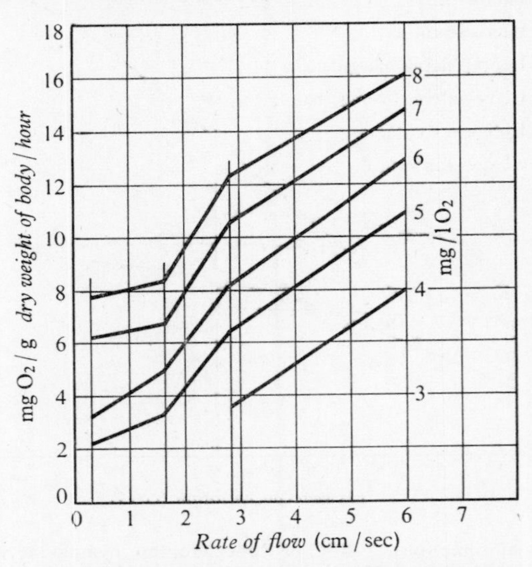

42. Oxygen consumption by *Rhithrogena semicolorata* at different rates of flow and different concentrations of oxygen (Ambühl 1959) (*Schweiz. Z. Hydrol.* **21**, p. 239, fig. 54).

which all the experiments were carried out, *Rhithrogena semicolorata* consumes just under 8 mg O_2 per g dry weight of body per hour at a current speed of about 0·3 cm/sec. An increase in the rate of flow is accompanied by a marked increase in the rate of oxygen consumption, and at 6 cm/sec 16 mg are being consumed. A reduction in the oxygen content of the water reduces the rate of oxygen consumption, but this increases with faster flow in the same way at all concentrations tested (fig. 42). It is no disparagement of this important work to regret that the tests were made at current speeds at the extreme low end of the range; 6 cm/sec. is a slow current compared with those in which *R. semicolorata* is usually found. It seems likely that there could come a point at which increased flow no longer increased oxygen consumption and that the incipient limiting point, not reached anywhere in fig. 42, depends on rate of flow as well as on oxygen concentration and temperature. The concentration lethal to *R. semicolorata* drops from just under 5 mg/l O_2 at a current speed of 0·3 cm/sec to about half that at 6 cm/sec, and from fig. 43 it appears likely that in swifter flow the lethal concentration would be lower still. The nymph of this species has adapted its gills to serve as a sucker for maintaining position on a flat surface in a current, and they cannot be used to waft a

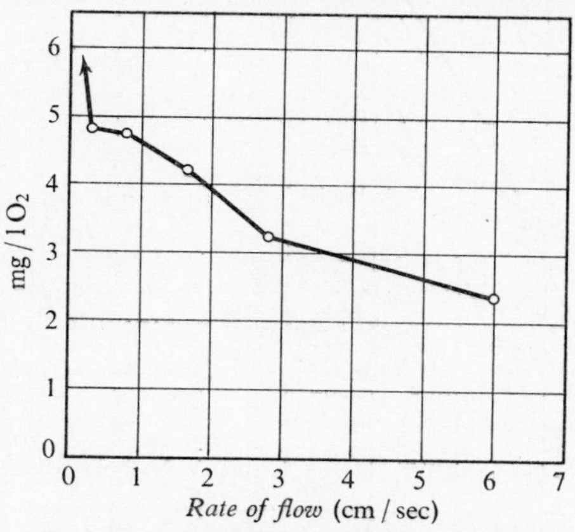

43. The concentration of oxygen lethal to *Rhithrogena semicolorata* at different rates of flow (Ambühl 1959) (*Schweiz. Z. Hydrol.* **21**, p. 240, Abb. 55).

flow of water over the body. *Ecdyonurus venosus* is very like *R. semicolorata* in bodily structure but the gills can be moved to maintain a respiratory flow. Increase in current speed affects its rate of oxygen consumption very little at 8 mg/l O_2 (fig. 44), but there is some increase when the concentration is quite low. The lethal concentration is little affected by change of current speed over the range used and remains fairly constant at just over 1 mg/l. *Ephemerella ignita*, which can also move its gills, is like *Ecdyonurus* but its rate of oxygen consumption hardly changes at all with increasing rate of flow even in water with little oxygen in solution. On the other hand the concentration lethal to it falls more (fig. 45). The three lines for *Baetis* in fig. 44 represent the results of experiments on three different days, twenty-five, seventeen and ten individuals having been held in the apparatus on the three successive occasions. That the results are somewhat different could have been due to a mixture of species, for the identification of none of the nymphs was certain, or to different proportions of nymphs in various stages of development. *Baetis* is like *Rhithrogena* in the way in which, with increasing current, oxygen consumption rises,

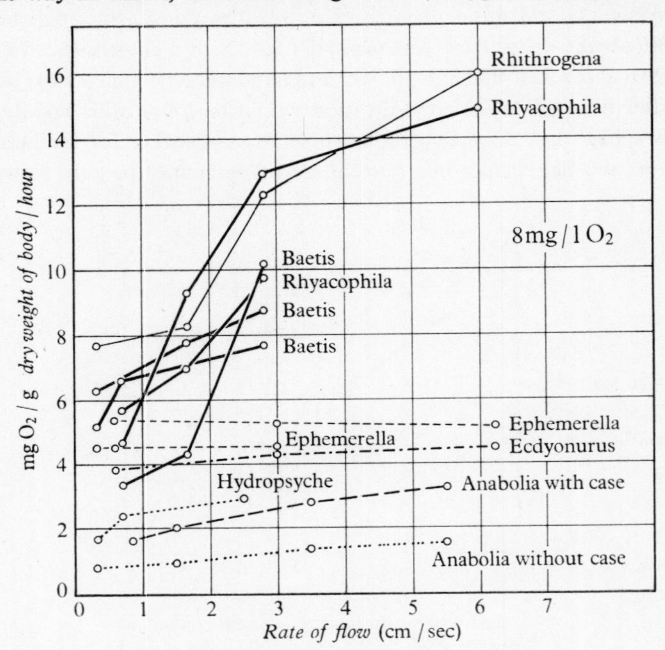

44. Rates of consumption of oxygen at different current speeds. Temperature 18-19°C. Oxygen concentration 8 mg/l (Ambühl 1959) (*Schweiz. Z. Hydrol.* **21**, p. 252, Abb. 67).

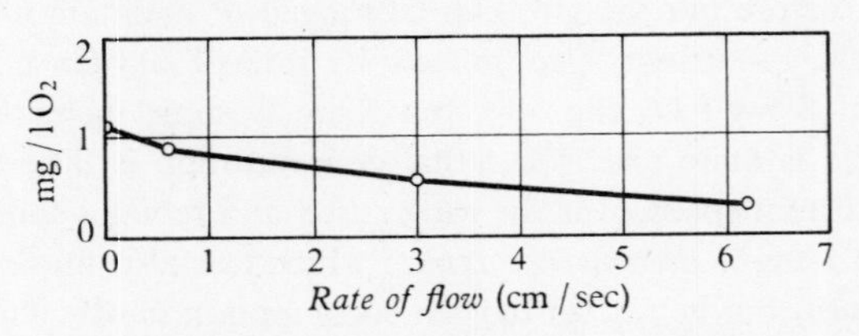

45. The concentration of oxygen lethal to *Ephemerella ignita* at different rates of flow (Ambühl 1959) (*Schweiz. Z. Hydrol.* **21**, p. 231, Abb. 46).

though the rise is not quite as steep, and also in the way in which lethal concentration falls.

The results for *Rhyacophila nubila*, a free-living caddis that rarely undulates the abdomen, resemble those for *Rhithrogena* at a concentration of 8 mg/l O_2, but differ from them in that, with falling concentration, the increase in consumption drops. In other words, whereas the lines in fig. 42 run parallel, in the corresponding figure for *Rhyacophila* they radiate.

Hydropsyche angustipennis, a net-spinning caddis, which undulates its abdomen at a rate which increases as oxygen concentration falls, shows some increase in consumption of oxygen as rate of flow increases at very low concentrations but not much when it is above about 1 cm/sec. The lethal concentration too falls steeply as current increases up to about this speed and remains rather constant at higher velocities.

The consumption by *Anabolia nervosa*, a case-bearing caddis, increases slightly but steadily with increasing rate of flow. The lethal concentration is rather constantly just below 1 mg/l O_2. The level is higher for larvae without a case and these also consume less oxygen.

The nymphs of most Odonata occur in still water, but those of *Calopteryx* (*Agrion*) *virgo* and *C.* (*A.*) *splendens* are found only in streams. They cling to roots projecting into the water from the bank and to plants such as *Ranunculus fluitans* and *Potamogeton crispus* but not to those that have flat leaves or a compact tufted habit. The current range is 2-80 cm/sec. They can move about in currents up to 60 cm/sec but in faster flow they remain stationary. Favourable temperature seems to be 13-18°C for *C. virgo* and 18-24°C for *C. splendens*. Zahner (1959) observed them in a box divided into channels in each of which current of a given speed could be maintained. Nymphs could wander from one channel to another but showed no signs of any preference even when different rates of flow were combined with different concentrations of oxygen. Low oxygen content made them restless, and this caused them to accumulate in the best oxygenated part of an aquarium divided by partitions over the tops of which they could pass. Zahner also suspended nymphs in cages in places with different amounts of oxygen, and, from observations on the mortality, made certain deductions about the minimum requirements. It proved impossible to determine the lethal concentration exactly because individuals varied so much; in the same experiment some would die soon, some would complete their aquatic lives but fail to complete metamorphosis, and some would apparently be unaffected. However, Zahner concludes that for both species in still water a concentration of 7-12 mg/l O_2 is unfavourable, which means that at all but extremely low temperatures the optimum concentration is above saturation. If there is some flow to renew the supply of oxygen at the surface of the nymph, these levels are lower; 4-5 mg/l O_2 is just unfavourable and over 7 mg/l optimal. A flow of 3-6 cm/sec brings *C. virgo* all the oxygen it requires and any beneficial effect of faster flow is cancelled by the need for more work holding on. A somewhat faster flow, 8-12 cm/sec, is most favourable for *C. splendens*.

These larvae cannot obtain from still water the oxygen they require to satisfy their needs. Their occurrence in streams is, however, largely brought about by the behaviour of male and female at the time of mating (chapter 5). Whether nymphs confined to running water by the reactions of their parents have never needed to develop tolerance to low oxygen content, or whether, the nymph being fundamentally exigent, selection has acted in favour of parents that would lodge it in suitable conditions is impossible to discover.

There can be no doubt that some species which have not the ability to create themselves a current of water over their bodies must rely on a natural current to bring them the oxygen which they require. Presumably there is an optimum which depends on the oxygen content of the water and also on the rate of flow, temperature playing a part as well. A full investigation of the influence of three variables takes a very long time, and it is not surprising that no statement is yet possible about the optimum for any species. The results of earlier work on stream animals in still water, for example the finding by Fox, Wingfield, and Simmonds (1937) that the incipient limiting level of *Baetis sp.* is well above the saturation concentration of oxygen, cannot be treated as reliable. The implication, however, that the incipient limiting point is always below saturation concentration, in other words that, under natural conditions the upper limit to activity is always set by the power of the animal to take up oxygen and never by the amount available, must be avoided. The amount of oxygen consumed by certain snails started to fall as soon as oxygen concentration fell below saturation concentration (Berg and Ockelmann 1959), and, though most showed a well marked incipient limiting point, some bottom-dwelling animals were found by Berg, Jónasson, and Ockelmann (1962) to respire in the same way.

RESPIRATION AND SALINITY

'Die Atmung ist in Salzwasser leichter als in Süsswasser.' Schlieper (1958a) attributes the formulation of that idea to Thienemann. *Mysis relicta* can tolerate as little as 1·6 cc/l O_2 in the Baltic, but dies if the concentration falls below 3-4 cc/l O_2 in fresh water. These figures are taken from the more recent work of von Ledebur (1939), which is a useful review of the subject. *Corixa stagnalis* is common in brackish water and rarely found outside it. Claus (1937) found its oxygen consumption to be least at 6 per mille salinity and higher in weaker and stronger concentrations. It reached a peak at 25 per mille and decreased in more saline water. The concentration in the blood of *C. stagnalis* is equivalent to 14 per mille salinity. Claus postulates that the permeability is naturally lowest at 6 per mille and that the extra oxygen consumed is used to prevent an increase in permeability and to maintain osmotic equilibrium. At 25 per mille control breaks down and oxygen consumption drops. Higher consumption of oxygen by a marine animal transferred to dilute sea water has been explained simply in terms of the extra amount needed to bale out the water drawn in by osmosis. If this be correct, the rate should remain constant, but Schlieper (1958a), in a critical discussion of the problem, points out that it does in some animals, e.g. *Carcinus*, but not in others,

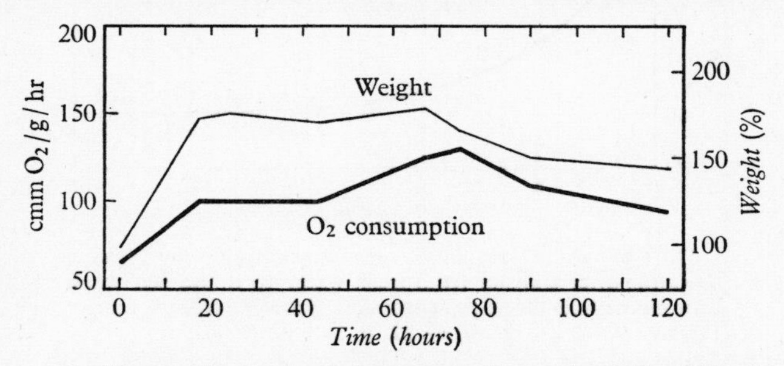

46. Change of rate of oxygen consumption and weight of *Nereis diversicolor* on transfer from sea water to 25 per cent sea water (Beadle from Schlieper 1958a) (*Die Binnengewässer* **22**, p. 288, Abb. 45).

e.g. *Nereis*. Furthermore changes in weight of *Nereis* transferred to dilute sea water, increase in weight being taken to be due to more water in the tissues, and changes in rate of oxygen consumption do not show the relation expected (fig. 46). Weight increases fairly steeply during the first twenty hours, remains more or less steady for fifty, and then starts to fall. Oxygen consumption should be at a maximum as long as there is excess water in the tissues according to the baling theory, but in fact it rises steadily for about seventy-two hours. Schlieper has calculated that the rate of oxygen consumption of *Carcinus* transferred to dilute sea water is greater than it should be if the only change was the extra work required to balance osmotic pressure.

There is therefore doubt about the interpretation that Claus put on his results. As Schlieper points out, an increased oxygen consumption in a different concentration may reflect no more than increased activity stimulated by change. *Corixa* is an animal whose activity is difficult to control. Another objection is that the experiments did not go on long enough. Time must always be considered, as the experiments on *Nereis* show. The rate of oxygen consumption of an *Astacus fluviatilis* transferred from fresh water to brackish water isotonic with its blood decreases 40 per cent, but nearly a month has elapsed before a steady rate is reached (fig. 47).

The permeable living membrane, therefore, has not fixed physico-chemical properties, and the idea that changes in the external medium do

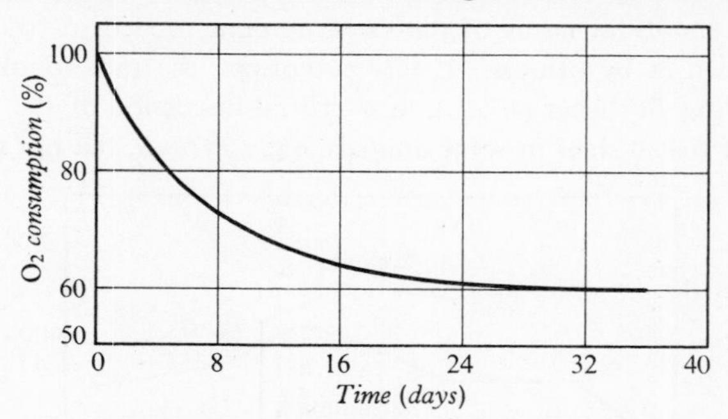

47. Change in rate of oxygen consumption by *Astacus fluviatilis* after transfer from fresh water to dilute sea water isotonic with its blood (15 per mille) (Schwabe from Schlieper 1958a) (*Die Binnengewässer* **22,** p. 287, Abb. 44).

nothing more than effect an alteration in the osmotic gradient that will be reflected in the oxygen consumed to maintain a steady state inside is too simple. The membrane itself changes and takes a comparatively long time to adjust itself to changes in the ambient medium.

Theodoxus fluviatilis has a similar rate of oxygen consumption in fresh and in brackish water at the same temperature, but the rate ceases to increase with increasing temperature at 29°C in fresh water and 35°C in brackish water. The reverse was true of *Potamopyrgus jenkinsi*, which also differed in that it consumed more oxygen in brackish water than in fresh water. Lumbye (1958), surprised at this, suggests that the lower consumption in fresh water was due to 'reduced growth'. The smaller size could, however, have had a genetic origin with which a physiological difference went too. The work of Berg on respiratory acclimatization has already been quoted to illustrate the pitfalls in comparing specimens of what appears to be one species from different places.

Whatever the physiological reason, oxygen demand may change with changing salinity, and the range of natural conditions which the species can inhabit expands or contracts. The work of McLeese (1956) has already been referred to. Hers (1942) showed that in fresh water without oxygen a *Chironomus* of the *plumosus* group was unable to maintain the normal concentration of chloride in the blood.

LIFE WITHOUT OXYGEN

Von Brand (1944) gives an account of the occurrence of anaerobiosis in the animal kingdom and (1945) discusses the processes involved. Energy is obtained by splitting carbohydrates into lactic acid, into fatty acids, or into various mixtures of the two, and at that point the process stops. Aerobic respiration is similar except that the splitting is taken a stage further to produce water and carbon dioxide. In frog's muscle the energy released by this process is sufficient to resynthesize 75 per cent of the lactic acid back to glycogen. Harnisch (1951) gives an account of anaerobiosis in freshwater animals and also discusses the physiology of the process. A number of animals can withstand absence of oxygen for a limited period but only a few can survive deprivation for long. Of them the chironomids are the only ones that have been studied in any detail. *Chironomus thummi* is found in shallow lakes where lack of oxygen may be acute but not prolonged. *C. anthracinus-bathophilus* and *C. plumosus* inhabit deeper lakes where oxygen may be absent for long periods.

Harnisch (1951, 1958) attaches particular importance to the rate of respiration of a species when it is placed in oxygenated water after a period without oxygen. It increases greatly in *C. thummi* and remains high for some time, whereas in the other species it is not higher than normal. His explanation is that *C. anthracinus-bathophilus* and *C. plumosus* can function satisfactorily without complete oxidation of the sources of energy, whereas *C. thummi* cannot.

Walshe's (1947a) results with *C. plumosus* are different, which may be due to the fact that her specimens came from the pond in Regent's Park and not from a deep lake. She found an increased rate of oxygen consumption, comparable to that which Harnisch found in *C. thummi*, when it was transferred from oxygen-free to oxygenated water. At 1°C this was certainly caused by repayment of an oxygen debt but at 17°C it was not possible to be certain that the whole increase was not due to sudden activity after quiescence and fasting in the absence of oxygen. The debt repaid was but 0·5 per cent of that which would have been incurred during the absence of oxygen had anaerobiosis not been possible and presumably represents the oxidation of a small residue of the products of anaerobiosis still left in the body. A species which differed only in being able to dispose of such products more rapidly would show less or no increase in the rate of oxygen consumption when oxygen became available again, and Walshe is inclined to attach less importance than Harnisch to this difference.

Haemoglobin is presumably important to animals that live in poorly oxygenated places because so many of them possess it. Some produce it only when oxygen is scarce. This has been shown for three pond species of *Daphnia* by Fox (1948) and Fox and Phear (1953), for *Artemia salina* by Gilchrist (1954), and for various other animals including ostracods, chironomid larvae, and young but not full-grown *Planorbis corneus* by Fox (1955). *Daphnia* that have been exposed to low concentrations of oxygen and have developed haemoglobin, compared with colourless specimens from oxygenated water, survive better in low concentrations (0·5-1·5 mg/l O_2) and also feed more quickly and produce more eggs (Fox, Gilchrist, and Phear 1951). Specimens of *Artemia* with haemoglobin consume more oxygen at low concentrations than those without (Gilchrist 1954).

Pause (1919) attributed a storage function to the haemoglobin of *Chironomus gregarius* which he found in a stream. He based this conclusion chiefly on observations in a sealed jar. Larvae survived only two hours in the first instar, and five hours at the beginning of the second instar, but by the end of this stage they lived for twenty-five and in the third instar thirty-four hours. Haemoglobin appeared during the second instar and was fully developed in the third. Pause's belief is that the larva lies safely in the bottom mud by day, and at night, when predators cannot see it, travels up to the surface to renew the store of oxygen. His survival figures certainly bear out the idea that enough oxygen can be stored to last a day, but Leitch (1916) had calculated that the volume of haemoglobin contained in a body the size of that of a chironomid could not store more oxygen than would last twelve minutes.

Ewer (1942) obtained from a commercial source larvae, some of which gave rise to adults that were identified as belonging to the *C. plumosus* group, probably *C. cingulatus* and *C. thummi*. Their incipient limiting point at 17°C was found to be 3 cc/l O_2. The oxygen consumption of larvae treated with CO was similar to that of normal larvae above the incipient limiting point but less below it, an indication that haemoglobin becomes effective at about that level. Walshe (1947b), using a similar technique, obtained a similar result with *Tanytarsus brunnipes*. It also led her (1947a) to the conclusion that haemoglobin played a part right at the beginning in the increased uptake of oxygen by *C. plumosus*, when it was transferred to oxygenated water from anaerobic conditions, but not later.

Last-instar larvae of *Chironomus plumosus* build tubes in mud by consolidating the walls with a salivary secretion. Walshe (1950) found that they could be induced to enter glass tubes, in which they could be watched, and that they appeared to be content in this substitute. Part of the time was spent in resting and some of the rests were so long that anaerobic respiration must have been taking place during them; spectroscopic examination showed that the haemoglobin lost its oxygen nine minutes after the animal stopped undulating. Part of the time was spent undulating the abdomen to keep a current of water flowing through the U-shaped tubes for respiratory purposes. The other activity was feeding; occasionally the larva browsed on the mud around its tube but the usual way was to spin a net across the tube, send a current through it by undulating, and then to eat the net and the particles that it had strained. Table 22 shows how time was divided

Table 22

Activity of *Chironomus plumosus* (Walshe 1950)

	Oxygen content as percentage of saturation	Percentage of total time devoted to: feeding	resting	respiratory irrigation
normal larvae	100-80	36	12	52
	37-22	32	16	52
	14-7	10	18	72
	7-5	0	12	88
	0	0	6	94
larvae treated with CO	90-75	22	14	16
	39-26	11	46	43
	11-7	0	74	26

between these different activities at different concentrations of oxygen and by larvae with active haemoglobin and with carboxyhaemoglobin. After a time with no oxygen, normal larvae cease to undulate and remain quiescent. Below 25 per cent saturation normal larvae are still feeding but the treated larvae are not. Evidently the presence of haemoglobin makes activity possible at an oxygen concentration at which otherwise it would not be. Moreover, recovery after anaerobiosis is quicker in normal larvae and starts at 7 per cent saturation compared with 15 per cent for treated larvae.

Walshe points out that *C. plumosus* is one of the few species that can survive in lakes 'where extensive summer and winter depletion occurs' but makes no comment on this in the light of her laboratory findings. Whether the larvae lie quiescent as long as there is no oxygen, or whether after a while they become active and rely on anaerobic respiration remains to be discovered.

EXCESS OXYGEN

Fox and Taylor (1955) kept various animals in water that was saturated with pure oxygen (100 per cent), saturated with air (21 per cent) or which was 4 per cent oxygenated. No *Artemia salina* died when kept for a fortnight at these concentrations but all the other species tested showed a mortality that varied according to the concentration. *Limnaea stagnalis* survived better at 21 per cent and 100 per cent than at 4 per cent. Young *Planorbis corneus* survived fairly well in 21 per cent, not so well in 4 per cent and badly in 100 per cent; older specimens survived well in 4 per cent and 21 per cent. *Tubifex* sp. out of its tubes succumbed quickly in 100 per cent but survived equally well in the other two concentrations. *Chironomus* spp. of the *plumosus* group, *Anatopynia varia* and *Heterocypris incongruens* showed best survival in 4 per cent. All these animals except the first two appear to have been adversely affected by abnormally high concentration of oxygen, and survival of the last three was better in a concentration well below air saturation.

Mortality of fish due to supersaturation with oxygen is mentioned by several authors. Woodbury (1942) attributes the death of thousands of fish in a Wisconsin lake to an oxygen supersaturation of 327 per cent caused by a dense growth of *Chlamydomonas*. Mortality in ponds supersaturated by photosynthesis, and also in fish containers fitted with oxygenating apparatus, is recorded by Mann (1952), Harnisch (1951), and Schäperclaus (1954). Wiebe and McGavock (1932) had a different experience when they placed fish in an aquarium, raised the oxygen content by means of compressed oxygen, and kept it high for twenty days. There were never less than 18·4 p.p.m. O_2 in solution, which was 183 per cent saturation; it remained above 300 per cent saturation for nine consecutive days. At the end of twenty days the oxygen concentration was allowed to sink back to its normal level. A number of different species survived these conditions without ill effect. Investigation is clearly required into the questions not only of when supersaturation kills fish but of how it does it. Some of the explanations offered are not in accord with known physical and physiological facts and it is advisable to reserve comment until more work has been done.

SUMMARY

Salvelinus fontinalis has a fairly high consumption of oxygen when resting (standard metabolism), and a comparatively very high consumption when active (active metabolism), the first increasing steadily, the latter increasing and then decreasing with rising temperature. The concentration of oxygen below which lack of it limits activity (incipient limiting point) is high. At the same temperature the standard metabolism of the pond species, *Carassius auratus*, is lower, the active metabolism much lower, and the incipient limiting point very much lower. *Carassius* can obtain the oxygen it requires for various levels of activity from a much lower concentration than *Salvelinus*. The significance of a concentration below which a species cannot live is obvious, but more common and less easy for the ecologist to evaluate is the concentration in which a species cannot exert itself fully. Similar relationships are described for invertebrates, some of which are confined to running water not only by high oxygen demand but also by lack of any mechanism to cause water to flow over the body and bring oxygen.

Chironomids can live in lakes from whose deeper layers oxygen disappears in summer because they can respire anaerobically. The haemoglobin possessed by many animals that live where oxygen may be scarce is an important transporting agent at low concentrations. Too much oxygen can be unfavourable.

SYNONYMS

Species	Group	Other generic names	Other species names
Acroloxus lacustris (Linnaeus)	Moll. Gast. Pulmonata	*Ancylus*	
Ancylus fluviatilis Müller	Moll. Gast. Pulmonata	*Ancylastrum*	
Astacus fluviatilis Fabricius	Crust. Malacostraca	*Potamobius*	*astacus* Linnaeus
Chironomus anthracinus Zetterstedt	Ins. Dipt. Chironomidae	*Tendipes*	
C. bathophilus Kieffer	Ins. Dipt. Chironomidae		*anthracinus* Zetterstedt
C. cingulatus Meigen	Ins. Dipt. Chironomidae	*Tendipes*	
C. gregarius Kieffer	Ins. Dipt. Chironomidae	*Tanytarsus*	

Species	Group	Other generic names	Other species names
C. plumosus (Linnaeus)	Ins. Dipt. Chironomidae	*Tendipes*	
C. riparius Meigen	Ins. Dipt. Chironomidae	*Tendipes*	
C. thummi Kieffer	Ins. Dipt. Chironomidae	*Tendipes*	*riparius* Meigen
Corixa stagnalis (Leach)	Ins. Heteropt. Corixidae	*Sigara*, *Halicorixa*	*lugubris* Fieber
Glossiphonia complanata (Linnaeus)	Hirud.	*Glossosiphonia*	
Limnaea stagnalis (Linnaeus)	Moll. Gast. Pulmonata	*Lymnea*	
Mysis relicta Lovén	Crust. Malacostraca		*oculata* (Fabricius)
Planaria alpina Dana	Platyhelminthes	*Crenobia*	
P. gonocephala Dugès	Platyhelminthes	*Dugesia*, *Euplanaria*	*subtentaculata* Draparnaud
Planorbis corneus (Linnaeus)	Moll. Gast. Pulmonata	*Planorbarius*, *Coretus*	
Potamopyrgus jenkinsi (Smith)	Moll. Gast. Prosobranchia	*Paludestrina* *Hydrobia*	*crystallinus carinatus* Marshall
Salmo trutta Linnaeus	Pisces	*Trutta*	*lacustris* Linnaeus *fario* Linnaeus
Theodoxus fluviatilis (Linnaeus)	Moll. Gast. Prosobranchia	*Neritina*	

CHAPTER 10

Salinity

MIXOHALINE WATER

INTRODUCTION AND NOTES ON THE FAUNA

Hitherto I have used the adjective brackish in a loose sense to include all waters that contain more salt than inland lakes, rivers, and ponds generally do. At a symposium held in Venice in 1958 (reported in *Arch. Oceanogr. Limnol. Roma*, 11, suppl. 1959) this usage was deplored on the grounds that, from a biological point of view, it is desirable to distinguish those waters in which the proportion of ions is roughly the same as in the sea from those in which it is not. The term mixohaline was recommended for brackish water that is dilute sea water.

There is a diversity of ways of expressing the amount of salt in saline water that may bewilder the biologist who is not in constant touch with work on that medium. Table 23 shows figures for a widely accepted analysis

Table 23

The major constituents of sea water
(Sverdrup, Johnson, and Fleming 1942)

	g/kg	%
Cl^-	18·980	55·04
$So_4^{=}$	2·649	7·69
Mg^{++}	1·272	3·69
Ca^{++}	0·400	1·16
K^+	0·380	1·10
Na^+	10·556	30·61
Others	0·245	0·72
	34·482	100

of sea water. There are 34·482 g of salt per kilogram of water, which is grams per litre at normal temperature and pressure, and also parts per thousand. The proportions of the various ions remain constant but the total figure varies; where there is strong evaporation it may be as high as 40, as for example in the Red Sea and parts of the Mediterranean. A round figure of 35 per mille is often taken as the salinity of sea water. If it be diluted with an equal volume of fresh water, the result may be said to have a salinity of 17·5 per mille or it may be called 50 per cent sea water. A complication arises from the practice of expressing the osmotic pressure of body fluids, which may incidentally be due to a mixture of salts and organic substances, as per cent NaCl. However, a rough conversion from one to the other is easy. If sea water be taken as having a salinity of 33·3 per mille, and if the salinity be treated as being wholly due to NaCl, not significant errors for many practical purposes, 100 per mille salinity is 100 g/l which is 10 per cent NaCl. Salinity can be measured simply by finding the specific gravity or the extent to which the freezing-point is depressed below that of distilled water. Table 24, taken from Schlieper (1958a) shows the relation

Table 24

Physical and chemical characteristics of mixohaline and sea water (Schlieper 1958*a*)

Salinity ‰	Specific gravity at 0°C compared with distilled water at 4°C	Freezing-point °C	mM/l
5	1·0040	– 0·266	79
10	1·0080	– 0·533	156
15	1·0121	– 0·795	233
20	1·0161	– 1·077	317
25	1·0201	– 1·350	396
30	1·0241	– 1·628	477
35	1·0281	– 1·907	559

between salinity and these two methods of measurement. As the composition is constant, salinity may be expressed in terms of the ion whose concentration can most easily be measured. A titration with silver nitrate will give the concentration of halides, and all except those making a very accurate analysis ignore bromine and iodine and treat the result as if only chlorine were present. It is sufficient for most purposes to approximate the figure in table 23 and regard a chlorinity of 19 per mille as representing sea water.

Fresh water has so little in solution that quantities are expressed, not as parts per thousand but as parts per million, which is mg/l. It is often convenient, particularly when calculations of how much of each of two substances combine or react are to be made, to express quantity not as weight but as milliequivalents, which is mg divided by equivalent weight, or as millimoles (mM) which is mg divided by atomic or molecular weight. The total amount of salts in solution may be ascertained roughly by determining the conductivity of the water. One milliequivalent gives a conductivity of 112 $\frac{1}{\text{megohms}}$ at 25°C.

Mixohaline waters have attracted a great deal of attention from physiologists, because they afford conditions in which faunistic differences can be correlated with one obvious variable, and this is reflected in the number of reviews. Krogh (1939), Harnisch (1951), Beadle (1957, 1959), and Remane and Schlieper (1958) (from here on referred to as Remane (1958) or Schlieper (1958a)) are the recent and important ones. In addition Segerstråle (1958) gives a valuable list of references to work in all parts of the world.

Where masses of fresh water and sea water adjoin, species from each invade the mixohaline region between them, but few penetrate far, and the number of species at a point midway between the two is small (fig. 48). Many of those that are found are known only from mixohaline water, though nearly all have close relatives in the sea or in fresh water. Although it is often convenient to make a distinction between mixohaline and fresh water (0·5 per mille salinity according to the Venice Symposium), any division must be arbitrary. Many species have over the years adapted themselves to water increasingly less saline than the sea in which their ancestors originated, and there is no reason to suppose that the process has ceased to-day even though they now inhabit what we choose to call fresh water. Accordingly although only mixohaline organisms are referred to in the descriptive part of this chapter, in the section on the physiological problems facing an animal invading dilute conditions, the lower limit is taken to be distilled water.

The classic stretch of mixohaline water is the Baltic, which offers a gentle gradient and unusually stable conditions. The salinity in the North Sea is 34 per mille, at the entrance to the strait between Denmark and Sweden it is 30 per mille, at the other end it is 8 per mille, and there is a gradual decrease to 2 per mille at the extreme tips of the two bays. The salinity is about 15 per mille off Kiel, a well-known centre for German research, and 5 per mille off Tvärminne where the Finns have a biological station (fig. 49).

Up to 1931 the Zuider Zee was a large inlet of the North Sea penetrating far into the Netherlands. Its salinity decreased progressively towards the mouths of the rivers. Early in 1932 it was cut off from the North Sea by means of a dam, and renamed the Ijsselmeer. Salinity fell rapidly at first, then more gradually and reached equilibrium at the end of 1935 after which it was about 0·3 per mille in the middle rising to 0·9 per mille near the dam; or limnetic (fresh water) in the middle and mixo-oligohaline near the dam to use the terms recommended by the Venice Symposium. The changes in the fauna were kept under observation by a number of workers,

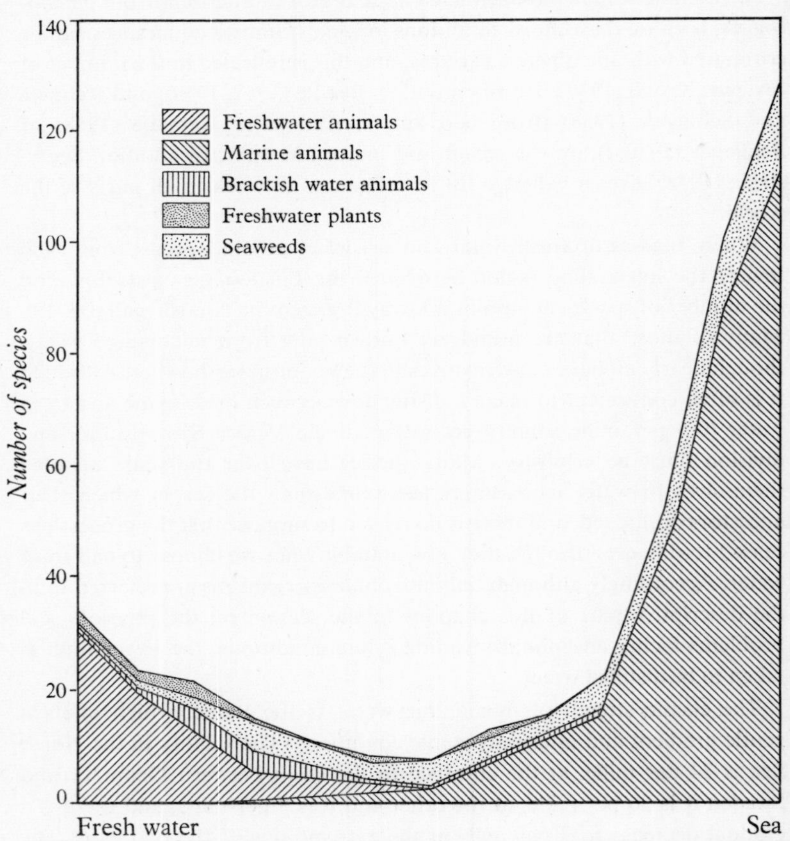

48. Fauna of the Tees estuary (Alexander, Southgate, and Bassindale from Macan and Worthington 1951) (*Life in Lakes and Rivers*, p. 116, fig. 13, London: Collins).

and their results were collated and edited by de Beaufort (1954). The publication is in Dutch, but there is an English summary of each chapter and a résumé of the whole in English at the end. In Britain estuaries have received particular attention, and there have also been some studies of small areas of mixohaline water in places where the coast is low and flat. In the Mediterranean, where the rise and fall of the tide is slight, brackish lagoons have been investigated, and reference may be made to the papers given by Petit and Schachter and by D'Ancona at the Venice Symposium. In the same volume may be found a report on the Black Sea by Băcescu and Mărgineanu.

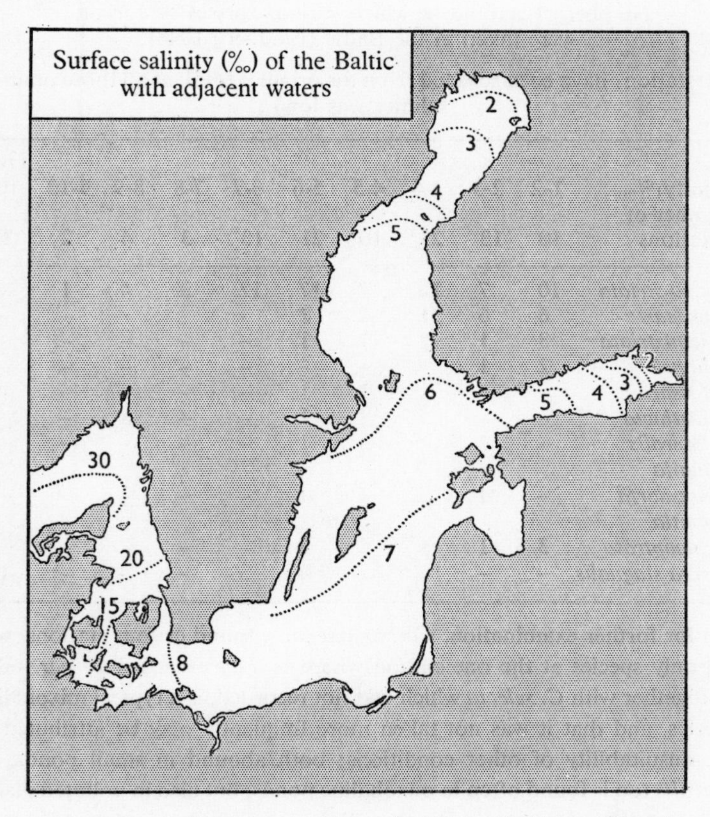

49. Surface salinity (per mille) of the Baltic Sea (Segerstråle 1949) (*Oikos* **1**, p. 131, fig. 2).

A selection from the findings in some of these places illustrates the descriptive side of the ecology of animals inhabiting mixohaline water. There is no space to give a full account or to name all the species that have been recorded in dilute sea water; Remane (1958) and Schmitz (1959) give lists. Segerstråle (1949) discusses briefly the mixohaline fauna off the Finnish coast and (1957) gives maps showing the range of various species in the Baltic. Lindberg (1937, 1948) reports on the Hemiptera and Coleoptera found in the Baltic, and, since much has been written about the habitats of the former, particularly the Corixidae, his findings (table 25)

Table 25

Number of stations at which various species of corixid were taken in the Baltic (Lindberg 1948)

(100 stations have been selected from the original total; at all those omitted salinity was low)

Salinity $^0/_{00}$	1-2	2-3	3-4	4-5	5-6	6-7	7-8	8-9	9-10	Over 10
Number of stations	10	13	23	10	21	13	3	4	2	1
Corixa striata	10	7	16	7	17	12	3	4	1	–
C. praeusta	6	6	1	1	2	–	–	–	–	–
C. semistriata	3	1	2	1	1	–	–	–	–	–
C. linnei	2	1	–	–	–	–	–	–	–	–
C. fossarum	–	–	1	1	1	–	–	–	–	–
C. distincta	–	–	1	1	1	–	–	–	–	–
C. lateralis	–	–	–	–	1	–	–	–	–	–
Cymatia bonsdorffi	–	1	2	1	1	–	–	–	–	–
Cymatia coleoptrata	3	1	1	2	2	–	–	–	–	–
Corixa stagnalis	–	–	–	–	–	3	–	–	–	1

call for further examination. *Corixa stagnalis*, found at four stations, was the only species at the one station where salinity exceeded 10 per mille. It, together with *C. selecta* which was not recorded, is a typical mixohaline species, and that it was not taken more frequently may be attributed to the unsuitability of other conditions; both abound in small ponds. *C. lateralis* too is found often in mixohaline ponds, but also in polluted places and, as will be seen later, in waters with high salinity but ionic composition unlike that of the sea. All but two of the remaining species fall into the lake-river group of Macan (1954a) and these two, *C. sahlbergi* and *C. semistriata*, have associations with species in that group. Four lake-river species were not found in the Baltic; two are rare and can probably be safely ignored but the other two, *C. falleni* and *Micronecta poweri*, are not and their absence may be attributable to inability to tolerate saline water. The frequency of *C. semistriata*, which, although associated with some of the lake-river species, is not often found in lakes, suggests that it, on the other hand, may be favoured by saline water. Apart from the mixohaline species, *C. striata* was the only one found at a salinity greater than 6 per mille. This might have been due to greater tolerance or to the greater exposure of the stations at which this concentration was exceeded, for it commonly occurs alone in lakes at the most exposed of the stations at which *Corixa* is found. Whether the second explanation is admissible cannot be determined from the data. The point to be stressed is that the corixid fauna of the Baltic, which is in effect a large lake, is almost exactly that which would be expected in a large lake, and it illustrates that conclusions about tolerance of salinity based on field observations must be guarded unless the other requirements of both the species found and the species not found are known too.

In rock pools with a salinity of 2·45-9·8 per mille Lindberg (1944) found *Corixa carinata*. They were on a latitude at which the species is otherwise recorded only from high altitudes; further north it occurs at sea-level in fresh water.

Lindberg (1948) records the occurrence of all the insects which he found in the Baltic. Remane (1958) gives figures for the maximum salinity tolerated by all the freshwater animals that have been taken in mixohaline water; all the main insect orders are represented except the Plecoptera, and they are scarcely to be expected since nearly all inhabit running water. The length of Remane's lists came as a surprise to the writer, whose experience is mainly of British waters. This no doubt is due to the ability of many species to tolerate the constant salinity in places such as the Baltic but not the more fluctuating conditions in smaller pieces of water, which are subject to greater influence from rainfall and evaporation.

When the Zuider Zee became the Ijsselmeer, twenty species of coelenterate disappeared, and the only survivor was *Cordylophora caspia*, which changed from the mixohaline form *typica* to the freshwater form *transiens* (= *lacustris*). *Pelmatohydra oligactis* was recorded in 1936 and the very rare medusa, *Craspedacusta sowerbii*, in 1937. The number of polychaetes dropped steadily from thirteen, when the dam was closed in 1932, to one, *Nereis diversicolor*, in 1938. Oligochaetes increased from four to eighteen; only one of the original four disappeared and it persisted till 1939. The amphipods *Gammarus zaddachi*, *Corophium volutator* and *C. lacustre* died out, *G. duebeni* persisted, and *G. pulex* and *Orchestia bottae* came in. Four marine isopods died out, and a mixohaline species, *Cyathura carinata*, first became more abundant and extended its range but later declined. *Asellus aquaticus* and *A. meridianus* were abundant when the survey was concluded but there was some doubt whether they were newcomers or whether they had always been present. Of seven mysids, only *Neomysis integer*, a mixohaline species, survived. Seven marine decapods, originally confined to the northern part where salinity was higher than elsewhere, disappeared quickly, as did *Crangon crangon* and *Carcinus maenas*, which leave mixohaline water to breed in the sea. The mixohaline species, *Palaemon longirostris* and *P. varians* were confined to the extreme north-west corner when the survey ended. *Eriocheir sinensis* and *Atyaephyra desmaresti* came in. A number of marine and two mixohaline (*Hydrobia ventrosa* and *Corambe batava*) molluscs were not found after three years, and there was then a period when this group was not represented. Later *Dreissena polymorpha* and *Hydrobia jenkinsi*, both of which inhabit mixohaline and fresh water, were recorded, and then freshwater species appeared.

Some species of fish that breed in the sea but used to enter the Zuider Zee to feed do not find their way into the Ijsselmeer. An exception is *Pleuronectes flesus*, the flounder, which is thought to enter when the sluice gates are opened. *Salmo salar* and *Anguilla anguilla* manage to pass through the Ijsselmeer, and special provision is made for the entry of the elvers of the latter. *Osmerus eperlanus* (smelt) is a survivor from the Zuider Zee but does not grow as large as it did before the dam was built. Other common species of the Ijsselmeer are *Leucioperca sandra* (pike-perch), *Perca fluviatilis* (perch), *Acerina cernua* (pope), *Abramis brama* (bream), and *Leuciscus rutilus* (roach).

Interest in estuarine faunas has centred round *Gammarus*, whose species show a zonation between sea and fresh water. There is divergence between the various accounts, due in part to misidentifications and in part to differences between estuaries; the nature of the bottom is important, and the salinity at one point at a given stage of the tide will vary with the shape of the estuary and the volume of fresh water flowing into it. There is general agreement that *G. pulex* occurs in rivers down to a point where it comes into contact with sea water. Also found in fresh water, but not known to breed in it, is *G. zaddachi zaddachi*, which extends down the estuary to a point where the salinity at mean high tide is 10-15 per mille. *G. zaddachi salinus*, separated from the preceding by Spooner in 1948, occurs below it in the middle reaches of the estuary, and can tolerate a salinity a little less than that of sea water. Below it again comes *G. locusta*, a common shore species that can tolerate dilution to 17 per mille. In Scotland, Denmark, and countries further north, *G. locusta* is replaced by *G. zaddachi oceanicus*, described by Segerstråle in 1948, though there is considerable overlap where the two meet (Spooner 1951). *G. z. oceanicus* is the main species in the Baltic except in the less saline areas where its place is taken by *G. z. zaddachi*. *G. z. salinus* and *G. locusta* occur as well (Segerstråle 1948). Spooner (1952) states that earlier records of *G. locusta* in America are *G. z. oceanicus* and that *G. z. zaddachi* in estuaries had often been misidentified as *G. duebeni*.

G. duebeni is rare in the main stream of an estuary and is typical of ditches and backwaters (Spooner 1948). Forsman (1952) found it mainly in rock-pools along the Swedish coast. It survived seventeen days when kept in distilled water and can tolerate water twice as saline as the sea. The temperature in small rock-pools can rise to more than 30°C on a summer day and this species has been found in Greenland in warm springs at a temperature of over 40°C, but these Greenland specimens must surely be a separate race, for Kinne (1954) has shown that it breeds at relatively low temperatures. Segerstråle (1946) has also found it mainly in rock-pools and showed experimentally that it is more tolerant of oxygen lack than *G. zaddachi*. It appears to be tolerant of a variety of unfavourable conditions but is notably unsuccessful in competition with other species (Kinne 1954). Hynes (1954) brings forward evidence that it can colonize fresh water when *G. pulex* is absent (chapter 6) and does not confirm Segerstråle's (1946) statement that the freshwater specimens belong to a different race, which will not breed in mixohaline water.

Pantin (1931) surveyed a small stream that ran over shingle at the top of the beach and over rock and boulders lower down. Above the level reached by high-water spring tides there was a freshwater fauna typical except for the mixohaline snail *Hydrobia ventrosa*; between the levels reached by high-water spring tides and high-water neap tides there was no fauna; and between high- and low-water marks of neap tides there was an estuarine fauna. *Procerodes* (=*Gunda*) *ulvae*, whose physiology was later investigated, an unidentified species in the same group, and the worm, *Protodrilus flavocapitatus*, occupied the whole of this stretch but did not extend beyond it. In the rocky section a few marine species occurred as well. Naylor and Slinn (1958) studied three rock-pools one above the other with a freshwater stream flowing through them. The lowermost was not covered at high tide but sea water entered it when there was a swell. Spray entered the other two. The fresh water tended to flow over the top of the sea water and the lower layers of the lowermost pool were generally pure sea water. Conditions in the other two were more variable. *Procerodes ulvae* and *Gammarus duebeni* occurred in all three pools and extended into fresh water above them.

In temporary collections of water mosquito larvae of the genus *Aëdes* are often to be found. Near the sea, but also in inland saline pools near

Droitwich (Marshall 1938), the species is *A. detritus*. It is impossible to state within what limits the larvae occur, because nobody has made sufficiently regular observations on the salinity of the breeding places. The range must be great because such places are often situated just within reach of exceptionally high tides. If dry weather follows the filling of one of these ponds, the sea water will be concentrated by evaporation, but, if the weather is rainy, it will be diluted. The larvae can tolerate salinities well above that of sea water and in the laboratory Beadle (1939) has successfully acclimatized them to distilled water.

More permanent ponds near the sea contain a varied fauna, and some of the species are not found inland, but the only ones to be mentioned here are *Corixa stagnalis* and *C. selecta*.

The physiology of certain species, some already mentioned, has been investigated and the habitats of these require brief mention. *Arenicola marina* burrows in sand between tide-marks. *Nereis diversicolor* occurs on the surface of estuarine mud-flats and is perhaps to be considered a mixohaline rather than a marine species. *Carcinus maenas*, the shore crab, is found on substrata of all kinds. *Mytilus edulis*, the mussel, requires a hard surface for attachment. *Asterias rubens*, the starfish, occurs on rocky bottoms not exposed at low tide. Both species of *Pleuronectes* are found in estuaries, and *P. flesus*, the flounder, penetrates far up them, but they breed in the sea. *P. platessa*, the plaice, is confined to the southern Baltic but *P. flesus* extends to the mouths of the gulfs (Segestråle 1957).

Lastly, there are the species that alternate between sea and fresh water. *Salmo salar* lays its eggs in fresh water and the young live in rivers for one, two or three years. Then they go down to the sea and spend one to five years feeding and growing, exactly where nobody knows. At the end of that time they return to fresh water, mate and lay eggs, after which most die, though a few survive to regain the sea and to return to fresh water for more reproduction. *Anguilla anguilla* starts life in the sea and spends two or three years drifting back to Europe as a typical marine planktonic organism. When these larvae reach the coasts and find fresh water, they swim upstream and eventually come to rest in some lake or pond where they may spend up to nineteen years before returning down-stream to embark on a voyage across the Atlantic Ocean to the Sargasso Sea. Tucker (1959) has recently suggested that they never get there, basing his hypothesis on two points. First he finds it difficult to imagine any direction-finding mechanism that would enable a fish that has drifted one way at the mercy of currents to return to a comparatively small area of ocean from starting points as far apart as the north of Norway and the Adriatic. Secondly, since there is evidence that the alimentary canal is degenerating by the time the fish enter the sea, it is difficult to see whence they derive the energy necessary for crossing an ocean and reproducing at the other side. He postulates that only the American eel, whose journey is so much shorter, succeeds in returning. The young are carried to America or to Europe according to which current they enter, and the different conditions during which larval development takes place cause the slight morphological differences on which the two species have been separated.

Eriocheir sinensis, the mitten crab, a native of China, was recorded in the River Weser in 1912. Since then it has spread throughout the Baltic and down the west coast of Europe to the South of France, where Hoestlandt (1959), who has given the most recent map of its range (fig. 50), believes it has reached water warm enough to check further advance. The young hatch from eggs laid in the sea or in brackish water and soon start to migrate upstream. They travel 700 km up the River Elbe but this is only half the distance they have been known to cover in China. The journey sometimes takes three to four years but the mature adults returning with the current may complete it in as many months.

Petromyzon fluviatilis adults live in brackish water. In the autumn mature individuals migrate into rivers where they fast throughout the winter. In the spring they spawn and die.

PHYSIOLOGICAL PROBLEMS CONFRONTING MARINE SPECIES ENTERING MIXOHALINE AND FRESH WATER

The body fluid of many marine animals has the same osmotic pressure as the ambient medium though the proportions of the various ions are not always identical (Beadle 1957, Schlieper 1958b). Presumably there must be a mechanism whereby some salts are taken up and others excreted. The compositions of the body fluid and of the tissue fluid are not identical either (Schlieper 1958a) but rather little is known about the relation between the two. Almost all marine animals respire in water and, as far as is known, no animal has evolved a membrane which is permeable to oxygen and carbon dioxide but impermeable to water and to salts.

A few inhabitants of the sea can penetrate mixohaline water because their tissues continue to function even though the concentration of salts in the body fluid falls to that of the outside medium. *Arenicola*, for example, can tolerate mixohaline water of salinity as low as 8 per mille (Krogh 1939). *Nereis diversicolor* can tolerate a salinity of 4 per mille, but it has evolved further than *Arenicola*, for at this low concentration it can maintain a higher concentration in the body fluid. Where the salinity is

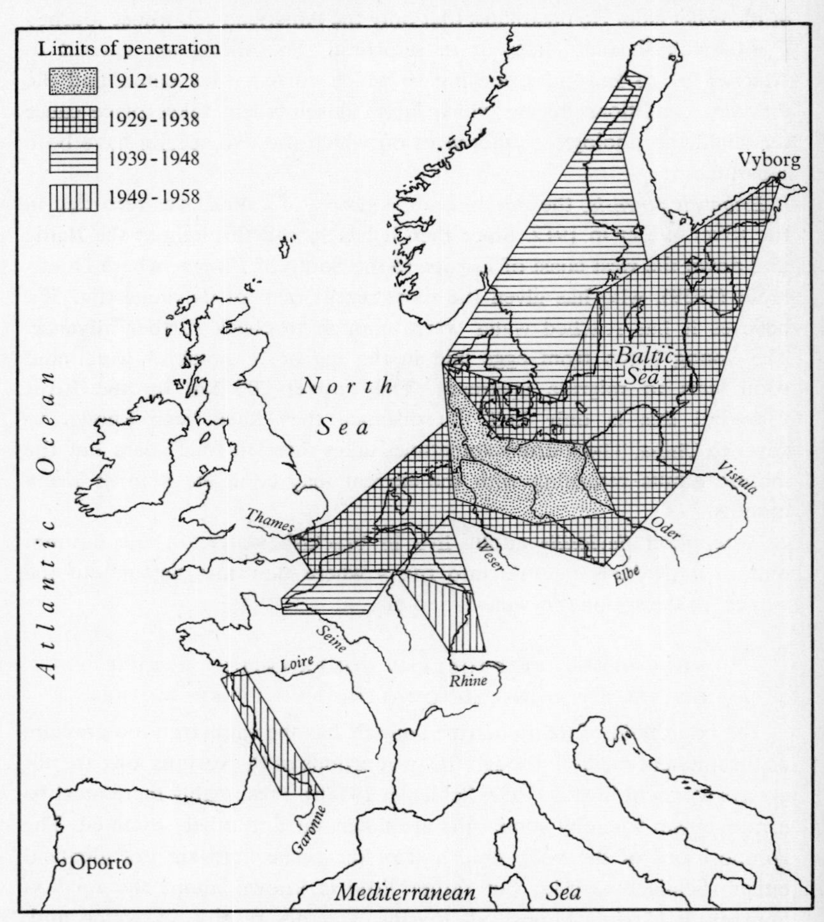

50. Distribution of *Eriocheir sinensis* (Hoestlandt 1959) (*Bull. franç. piscic.* No. 194, p. 12, fig. 3).

constant, as in the Baltic, inability to maintain activity or to reproduce may limit range before the salinity becomes intolerable. Schlieper (1958a) gives the following examples. At a temperature of 18-21·5°C *Ostraea virginica* is affected by salinity thus:

35-25 per mille development is normal;
23-21 per mille development is normal but slow;
19·3 per mille development is slow and there is some mortality;
17·5 per mille some eggs die though there is a fair production of free-swimming larvae not many of which, however, live to produce shell;
15·8 per mille development is very slow and very few young reach the shell-bearing stage.

Specimens of *Asterias rubens* living in the Baltic at 15 per mille salinity, which is near the limit of their range (Segerstråle 1957, fig. 16), take twice as long to turn over when placed upside down as specimens living in the North Sea. The process takes longer still after it has been performed seven times one after the other, whereas a North Sea specimen can right itself fifteen times before showing signs of fatigue (Schlieper 1958b). A temperature of 15°C also affects the rate adversely compared with 10°C, whereas North Sea specimens turn over at the same speed at both temperatures. Eggs are ripe in January, February, and March in the North Sea but not till May in the Baltic, where they are also smaller.

The rate at which an object was passed across the surface of the gill of *Mytilus edulis* was:

North Sea	40 mm/min
West Baltic (16 per mille salinity)	30 mm/min
East Baltic (6 per mille salinity)	14 mm/min

When east Baltic specimens were acclimatized to west Baltic water their rate went up to 25 mm/min.

Other species have developed varying degrees of control, and it is convenient to consider next the primitive system of *Procerodes ulvae*, although this is an estuarine species, not a marine one that can invade dilute arms of the sea. In dilute sea water the worms swell owing to the entry of water, which accumulates in the cells of the parenchyma, and lose salts, but, when the concentration of these is reduced to between 6 and 10 per cent of what it was in sea water, the loss stops. It is thought that permeability is lowered. The arrest of salt loss takes place only when there is a fair amount of calcium present. Oxygen consumption rises in dilute sea water, and this is attributed to the work that is being done to take salts in against the osmotic gradient and to remove water, but the seat of neither of these activities has been discovered (Pantin 1931, Beadle 1934).

In the sea the blood of *Carcinus maenas* is isotonic with the medium, but contains more sodium, potassium, calcium, and chloride and less magnesium and sulphate. There is an active uptake of the ions in the first group at the gill surface, and a removal of those in the second by the antennary glands, in which also potassium is resorbed. In brackish water the crab has the advantage of the relative impermeability, due to the hard exoskeleton, of all of its body except the gills. The concentration of the salts in the body fluid sinks (fig. 51) but remains above that of the external medium down to a salinity of 14 per mille, below which the crab cannot live. The urine is always isotonic with the body fluids. The rate of oxygen consumption is higher in mixohaline than in sea water but the extra amount used is more than is required for the work of maintaining the state

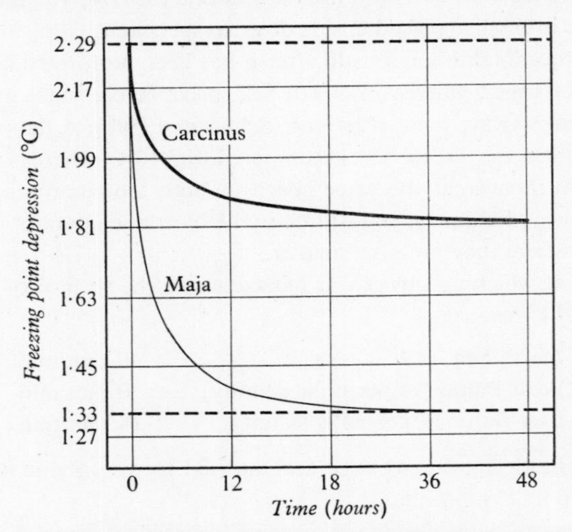

51. Change in concentration of the body fluids of *Carcinus* and *Maja* on transfer from salinity of △ 2·29–1·33°C (Schwabe from Schlieper 1958a) (*Die Binnengewässer* **22,** p. 262, fig. 27).

of disequilibrium (Schlieper 1958a). When placed on their backs, specimens of *C. maenas* from the North Sea and the Baltic do not show the difference which specimens of *Asterias* do. Beadle (1957) postulates that, as there is no osmotic gradient in sea water, there must be an active intake of water to compensate for that passed out by the antennary glands.

Protozoa with contractile vacuoles are animals in which the water exchange can be observed easily, though their small size has so far prevented any investigation of permeability or the difference between the concentration of ions inside and out. Certain marine peritrich ciliates possess contractile vacuoles whose function is to remove water taken in with the food (Kitching 1939). Some of them tolerated dilute sea water in experiments and increased in size when placed in it. The new volume was generally reached quickly and then maintained, but in *Cothurnia curvula* (fig. 52) it fell slowly. In solutions more dilute than 75 per cent sea water the rate of output by the contractile vacuole increased and, having dropped after an initial rise, remained fairly steadily well above the level in sea water.

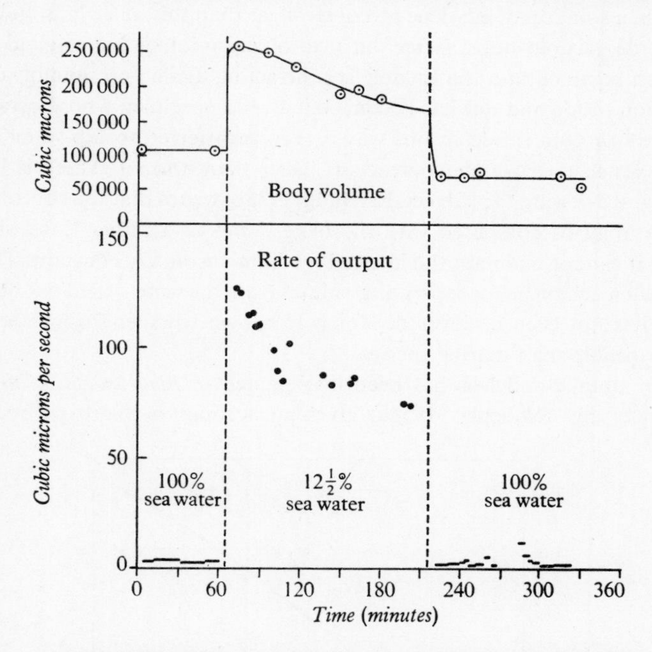

52. The effect of dilute sea water on the body volume and the rate of output by the contractile vacuole of *Cothurnia curvula* (Kitching 1934) (*J. exp. Biol.* **11**, p. 373, fig. 4).

Under these conditions the contractile vacuole is serving also to protect the body against excessive swelling by removing water drawn in by osmosis (Kitching 1934). It is not recorded whether any of the species used in the experiments occur naturally in mixohaline water, but others do (Remane 1958). Freshwater species presumably passed through mixohaline water. If their contractile vacuoles are stopped by means of cyanide, the bodies swell. The original size is regained when an animal treated in this way is returned to water with no cyanide, and during this process the contractile vacuole removes more water than usual. In a 0·05 M sucrose solution there is no swelling as a result of cyanide treatment and Kitching (1938a) concludes that the contractile vacuole normally maintains a difference in concentration equivalent to this. In a recent personal communication, however, he has stated that there could be an alternative explanation of the results obtained with cyanide. Less ambiguous are his conclusions derived from work on a suctorian (Kitching 1951). Its internal osmotic pressure is approximately that of a 0·04 M solution of non-electrolyte. If an animal is placed in a solution of ethylene glycol stronger than this, it shrinks and the contractile vacuole stops. After thirty to ninety minutes the original size has been regained and the vacuole is contracting again; presumably concentration inside and out has become equal. If a specimen whose osmotic pressure has been raised in this way is then transferred to tap-water, the rate at which the vacuole contracts is faster than normal. There is little swelling. Reviewing the subject, Kitching (1938b) writes that the contractile vacuole 'may be compared with the bilge pump of a rather leaky ship', though it cannot maintain the internal medium absolutely constant. There must be a mechanism for separating solutes from the water baled out but its nature has not been discovered. Nor is it known whether freshwater are less permeable than marine species.

Water uptake and loss has been investigated in *Pleuronectes platessa*, the plaice, and Schlieper (1958a) gives an account of the experiments,

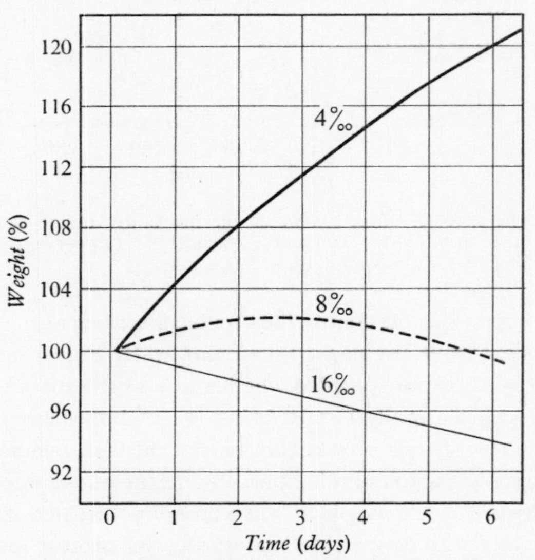

53. Change of weight in unfed *Pleuronectes platessa* in three different salinities (Henschel from Schlieper 1958a) (*Die Binnengewässer* **22**, p. 257, fig. 24).

rather drastic experiments, which cause some zoologists to ask whether the functioning of a normal animal can really be deduced from them. The fish is found in mixohaline water as well as in the sea and has been taken in a salinity as low as 5 per mille. The concentration of the blood is 10 per mille, which is the same as that of the freshwater fish investigated (table

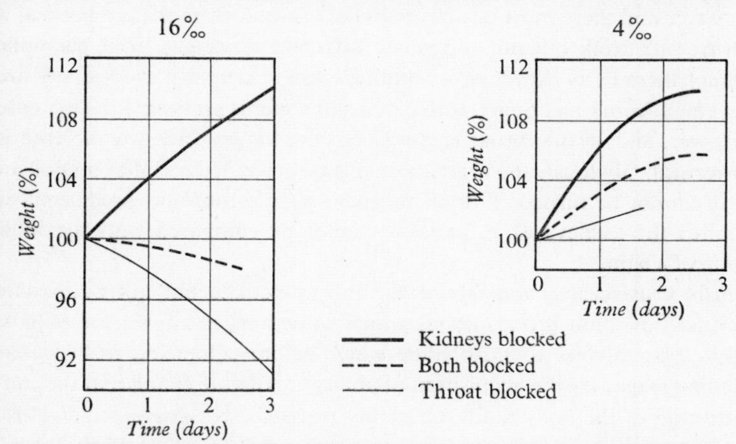

54. Alteration in weight of *Pleuronectes platessa* at two salinities and different experimental conditions (Henschel from Schlieper 1958a) (*Die Binnengewässer* **22**, p. 258, fig. 23).

26). A starving plaice slowly loses weight in a salinity of 16 per mille (fig 53). If its throat is bound so that it cannot take water into the alimentary

Table 26

Total osmotic concentration in body fluids of some mixohaline and freshwater animals (data from various authors compiled by Wikgren 1953)

	mM
Anodonta cygnea	16
Paramecium caudatum	*c*.25
Daphnia magna	68
Limnaea pereger	73
Hirudo medicinalis	106
Ephemera vulgata nymphs	128
Petromyzon fluviatilis	138
Leuciscus erythrophthalmus	145
Perca fluviatilis	149
Anguilla anguilla	185
Astacus fluviatilis	232

canal, the decrease in weight is comparatively rapid (fig. 54), evidently owing to the continued activity of the kidneys, since the decrease is slower if they are blocked as well. Under these conditions the loss in weight is presumably due to diffusion of water from the body to the more concentrated external medium. The normal fish makes good this loss by drinking water and there must be active uptake of water from the gut because a fish free to drink but not to urinate increases in weight. In 4 per mille salinity there is, as expected, a rapid increase in weight if the kidneys are prevented from functioning. If the intake of water is prevented the increase is slower, and if the throat is closed but the kidneys free the increase is slower still. Obviously the whole system is co-ordinated and the component parts cannot be turned on or off independently as osmotic circumstances require; the kidneys of *P. platessa* cannot be compared with the bilge pump of a ship.

Fish, crustaceans, lamellibranchs, and operculate but not pulmonate snails, to mention groups about which something is known, must have adapted themselves to mixohaline water before colonizing fresh water. The first stage was the acquisition of ability to tolerate changes in the concentration of the body fluids, an ability possessed by *Asterias* and *Arenicola* for example, but further progress depended on other physiological changes, some of which have been seen in the animals already mentioned. These changes are:

1. A permanent reduction in the concentration of the body fluids. Potts (1954) has calculated that this is the most important change in order to conserve energy but Shaw (1959a) does not agree with him.
2. A reduction in permeability;
3. The ability to take up ions
   (a) from the external medium, an ability already present in some marine animals,
   (b) from the urine. Potts (1954) calculates that the production of a urine more dilute than the body fluids is the most important change to make life in fresh water feasible.

The extent to which each of these changes has been carried varies in different species.

Table 26, showing the concentration of the body fluids of various animals, is a selection from the data which Wikgren (1953) has compiled. The body fluid of *Anodonta cygnea* has a concentration equivalent to about 1 per mille salinity, which is just under 3 per cent sea water. The body fluid of *Limnaea pereger* is equivalent to about 5 per mille salinity, that of the fish to about 10 per mille salinity. The salinity of the body fluids of *Astacus fluviatilis*, 15 per mille or about 40 per cent sea water, is stated by Wikgren probably to be higher than that of any other Finnish fresh-water animal.

Table 27

Permeability of some species to water (compiled by Wikgren 1953). Time taken for 1 cc water to pass through 1 sq. cm of membrane under a pressure of 1 atmosphere

Petromyzon fluviatilis	91 days
Carassius auratus	158 days
Astacus fluviatilis	*c*.2 years
Anguilla anguilla	*c*.5 years

The figures for the permeability of four species (table 27) were obtained by a measurement either of the output of urine or of the increase in weight after occlusion of the excretory ducts. Their reliability does depend on the assumptions that the excretion of water takes place only through the excretory organs, that these serve no purpose except to get rid of surplus water drawn in by osmosis, and that the animals are not affected by the rough treatment unavoidable in experiments of this nature. All three assumptions are open to challenge. As, furthermore, data are not yet available which would permit allowance being made for the different permeability of the various parts of the body surface, the figures quoted are the roughest of approximations, but the differences between them are so great that they are not wholly valueless. Schlieper (1958a) quotes figures for the uptake of sodium iodide as an indication of permeability. It passes quickly into the stenohaline *Hyas arenea*, less quickly into *Carcinus maenas*, which is active in mixohaline water, and less quickly still into the fresh-water *Astacus fluviatilis. Eriocheir sinensis*, which passes from fresh water to the sea is the least permeable (fig. 55). As will be noted later, this crab produces urine isotonic with the blood in contrast to the crayfish, which produces dilute urine.

Increase in the concentration of the outside medium produces a small rise in the concentration of the blood of a normal eel, but a new state of equilibrium is soon reached. If the mucous has been wiped off and the sensitive epidermis damaged, similar treatment produces a faster rise in the internal medium and death if the outside concentration is high enough (Schlieper 1958a) (fig. 56).

The concentration of ions below which a species can no longer take them up from the outside medium is likely to be an important factor limiting range in fresh water, and, since in dilute media ions diffuse out, the rate of uptake (table 28) must be taken into consideration too. Wikgren (1953) points out that some of the species occur in water with a chloride concentration below that shown as the limiting concentration in table 29,

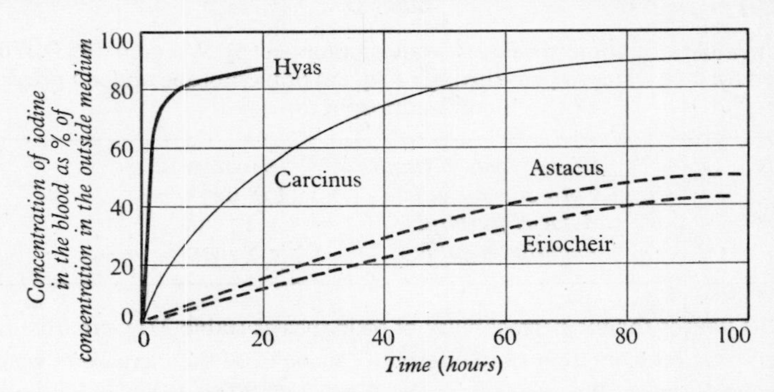

55. Rate of uptake of iodine by *Hyas*, *Carcinus*, *Astacus*, and *Eriocheir* (Nagel from Schlieper 1958a) (*Die Binnengewässer* **22,** p. 296, fig. 51).

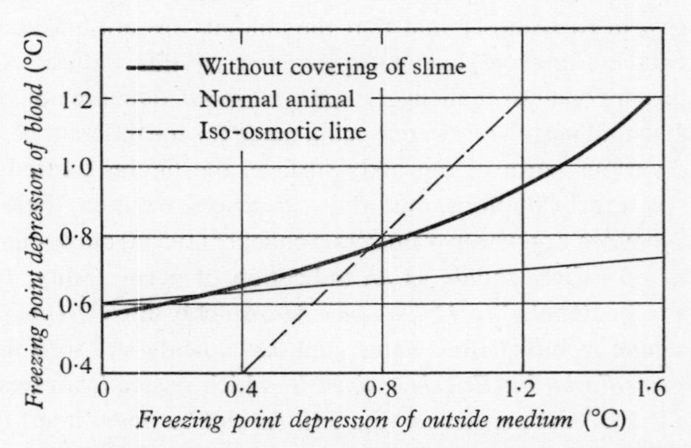

56. Change in concentration in the blood of *Anguilla anguilla* with increasing concentration in the medium (*Die Binnengewässer* **22,** p. 270, fig. 35).

and puts forward two possible explanations. Either there are physiological races with a lower threshold than the animals used in the investigations, a possibility that has not been explored, or the technique was faulty. Krogh discovered the threshold by keeping the animals for some time in running distilled water and then transferring them to solutions of known concentration, and there is a possibility that the long exposure to distilled water damaged the chloride-uptake mechanism. There is evidently still work to be done on this point. It will be noted, moreover, that the workers whose researches have so far been considered have confined themselves to chloride, although there are other ions that may be more important. Some of these have been studied by more recent investigators whose results will be noted presently.

Table 28

Rate of absorption of chloride (data from Krogh and Wikgren, compiled by Wikgren 1953)

	μM/100 g/hr
Petromyzon fluviatilis	90
Carassius auratus	27
Astacus fluviatilis	20
Limnaea stagnalis	17
	4 (Krogh)
Leuciscus rutilus	16 (Wikgren)

Table 29

Lowest concentration in the outside medium from which chloride can be taken up (Wikgren 1953)

	Concentration mM/l Cl	mg/l Cl	Authority
Petromyzon fluviatilis	0·00–0·05	0·0–1·75	Wikgren
Astacus fluviatilis	0·00–0·05	0·0–1·75	Krogh
Carassius auratus	0·00–0·05	0·0–1·75	Meyer
Carassius auratus	0·1 –0·2	3·5–7·0	Krogh
Pleuronectes flesus	0·2 –0·4	7·0–14·0	Wikgren
Eriocheir sinensis	0·2 –0·4	7·0–14·0	Krogh
Limnaea stagnalis	0·4 –0·8	28·0–56·0	Krogh

Petromyzon fluviatilis, the lamprey, fasts during the time it spends in fresh water. Its permeability to water (table 27) and to salts (0·1 μM/sq. cm/hr) is not high, but there is a fairly large loss of salts owing to the large surface area of the branchial region. Its uptake capacity is good, the threshold being low (table 29) and the rate rapid (table 28). It produces large quantities of dilute urine. *Cyprinus carassius*, the Crucian carp, a freshwater fish, differs mainly in that the loss of salts is less because the branchial area relative to the total body surface is only one-sixth that of the lamprey.

The crayfish, *Astacus fluviatilis*, another freshwater organism, differs in that it has a higher body-fluid concentration (table 26) and a very low permeability (table 27). The threshold for chloride uptake is low (table 29) but the rate is not fast (table 28). The urine is dilute but not great in amount (Wikgren 1953).

Sialis lutaria larvae, widespread in fresh water, appear to have no uptake mechanism and obtain salts with their food. They can survive washing with distilled water for several weeks owing to the low permeability of their cuticle and the large part that amino acids play in maintaining the osmotic pressure of the body fluid (Beadle and Shaw 1950).

During the last few years the investigation of loss and uptake has been studied with the aid of radioactive ions. The osmotic pressure of the body fluid of *Asellus aquaticus* is due largely to sodium and chloride, and experiments with heavy water and labelled sodium indicate that the animal is readily permeable to both. It can take up both sodium and chloride from a dilute medium. In distilled water a high osmotic pressure in the body fluids is maintained by absorption of water from them by the tissues, inside which the osmotic pressure is kept up by organic substances. Twenty to thirty per cent of the sodium in the body is in Zenker's organ, but the significance of this is obscure (Lockwood 1959a, b, c).

Bryan (1960) finds the following concentrations in the body fluid of *Astacus fluviatilis*: Na 203, K 4·6, Mg 2, Ca 15 and Cl 200 mM/l. Analyses by other authors show some variation but whether this represents a real difference or experimental error is not certain. In distilled water the concentration drops to 140 mM/l. When specimens are transferred from distilled water to a solution containing labelled sodium, the rate of both uptake and loss increases, the former preponderating.

According to Shaw (1959b) the maximum rate of influx of sodium into *Astacus pallipes* is about 100 μM/100 g/hr and requires a concentration of at least 1 mM/l NaCl. At a concentration of 0·3 mM/l the rate of uptake is reduced to one-sixth of the maximum. The threshold for uptake is 0·04 mM/l NaCl. Sodium is taken up as fast from Na_2SO_4 as from NaCl but all the sulphate remains outside. As there is no increase in the rate of excretion of NH_3, it is assumed that when uptake of Na is rapid some of the excess SO_4 is balanced by H. The rate of uptake of Na falls as the concentration of NH_3 rises and is stimulated by the presence of Mg but the amount of increase is not proportional to the amount of magnesium. Calcium up to 1 mM/l and potassium up to a concentration of 4 mM/l do not affect sodium uptake. Below pH 6 sodium uptake falls steadily and at pH 4 is but 20-30 per cent of what it was above pH 6 (fig. 57). Neither it nor NH_3 alter the rate of sodium loss (Shaw 1960a, b, c). The rate of sodium uptake varies with the total internal salt concentration; that of chloride, which is independent, varies with the concentration of sodium. The threshold is 0·028 mM/l a figure similar to that quoted by Wikgren for *A. fluviatilis* (table 29).

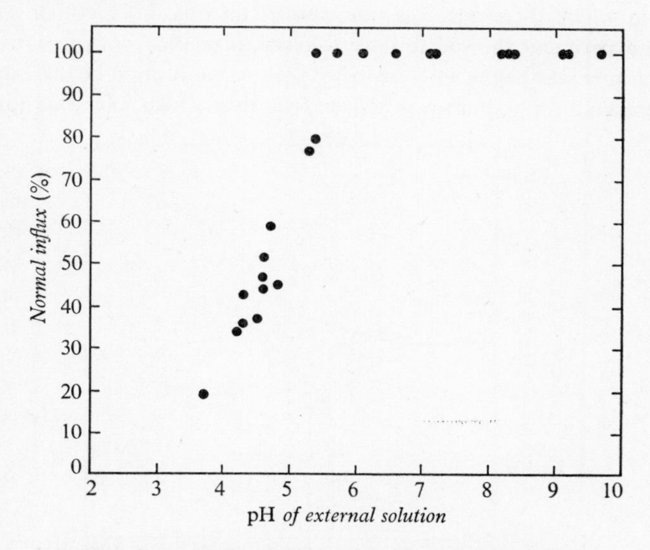

57. The effect of external pH on the sodium influx of *Astacus pallipes* (Shaw 1960) (*J. exp. Biol.* **37**, p. 554, fig. 6).

The work of Shaw (1959a) on the African freshwater crab, *Potamon niloticus*, is notable because physiological and ecological findings are brought together, though he emphasizes that time did not permit a careful ecological survey, and other possible limiting factors were perforce ignored. *Eriocheir sinensis*, *P. niloticus*, and *Astacus fluviatilis* have respectively blood concentrations equivalent to 342, 271 and 228 mM NaCl. *E. sinensis* does not produce a dilute urine, as freshwater animals generally do, but it is not completely adapted to fresh water because it has to return to brackish water to breed. *A. fluviatilis* does produce a dilute urine. *P. niloticus* resembles *E. sinensis* in that the urine is iso-osmotic with the blood, though it contains less potassium. *P. niloticus* is also more permeable to salts than *Astacus*, the rate of sodium loss being 20-40, 8 and 1·5 μM/10 g/hr in *Eriocheir*, *Potamon* and *Astacus* respectively. On these two counts it would appear to be ill-adapted to life in fresh water, but is nevertheless obviously successful in that medium. It appears to have evolved along lines different from those followed by *Astacus* and can live in fresh water without undue exertion to preserve osmotic imbalance (fig. 58) by virtue of greatly reduced permeability to water.

The lowest concentration from which it can take up sodium is 0·05 mM/l and the threshold for potassium is 0·07 mM/l. It was absent from a river in which there was less potassium than this, but another species, which experiment showed to have a lower threshold, was present. If the temperature was below 14°C or the oxygen concentration below 1·2 mg/l, the threshold for sodium was higher. In a river which, rising on top of a

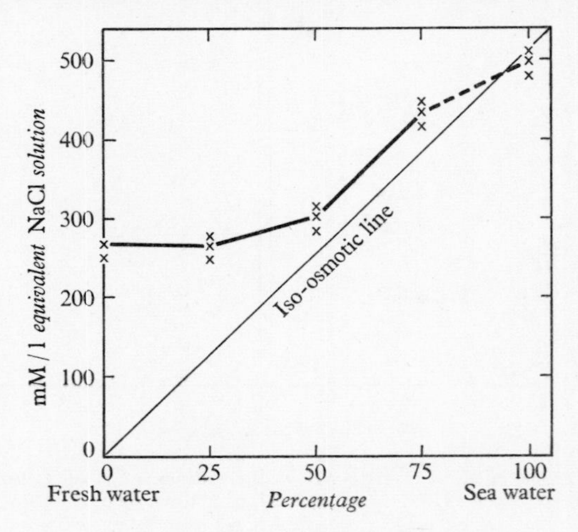

58. Concentration of salts in the blood of *Potamon niloticus* with increasing concentration in the medium (Shaw 1959) (*J. exp. Biol.* **36**, p. 161, fig. 1).

mountain and flowing into Lake Victoria, had a temperature range from 0° to 30°C, the crab occurred up to a point where the midday temperature was 13°C. The composition of the river water was constant all the way down. *P. niloticus* was also absent from a place where uptake of sodium could have been adversely affected by low oxygen concentration.

All freshwater organisms probably have a salinity optimum, but this is difficult to study, partly because the lower limits are low and partly because composition may vary for different species. It is, therefore, much easier to study in mixohaline water and most of the information about it comes from there.

Claus (1937) placed specimens of the freshwater *Corixa fossarum* and *C. distincta* and the mixohaline *C. lugubris* in distilled water and moved them every two hours into a solution in which the salinity was 2 per mille higher than in the one from which they had come. The weight of *C. lugubris* fell steadily at first, but remained unchanged between 8 per mille and 18 per mille, the range in which it is generally found in nature, and

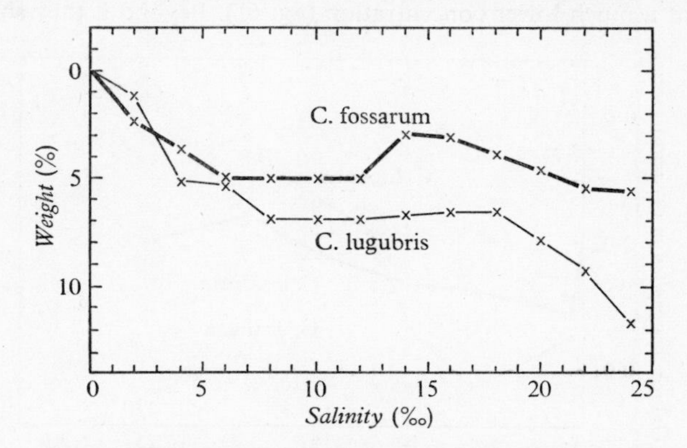

59. Change in weight of two species of *Corixa* with change of salinity in the external medium (Claus 1937) (*Zool. Jahrb.* **58**).

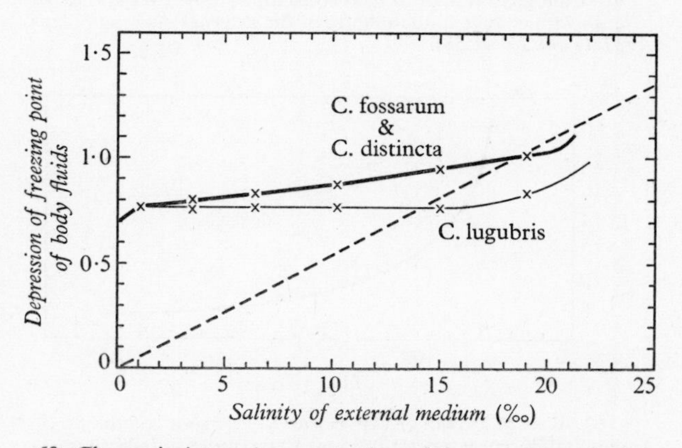

60. Change in internal concentration of three species of *Corixa* with changes in the external concentration (Claus 1937) (*Zool. Jahrb.* **58**).

decreased again at higher salinities. The weight of the other two species fell less and in a less regular way (fig. 59). Determination of the freezing-point of the body fluid showed that that of *C. lugubris* became more concentrated as the specimens were moved into more saline water, but maintained a steady level between about 0·5 and 17 per mille (fig. 60). The freshwater species were unable to prevent the internal concentration rising steadily in pace with that outside (fig. 60) which was thought to be the reason for their less marked decrease in weight. The oxygen demand of the mixohaline species is least at 6 per mille salinity, that of the freshwater species at a much lower concentration (fig. 61). Beyond it they showed a

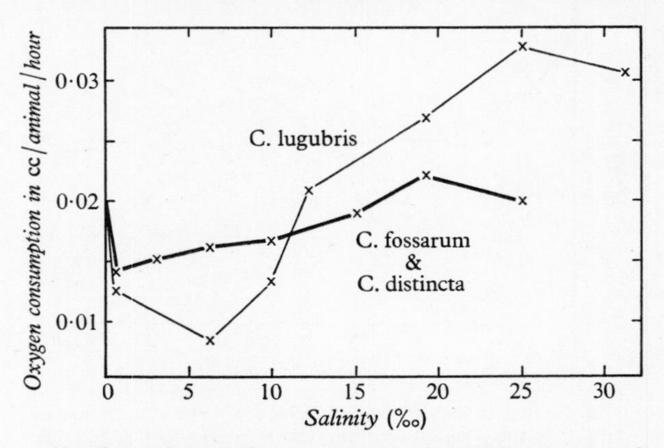

61. Change in rate of oxygen consumption by three species of *Corixa* with change in the salinity of the external medium (Claus 1937) (*Zool. Jahrb.* **58**).

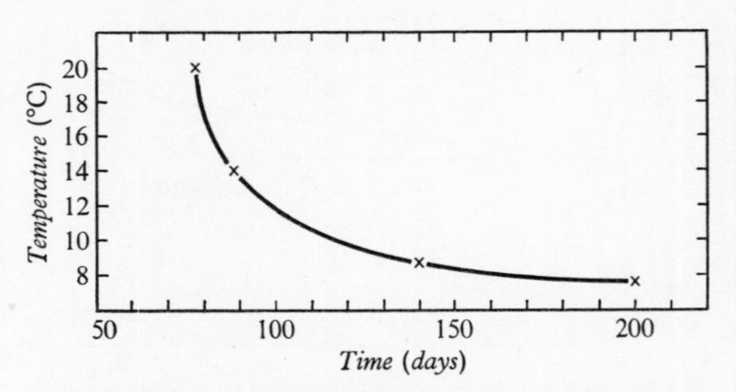

62. Rate of growth or *Gammarus duebeni* to a length of 9·3 mm at different temperatures and 10 per mille salinity (Kinne 1954) (*Z. wiss. Zool.* **157**, p. 444, Abb. 2).

much slower rate of increase, which Claus takes to be an indication of their lesser ability to maintain a state of disequilibrium.

The rate of growth of young *Gammarus duebeni* at 10 per mille salinity increases exponentially as the temperature rises from 8 to 20°C (fig. 62). Growth is fastest at this salinity but the rate is not greatly retarded till salinity passes the limits 5 per mille and 20 per mille (fig. 63). When young were transferred to one of three salinities as soon as they left the brood-pouch, fewest died at 10 per mille even when the mothers had been in a higher or lower salinity while they were in the brood-pouch (table 30). Death of eggs in the marsupium was also least at 10 per mille when measured in these three salinities. Temperatures of 8°C, 16°C and 20°C were increasingly unfavourable except at 10 per mille salinity at which mortality at the two lower temperatures was the same (table 31). The temperature above which there is no development is about 22°C at 10 per mille salinity and lower in waters more concentrated or more dilute (table 32) (Kinne 1954).

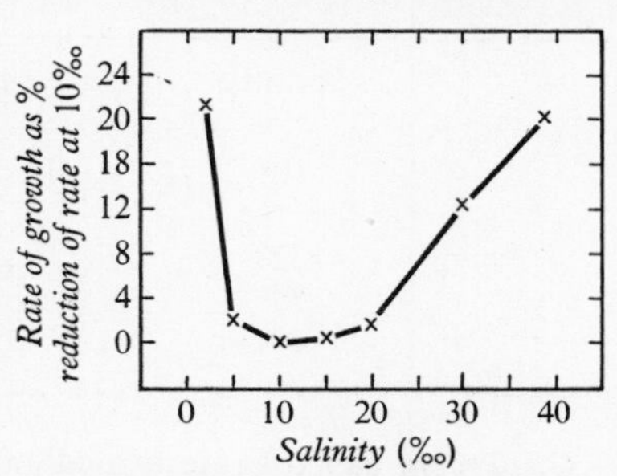

63. Rate of growth in six months at 18°C of *Gammarus duebeni* at different salinities (Kinne 1954) (*Z. wiss. Zool.* **157**, p. 448, Abb. 4).

Table 30

Percentage mortality of *Gammarus duebeni* kept at one salinity while in the brood-pouch and transferred to another on emergence (Kinne 1954)

Salinity during brood-pouch phase	2‰ 19°C	2‰ 9°C	10‰ 19°C	10‰ 9°C	30‰ 19°C	30‰ 9°C
2‰	20	11	8	6	30	?
30‰	24	21	6	7	20	?
37-40‰	79	29	10	8	?	59

Table 31

Percentage mortality of eggs in the marsupium of *G. duebeni* at different temperatures and salinities

Salinity	Temperature 8°C	16°C	20°C
10‰	5-15	5-15	20-50%
2‰	10-15	40	100%
30‰	15	40-50	100%

Table 32

Upper temperature limits for the development of the eggs of *G. duebeni* at different salinities (Kinne 1954)

Salinity	Upper temperature limit
10°/₀₀	about 22°C
2°/₀₀	16-17°C
22°/₀₀	about 21°C
30°/₀₀	18-19°C
40-50°/₀₀	under 13°C

G. duebeni can tolerate a medium between fresh water and about 50 per mille salinity but its optimum range is much narrower than this. Its temperature optimum is low and unfavourable temperatures reduce salinity tolerance. Panikkar (1940) found that certain prawns tolerated unfavourable salinities better at high temperatures, and concluded that this is a general rule which explains the rich decapod fauna of mixohaline and fresh water in the tropics. Kinne (1954) maintains that the optimum temperature may be high or low according to the species and that, in temperate latitudes, the salinity range of any mixohaline animal will be restricted at some time of year by unfavourable temperature. Colonization of tropical mixohaline waters is easier, not because of the higher temperature, but because of the smaller annual range.

Many species have been mentioned in this section. Not all have been studied in the same way or from the same point of view, and some excited the interest of the physiologists before they had received proper attention from the field workers. Another reason for the disconnected nature of the account springs from the acquisition by later workers of a tool, the radioactive ion, not available to the pioneers.

Marine species may possess a mechanism for removing water and for maintaining within a concentration of ions not the same as that without. To do this they must have a surface that is not completely permeable. They are therefore endowed with the abilities necessary for the invasion of dilute water. Much has been deduced about this invasion from physiological work, and one of the noteworthy conclusions is that these latent potentialities have been developed to a different degree by different species. There is no General Principle of the type which biologists so assiduously seek and organisms so rarely consent to be guided by.

Some of the present-day inhabitants of mixohaline and fresh water had a new environment thrust upon them. Large areas of sea water were isolated during the Ice Age and gradually became fresh, a change to which some of the original inhabitants, such as *Mysis relicta*, adapted themselves, to judge from their modern distribution. They would also have had to adjust themselves to other changes, such as the disappearance of a sandy shore when the tide ceased to rise and fall, and the arrival of strange animals as salinity reached a level where freshwater species of long standing could come in. Some of these could have been isolated by similar events in earlier epochs.

It may be supposed also that some species have achieved a new environment by gradual penetration of estuaries. Since few species inhabit both fresh water and the sea to-day, there has probably been isolation leading to the splitting of the original invader into several species. The three subspecies of *Gammarus zaddachi* may illustrate the process taking place now. Species might become isolated because they required certain physical conditions, and the splitting followed a course similar to that postulated by Fryer for the rockface fish in Lake Nyasa (chapter 6). There might, alternatively, have been a genetic differentiation comparable to that which Moore found in frogs. *Rana pipiens*, it will be remembered, is made up of a series of races each with different temperature requirements and some of these races are intersterile (chapter 8). This, however, is a realm in which only speculation is possible.

PHYSIOLOGICAL PROBLEMS CONFRONTING FRESHWATER SPECIES ENTERING MIXOHALINE WATER

Most of the work has been done with the larvae of the mosquitoes, *Aëdes aegypti*, which cannot tolerate mixohaline water, and *A. detritus*, which in nature is found only in that medium. Wigglesworth (1933a, b) showed by means of the flagellate technique invented by Fox that respiration in *A. aegypti* takes place all over the body surface and not particularly through the anal papillae, which the morphologists were wont to refer to as gills. These organs are much more permeable to water than any other part of the body, and there is through them a constant inflow which is excreted by the Malpighian tubules. The osmotic pressure of the body fluid in larvae reared in tap-water is equivalent to between 7·5 and 8·9 per mille salinity, though in fact more than half is due to organic solutes, not to salts. In distilled water these solutes account for a greater share of the osmotic pressure. When the external exceeds internal concentration, the latter begins to rise and at 16 per mille salinity death is rapid (cf.

fig. 58). Chloride can be absorbed from a dilute solution, though the threshold has not been determined, probably through the anal papillae (Wigglesworth 1938), and Treherne (1954) has shown that these are the main site of absorption of labelled sodium ions, though some reach the body fluid through the gut and the body surface.

Aëdes detritus is an astonishingly tolerant larva and Beadle (1939) was able to acclimatize it to distilled water. Since it cannot take up chloride from a solution much more dilute than the body fluid, it must, presumably, rely on its food as a source of salts under these conditions. In distilled water the osmotic concentration of the blood is similar to that of *A. aegypti* and, as in that species, the first rise in the salinity of the medium leads to a sharp increase in the proportion of the internal osmotic pressure due to chloride. At higher salinities the proportion remains constant and total osmotic pressure rises inside but far more slowly than it does outside. In 70 per mille salinity, which is 200 per cent sea water, the body fluid concentration is equivalent to a salinity of 14 per mille. The functioning of the tissues of *A. detritus*, like those of some marine animals that can live in dilute sea water, is not disturbed by quite large changes in the composition of the body fluid. Its tolerance of a wide salinity range is due also to its ability to get rid of excess salts. Ramsay (1950) showed that in dilute solutions *A. aegypti* absorbs salts from the fluid in the rectum, and brings forward some evidence that *A. detritus* has this power too. In strong solutions *A. detritus* can concentrate the rectal fluid but *A. aegypti* cannot.

Ramsay (1951) showed that in distilled water the osmotic pressure of the fluid in the intestine, which seems to be derived from the Malpighian tubules, is a little more dilute than the body fluid but contains distinctly less sodium. As the sodium concentration in the medium rises, the amounts in body fluid and Malpighian tubules approach each other but there is always less in the Malpighian tubules. The results of a further series of ingenious manipulations by Ramsay (1953) are not relevant to the present theme, as far as can be seen in the light of present knowledge, but they give an insight into the complexity of ionic relations of which ecologists must certainly take note. There is about as much potassium as sodium in the body of a larva in KCl solution, and about two and a half times as much sodium as potassium in that of one in NaCl solution. In the haemolymph there is about twenty-five times as much sodium as potassium, and evidently a high proportion of the latter ion is in the tissues. There is a constant removal by the Malpighian tubules of potassium which is then resorbed in the midgut.

OTHER WATERS OF HIGH SALINITY

Hedgpeth (1959) has put forward a provisional scheme for classifying saline waters other than those which are mixohaline:

1. Hypersaline lagoons have permanent or intermittent connexion with the sea and most are in arid regions. They generally hold an attenuated marine fauna with the addition of *Artemia salina* and ephydrid fly larvae.

2. Relict waters have lost all connexion with the sea, but had one once to judge from some of the species found and from the proportions of the ions.

3. Salterns and brines are ponds in which sea water is evaporated to obtain salt. The organisms which survive the increasing concentration longest are *Artemia* and the flagellate *Dunaliella viridis*.

4. Carbonate and sulphate waters. These are referred to as 'athalassohaline waters' by Bond (1935) because the ions are present in proportions unlike those in sea water. Hedgpeth devotes six pages to lists of animals that have been found in water with a salinity exceeding 45 per mille, but omits Protozoa and Insecta. His paper is also a valuable source of references.

Many of these waters lie in remote places, and whoever wishes to study them must often devote more effort to getting there and back than to scientific work. Consequently few if any of the lists are complete and some are obviously fragmentary. Furthermore little physiological work has been done. The rest of this chapter is, therefore, without long lists but otherwise mainly descriptive; for reasons of space only athalassohaline waters are considered.

Beadle (1943) spent five weeks investigating sixteen bodies of water in Algeria. They ranged in salinity from 4 per mille to 261 per mille, and the composition resembled sea water in that Na and Cl were the predominant ions, though there was distinctly more sulphate than in the sea. *Artemia salina* was the only animal found exclusively in salinities well above that of sea water. Species with a range from a salinity well below to one well above that of sea water were:

Rotifera	*Brachionus plicatilis*
	Pedalia fennica
Copepoda	*Diaptomus salinus*
	Cletocamptus retrogressus
	Nitocra lacustris
	Cyclops ? viridis
Insecta	*Aëdes detritus*
	Ephydra macellaria

One fish, *Cyprinodon fasciatus*, extended to water only a little more saline than the sea. Algae with the same range as the animals listed were:

Cladophora fracta	*Lyngbya aestuarii*
Spirulina subsalsa	*Schizothrix lardacea*

as well as several unidentified Chlamydomonadales, *Cladophora* and *Chroococcus* species.

Hutchinson (1937), investigating nine lakes in Tibet, was more interested in limnological than in faunistic and floristic studies, and did not attempt to make complete lists. Only one lake was more saline than the sea. It contained considerable amounts of sodium, potassium, sulphate, and chloride with sodium and sulphate preponderating (table 33). Phytoplankton was limited to a few colonies of *Merismopedia* sp., and algae covered the bottom of the lake. The commonest animals of the lake were *Artemia salina* and *Brachionus plicatilis*, both in Beadle's list also, and the only others found were ephydrid larvae; adults of *Ephydra glauca* and *Halmopota hutchinsoni* were taken beside the lake.

Little Manitou Lake in Canada is not dissimilar, though it is more saline (120 per mille) and has more magnesium (table 33). It was four times more saline than the next of sixty lakes, which ranged between that figure and fresh water. Algae recorded were:

Lyngbya Birgei	*Surirella ovalis*
Pediastrum Boryanum	*Cladophora glomerata*
P. duplex	*Oedogonium* sp.
Ceratium hirundinella	*Enteromorpha prolifera tubulosa*
Cyclotella Meneghiniana	*Amphora commutata*
Cymatopleura solea	*Cymbella tumida*

Higher plants, absent in the more saline lakes, first appeared at 20 per mille and increased in number of species with decreasing salinity below that concentration. *Brachionus plicatilis, B. p. spatuosus, B. satanicus, Conochilus unicornis*, and *Pedalia* sp. near *fennica* were taken in lakes with salinities ranging from 10-30 per mille, but whether any were taken in Little Manitou Lake is not clear. A table shows how the number of species increases with decreasing salinity. There is a similar table for Entomostraca, from which it may be seen that *Artemia salina* occurred in Little Manitou Lake and in no other; that *Daphnia longispina* covered the whole range of salinity; and that, of the Copepoda, *Cletocamptus albuquerquensis* occurred over the range 20-120 per mille and *Diaptomus siciloides* over the range 10-120 per mille. The only species not in Little Manitou Lake but in the

Table 33

Dissolved ions in some saline lakes (mg/l)

Country	Lake	Na	K	Ca	Mg	CO_3	HCO_3	SO_4	Cl	Reference
Tibet	Tso Kar	16346	5478	406	2716	–	2141	35075	11662	Hutchinson 1937
Canada	Little Manitou	17950	1017	514	11160	236	554	51720	23295	Rawson and Moore 1944
U.S.A.	Hot	16790	1564	720	53619	0	3062	243552	1882	Anderson 1958
Kenya	Nakuru	5550	256	10	0	6150		253	1375	Beadle 1932
Turkey	Van	8100	400	9	107	3492	2428	2447	5900	Gessner 1957

next most saline ones was *Cyclops viridis*, which otherwise spanned the entire range. The distribution of the remaining animals is shown by group only and there are but three, Coleoptera and Ephydridae and Dolichopodidae, which were found in Little Manitou Lake. The hydrophilid beetle *Enochrus diffusus* is mentioned in the text. Ephydrid larvae were thought probably to be *Ephydra subopaca*, and adults of *E. hians* were collected in the saline-lake region. *Trichocorixa verticalis* was found only in the more saline lakes, but not in Little Manitou. Twenty-four species of fish were recorded in the lakes. Four reached a salinity of 10 per mille and one, *Pungitius pungitius* (nine-spined stickleback), reached 20 per mille.

In Hot Lake also sulphate is the main anion but magnesium is the predominant cation. The lake is peculiar in that stratification leads to accumulation of heat in the hypolimnion, which may reach a temperature of 50°C. *Chara* sp. was abundant in shallow water, and below it, starting at a depth of about 1 m. was a mat made up of the blue-green algae: *Plectonema nostocorum*, *Oscillatoria chlorina*, *Anacystis thermalis*, and *Gomphosphaeria aponina*. This extended down into the hot anaerobic layers of the lake. The zooplankton consisted mainly of *Artemia salina* and the only other species of note was the rotifer, *Brachionus angularis*.

There were occasional records of the rotifer *Keratella quadrata*, of calanoid and cyclopoid copepods, and of an ostracod, but no species names are given and nothing is mentioned about a littoral fauna.

The last two lakes in table 33 differ from the rest in that sodium and carbonate are the commonest ions. They differ from each other in the quantity and proportion of the other ions. Lake Nakuru was analysed by Beadle but the biological survey, one of the most detailed of any made in lakes of this type, was by Jenkin (1932, 1936). Five lakes were visited. In Lake Naivasha, with 210 mg/l Na_2Co_3, over seventy species of animal were recorded, but in the remaining lakes a higher concentration was accompanied by a much less varied fauna (Jenkin 1936, table 9). In Lake Nakuru only five species were found: the rotifers *Brachionus pala* and *B. plicatilis* and the corixids *Micronecta scutellaris*, *M. jenkinae*, and *Corixa lateralis kilimandjaronis*. The amount of the alga *Arthrospira platensis* was so great that it formed the main diet of the flamingoes. Other algae named were *Melosira nyassensis*, *Cyclotella meneghiniana*, *Navicula sphaerophora*, *Spirulina laxissima*, *S. subtilissima*, and *Anabaenopsis circularis* var. *javanica*.

Gessner (1957) was able to spend only a week beside the remote Turkish Lake Van, but, like Miss Jenkin, he endeavoured to obtain as complete a picture of the fauna and flora as possible. *Chaetoceros orientalis* was very abundant in the plankton and *Surirella armenica, Oocystis Borgei,* and a cryptomonad were also taken. There were no higher plants but twenty-five species of algae were recorded from the littoral region. The zooplankton consisted of six rotifers, two copepods and one cladoceran: *Pedalia fennica* var. *polydonta, Brachionus angularis, B. plicatilis, Filinia major, Colurella adriatica, Trichocerca taurocephala, Diaptomus spinosus, Cyclops viridis,* and *Moina* sp. (*macrocarpa* group). Animals taken in the littoral region were *Corixa lateralis,* two chironomids (*Eucricotopus atritarsis* and *Microchironomus* sp.), an enchytraeid worm, and *Cyclops viridis.* There was a fish, *Alburnus tarihi,* in the lake. It was said to spawn in the inflowing rivers.

Hutchinson (1937) visited a series of saline lakes in Nevada, U.S.A., but did not make analyses comparable with those in table 33. In Winemucca Lake, in which there was 23·7 g/l Cl, he recorded: a narrow-leaved *Potamogeton,* the rotifers *Brachionus plicatilis* and *Pedalia jenkinae,* the cladoceran *Moina hutchinsoni,* the copepod *Cletocamptus albuquerquensis,* an ephydrid larva, and *Trichocorixa verticalis,* which was very abundant.

Decksbach (1924) gives lists of animals found in lakes on the Kirghiz steppes in Russia, but does not indicate in which lakes each one was found. Nor is there a complete chemical analysis for each lake.

Bond (1935) investigated one athalassohaline and two thalassohaline lakes in the Caribbean, and was the inventor of those two terms. He gives neither complete analyses nor complete lists of organisms. The athalassohaline lake had a salinity of 40 per mille and ten times as much sulphate as there would have been in sea water of the same strength. He records *Brachionus plicatilis, Trichocorixa verticalis,* a copepod, *Chara Hornemannii,* blue-green algae, and *Dunaliella* sp.

The similarities between these lakes, scattered over America, Africa and Asia and containing widely different amounts and proportions of the main ions, are startling. *Artemia salina,* Rotifera, Cladocera, Copepoda, Corixidae, and Ephydridae constitute the fauna. Within those groups certain species occur with remarkable constancy; the rotifer *Brachionus plicatilis* appears in seven out of the eight lists, and in the eighth there is another member of the same genus, *B. angularis. Pedalia fennica* and *Cyclops viridis* both appear in three lists, one in Africa, one in Canada, and one in Turkey. These, being the three longest lists, are probably the most complete and it is far from certain that a species missing from any one of them is really absent. *Trichocorixa verticalis* was not recorded in Hot Lake but was found in the other three American lakes. *Corixa lateralis* is in both the soda lakes, though the African specimens are assigned to a distinct subspecies. It is one of the few species common to the Ethiopian and the Palaearctic Region. *Artemia salina* is listed four times, and was not taken in the soda lakes.

The other striking feature of these athalassohaline lakes is that the affinities of the fauna are all with fresh water and not with the sea; the name is as applicable to them as it is to the ions. Branchiopoda, Cladocera, Rotifera and Insecta are all predominantly freshwater groups. Evidently the characteristics that enable an animal to tolerate extreme dilution are also those required for the toleration of extreme concentration; low permeability is no doubt of the first importance.

Pools of waste oil can scarcely be considered fresh water, but are mentioned here because they have been colonized by a typical saline water group, the Ephydridae (Thorpe 1930). The oil is highly toxic if it comes in contact with the tissues, but is kept from them by the properties of the cuticle and of the perithrophic membrane. The larvae feed on organisms which fall into the pools and become trapped in the oil, but the details of the process of digestion have not been investigated.

Artemia salina is the only one of these species about whose physiology something is known. Boone and Baas Becking (1931) and Jacobi and Baas Becking (1933) have shown that, although it thrives over a very wide range, the salinity must be due mainly to sodium chloride or sodium bromide. The concentration of sodium must exceed that of the other cations by an amount that becomes proportionately larger as total concentration increases.

Internal and external osmotic concentrations are the same at about 9 per mille salinity. As is customary in animals that can tolerate a big fluctuation in the outside medium, the tissues of *Artemia* are not affected by considerable changes in the concentration of the body fluid. This ranges from the equivalent of about 5-25 per mille, while the outside medium passes from a salinity of 3 per mille to one of 300 per mille. Since the first ten branchiae stain with silver chloride, it is deduced that they are the seat of the controlling mechanism. Radioactive sodium and halides passed in and out of the body with remarkable rapidity. Croghan, who made these observations (1958a-e), suggests that there is some mechanism in the epithelium by means of which ions can be seized and passed inwards or outwards according to circumstances. In the concentrated solutions in which it is generally found, the animal is constantly drinking the medium, whether it is feeding or not. The fluid in the gut, however, is generally less concentrated than the body fluid. Croghan attributes this to a rapid absorption of sodium and chloride through the gut wall and their return to the external medium by the branchiae. When the fluid in the gut has been made more dilute than that in the body, water can diffuse inwards through the gut wall, which is the part of the body most permeable to it. Active uptake of water against an osmotic gradient appears to be difficult for an animal. The method evolved by *Artemia* is similar to that found in the marine teleosts, which, accordingly, are believed to be descended from freshwater ancestors.

SUMMARY

Some animals can penetrate mixohaline water from the sea because their tissues can continue to function in body fluids whose concentration falls with that of the medium. Others are able to maintain the body fluid concentration above that of the medium by means of an active uptake of salts from the medium or from the urine. Invasion of fresh water has generally been accompanied as well by decrease in permeability and a lowering of the concentration of the body fluids. The lowest concentration at which some ion can be taken from the medium determines the range of some species. Adaptation to brackish water by a freshwater organism involves toleration by the tissues of a changing body fluid and the ability to get rid of unwanted ions. *Artemia*, an inhabitant of water generally more concentrated than the sea, drinks the medium and renders it more dilute than the body fluid by taking salts through the gut wall and passing them out through the branchiae. It can then make good by diffusion through the gut wall losses due to exosmosis. Tolerance of a salt concentration outside the optimum range is easier at a particular temperature.

Waters with a high concentration of salts in proportions unlike those in the sea are inhabited almost exclusively by animals of freshwater origin.

SYNONYMS

Species	Group	Other generic names	Other species names
Aëdes aegypti	Ins. Dipt. Culicidae	*Stegomyia*	*fasciata* Meigen *calopus* Meigen
A. detritus Haliday	Ins. Dipt. Culicidae	*Ochlerotatus*	
Anacystis thermalis (Menegh)	Cyanophyta	*Chroococcus*	
Anguilla anguilla (Linnaeus)	Pisces		*vulgaris* Linnaeus
Arthrospira platensis (Nordst.) Gomont	Cyanophyta	*Spirulina*	*jenneri* Hassal

Species	Group	Other generic names	Other species names
Astacus fluviatilis Fabricius	Crust. Malacostraca	*Potamobius*	*astacus* Linnaeus
Brachionus plicatilis Müller	Rotifera		*mülleri* Ehrenberg
Cletocamptus retrogressus Schmank	Copepoda	*Wolterstorffia*	*blanchardi* Richard
Corixa carinata (Sahlberg)	Inst. Heteropt. Corixidae	*Sigara, Arctocorisa*	
C. distincta (Fieber)	Inst. Heteropt. Corixidae	*Sigara, Subsigara*	
C. falleni (Fieber)	Inst. Heteropt. Corixidae	*Sigara, Subsigara*	
C. fossarum (Leach)	Inst. Heteropt. Corixidae	*Sigara, Subsigara*	
C. lateralis (Leach)	Inst. Heteropt. Corixidae	*Sigara, Vermicorixa*	*hieroglyphica* Dufour
C. lateralis kilimandjaronis Kirkaldy	Inst. Heteropt. Corixidae	*Sigara, Vermicorixa*	*kilimandjaronis*
C. sahlbergi (Fieber)	Inst. Heteropt. Corixidae	*Sigara, Anticorisa, Hesperocorixa*	
C. selecta (Fieber)	Inst. Heteropt. Corixidae	*Sigara, Halicorixa*	
C. semistriata (Fieber)	Inst. Heteropt. Corixidae	*Sigara, Retrocorixa*	
C. stagnalis (Leach)	Inst. Heteropt. Corixidae	*Sigara, Halicorixa*	*lugubris* Fieber
C. striata (Linnaeus)	Inst. Heteropt. Corixidae	*Sigara*	
Cyprinus carassius Linnaeus	Pisces	*Carassius*	*vulgaris* Nilsson
Diaptomus salinus Daday	Copepoda	*Arctodiaptomus*	
Dreissena polymorpha (Pallas)	Moll. Bivalvia	*Dreissensia*	
Hydrobia ventrosa (Mont.)	Moll. Gastropoda	*Paludestrina*	*stagnalis* (Baster)

Species	Group	Other generic names	Other species names
Keratella quadrata (Müller)	Rotifera		*divergens* (Voight)
Leucioperca sandra Linnaeus	Pisces		*leucioperca* Linnaeus
Leuciscus erythrophthalmus Linnaeus	Pisces	*Scardinius*	
Limnaea pereger (Müller)	Moll. Gast. Pulmonata	*Lymnaea, Radix*	*ovata* Drap.
Melosira nyassensis (Müller)	Chrysophyta (diatom)		*deuriesii* Müller *minor* Müller
Micronecta poweri (D and S)	Inst. Heteropt. Corixidae	*Sigara*	
Navicula sphaerophora (Kützing) Pfitzer.	Chrysophyta (diatom)	*Anomoeoneis*	
Neomysis integer (Leach)	Crust. Malacostraca		*vulgaris* Thompson
Petromyzon fluviatilis Linnaeus	Pisces	*Lampetra*	
Pleuronectes flesus Linnaeus	Pisces	*Platichthys*	
Procerodes ulvae (Oersted)	Platyhelminthes	*Gunda*	
Salmo salar Linnaeus	Pisces	*Trutta*	

CHAPTER 11

Calcium

Calcium is probably more variable in amount than any other ion in the general run of fresh waters; there may be less than 1 mg/l in soft waters derived from hard rocks and over 100 mg/l in hard water that is derived from chalk or limestone. There is scarcely an animal group in which the distribution of at least some species has not been related to the calcium concentration, though the significance of the relationship, which will be discussed later in the chapter, is not known, and there are several unexplained anomalies.

It is possible to make a rough division into:

1. Groups in which the presence of calcium appears to be favourable to all or nearly all the species, and
2. groups in which species characteristic of lime-poor water contrast with species characteristic of calcareous water,

but the distinction cannot be pushed too far.

The Mollusca, typical of the first division, were studied by Boycott, who summarized the results of a lifetime's observations shortly before his death (Boycott 1936). Fifty per cent of the species are hardly ever found in water with less than 20 mg/l Ca, and Boycott treats this as the lower limit of hard water, though it would not be so regarded by a waterworks engineer. Other authors too set the line between hard and soft according to the occurrence of the species they are studying, and there is no definition generally accepted by biologists. With decreasing amounts of calcium below this concentration the number of species becomes progressively fewer. A large volume of water seems to be one factor that compensates for low concentration, and *Planorbis carinatus*, for example, though generally a typical hard-water species, is common in Windermere in whose waters there are but 5 mg/l Ca. Very few species appear to be confined to soft water. *Limnaea glabra*, characteristic of what Boycott calls 'poor' places, has not been taken in water which might be presumed to be hard, but Boycott had no difficulty in rearing it in hard water in an aquarium. Nearly all the records for the large bivalve, *Margaritifera margaritifera*, are from soft water, and Hendelberg (1960), in a recent comprehensive study, found that its distribution in Sweden coincided with the non-calcareous areas, but there are a few records of it in hard water.

Hubendick's (1947) findings in Sweden bear out those of Boycott in Britain, but a few species appear to be more and a few to be less exigent in their calcium requirements when the two countries are compared. *Limnaea stagnalis* is one species which, confined to hard water in Britain, occurs in soft water in Sweden, and I can testify to my own surprise when I took it in a peaty lake in the north of Finland, and a Finn assured me that the water was as poor in lime as it looked.

Mann (1955a, b) collected Hirudinea in fifty-eight pieces of water, twenty-nine in Berkshire and twenty-nine in the Lake District. He designates 'soft' those with less than 7 mg/l Ca, 'intermediate' those with 7-24 mg/l Ca, and 'hard' those with over 24 mg/l Ca. Eight species were taken fairly often, and all except *Piscicola geometra*, which is found in running water and on the stony shores of lakes, occurred more frequently in hard than in soft or intermediate water. A large volume of water appeared to be a favourable factor for leeches in addition to hardness. *Hemiclepsis marginata* was not taken in soft water. *Helobdella stagnalis* was the commonest species in twelve out of sixteen places with hard water, *Erpobdella testacea* in three out of the remaining four, all of which were small, and *Glossiphonia complanata* and *Dina lineata* in the last, which was temporary (Mann 1955b, table 2). In intermediate waters that were large the most abundant species was generally *H. stagnalis*, in those that were small it varied. In those soft waters in which leeches were taken at all, *E. octoculata* was usually the most numerous species. Boisen Bennike (1943) also finds fewer species of leech in waters with little calcium.

Reynoldson (1958a, b) searched for planarians in 122 lakes with a wide range of calcium concentration, and the eight in which he failed to find any all had less than about 7 mg Ca/l. One species, *Planaria vitta*, is confined to soft water; in the six lakes in which Reynoldson found it, there was never more than 2·5 mg Ca/l, and all were high up and fringed with *Sphagnum*. *Polycelis nigra* is also found in soft water but its range includes the hardest waters as well. Three other species come in, one after the other, with increasing concentration of calcium (fig. 64). Collections were not made from the kind of place where Mann (personal communication) finds that small size begins to exert a noticeable effect on the occurrence of leeches. Reynoldson observes that each species tolerates less calcium in the more southerly of two places; he assumes that there is a temperature difference and that warmer water is more easily colonized than colder.

Conclusions about the relation between some Crustacea and calcium is conflicting. Couégnas (1920), collecting in a granite area, found crayfish only in streams that flowed over amphibolite, a rock containing 13·23 per cent Ca, and their presence at one station led to the discovery of an

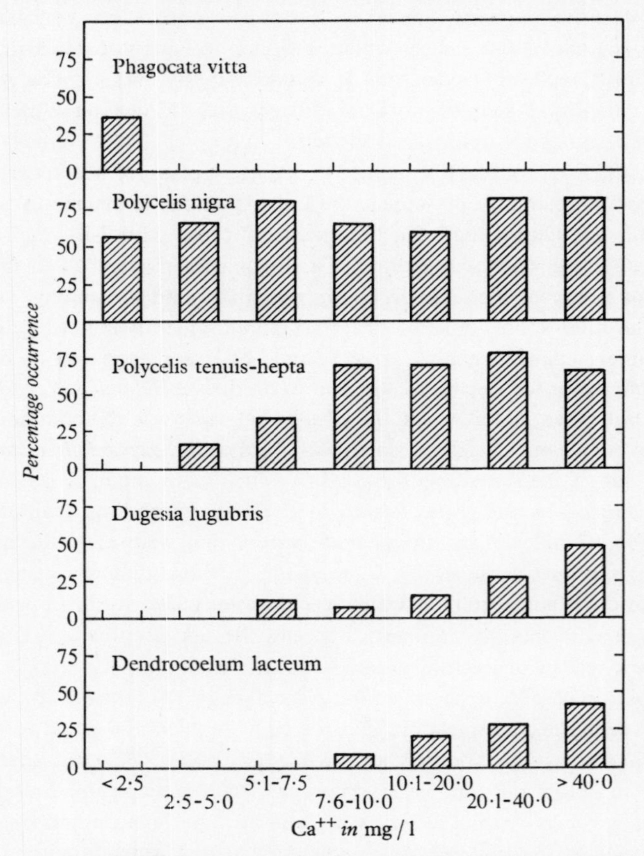

64. Percentage occurrence of six species of triclad in waters with various concentrations of calcium (Reynoldson 1958) (*Verh. int. Ver. Limnol.* **13,** p. 324, fig. 3).

outcrop not shown on the geological map. On the other hand Nygaard (1955) records a thriving and breeding population of *Astacus fluviatilis* in a lake with only 1·8 mg Ca/l. Had the French writer given a scientific name to his crayfish, the comparison might have been more profitable.

It is not necessary to mention here all the papers which have purported to explain the range of *Gammarus pulex*. Suffice it to quote Hynes (1954, p. 72) who, after extensive investigation, failed to substantiate earlier claims that deficiency of calcium or excess of magnesium relative to calcium limited the range of this amphipod. The factor concerned remains elusive. As was noted in chapter 3, the absence of *G. pulex* in Welsh streams contrasts strongly with its abundance in many Lake District streams. There appear, however, to be areas in the Lake District where it does not occur, and, as the streams seem to be similar in all other respects, a chemical difference is the most probable explanation; at the moment the identity of the ion or ions involved remains undiscovered.

The controversy over *Asellus* was discussed in chapter 4, and the conclusion reached was that it is found in most places where there is more than 12·5 mg/l Ca and in few places where there is less than 5 mg/l Ca. In places with an intermediate concentration it may or may not occur.

Tucker (1958) lumps all these groups together and records that numbers of both species and individuals tended to increase when fifteen ponds and one lake were arranged in ascending order of calcium concentration. Perusal of his data reveals possibly significant correlations with factors other than chemical ones. For example, only four of the ponds have an outflow, and *Gammarus pulex* is found in three of these and nowhere else. *Limnaea pereger* is absent from five out of the six described as being covered by floating vegetation. *Limnaea stagnalis*, *Planorbis corneus*, and *P. vortex* are recorded only from the lake. Nevertheless the main conclusion cannot be doubted. It is the same as one reached in chapter 3.

Of these animal groups in which all or nearly all species appear to be favoured by calcium, none belong to the Insecta. Many of the groups in which some species are typical of soft waters do. Perhaps the most striking example is the genus *Heptagenia* (Ephemeroptera). *H. sulphurea* and *H. fuscogrisea* abound on stony shores of lakes in the limestone areas of Ireland, whereas, on a similar substratum in softwater lakes, the only species is *H. lateralis* (Harris 1952). The same kind of difference but in ponds is seen in the family Corixidae. The main species of *Corixa* in dewponds on the chalk downs are *C. punctata*, *C. lateralis*, *C. limitata*, and *C. nigrolineata*, and Macan and Macfadyen (1941) remark that all but the last of these four were found by Macan (1938) to be rare in the soft waters of the Lake District. Similarly, among the mosquitoes, *Aëdes punctor* is the common species in temporary pools in the Lake District (Macan 1951), whereas such places on the calcareous soils of Cambridgeshire are inhabited by *A. rusticus* and, in woods, by *A. cantans* (Macan 1939).

A well-known cladoceran that is regarded as typical of soft waters is *Holopedium gibberum* (Thienemann 1950). Hamilton (1958) states that it has not been found in water with more than 20 mg Ca/l. Pacaud (1939) mentions *Acantholeberis curvirostris* and *Streblocerus serricaudatus* as two species taken only in soft waters, and *Simocephalus vetulus* as one that flourishes only in hard. Smyly (1958a) finds that *Streblocerus serricaudatus* has a wider range but agrees about the other two. *Alanopsis elongata* and *Bosmina coregoni* var. *lilljeborgii* are typical of soft-water tarns in the mountains.

Jewell (1939) collected sponges from many lakes. Three species were found only in lakes with little calcium, two were absent both from those with the highest and those with the lowest amounts of calcium, and one had a wide range. Most of the species did not survive in cultures which had calcium concentration above or below the limits of the range in which they occurred wild.

This is but a selection from the data which could be brought forward to demonstrate a relation between the range of certain species and the concentration of calcium. The unanswered question is how far calcium affects the animals directly and how far indirectly. The indirect effect is certainly important. Tansley (1939, p. 82) writes: 'A good supply of "exchangeable bases" (i.e. chemically active basic ions), particularly calcium, is one of the primary conditions of a "good" soil, i.e. a soil favourable to plants apart from the extreme oxiphilous species.' Where these bases are plentiful the organic matter derived from dead plants is broken down rapidly; where they are in short supply decomposition is slow. Nor is it entirely a question of speed or quantity; there are qualitative differences producing different organic and inorganic compounds and determining what small organisms are present, but, in fresh water at least, not much is known about the processes. Reynoldson (1958a) writes that, as lakes become more productive, the number of triclads which they contain increases, and he relates the presence and absence of *Asellus* to the total amount of dissolved matter as well as to calcium. Tucker (1958) states that the amount of dissolved organic matter tends to be higher in the more calcareous waters. It is obviously relevant, therefore, to consider what is found in lakes where the water is rich in lime but where there is little decomposable organic matter. Three lochs on the Isle of Lismore in Scotland are among the few of this type that have been studied. Reynoldson (1958a) found in them only *Polycelis nigra* of the species in fig. 64. Hunter (1957) recorded almost exactly the same snails as occur in Windermere and Loch Lomond (2-3 mg/l Ca) except for *Limnaea stagnalis*, which has not been found in either, and *Planorbis carinatus*, which has not been found in Loch Lomond. Reynoldson's record of *Planaria alpina* suggests that these lochs are colder than most. This could be partly responsible for the poor fauna, and further study of lakes of this type is required. It does seem probable, however, that the direct effect of calcium is not great. The relation between various species and the productivity of a lake, which is due to organic matter, to the products of the decomposition of organic matter, and to the inorganic substances derived directly from rock and also released during the decomposition of organic matter, is discussed in the next chapter.

More direct effects of calcium have been noticed in earlier chapters. Without a certain amount of it, *Procerodes ulvae* cannot tolerate extreme dilution (Pantin 1931), and *Nereis diversicolor* swells more at first and reduces weight less rapidly later when transferred to mixohaline water. The gill of *Mytilus* is less resistant to high temperature in 15 per mille salinity than in sea water, but this effect can be expunged by the addition of calcium to the mixohaline water (Schlieper 1958a). The rate of respiration of both *Salmo trutta* and *Planaria alpina* increases rather rapidly with rising temperature but the increase is reduced if much calcium is added to the medium. Corresponding with this reduction is longer survival at lethal temperature (Schlieper, Bläsing, and Halsband 1952).

The exact significance of these observations to the present discussion is, however, still obscure.

Bates (1939) experimented with culture media in which various species of *Anopheles*, and various races of *Anopheles maculipennis*, could be reared successfully. All required calcium, which was added in the form of $CaCl^2$, but the amount necessary varied. *A. superpictus* required most and *A. m. atroparvus* least.

Snails produce a heavier shell the more calcium there is in the water (Hubendick 1947) but it is not obvious that a thicker shell confers any advantage on the animal, nor has anyone explained why some species should require more calcium in the water than others. Frömming (1936, 1938) maintains unequivocally that the scarcity of individuals and species in waters poor in lime is related to a meagre food supply, not to a low concentration of calcium. This, however, is likely to be disputed until something is known, not only about what snails eat, but about what they derive most nourishment from.

Waters that emerge from the earth highly charged with calcium bicarbonate and flow away as streams may deposit calcium carbonate. Geijskes (1935) believes that the resulting incrustation makes the substratum suitable for some species, notably *Lithotanytarsus* which makes calcareous tubes, and species of *Pericoma* (Psychodidae) and *Riola*, and unsuitable for many of the usual stream-dwelling Ephemeroptera, Plecoptera and Trichoptera.

SUMMARY

In some animal groups calcium appears to favour all the species and the number drops as calcium concentration decreases. In others there are species typical of soft water and these often have close relatives typical of hard water. Calcium enables some mixohaline-water animals to tolerate extreme dilution. It also makes certain sensitive freshwater species more tolerant of unfavourable conditions. Apart from this, little is known about how it affects distribution and particularly about the importance of direct effects and indirect effects through its facilitation of decomposition in the soil.

Deposition of calcium carbonate may affect the fauna by altering the substratum.

SYNONYMS

Species	Group	Other generic names	Other species names
Aedes cantans (Meigen)	Ins. Dipt. Culicidae	*Ochlerotatus*	*maculatus* Meigen *waterhousei* Theobald
A. punctor Kirby	Ins. Dipt. Culicidae	*Ochlerotatus*	*nemorosus* Meigen
A. rusticus Rossi	Ins. Dipt. Culicidae	*Ochlerotatus*	
Anopheles atroparvus Van Thiel	Ins. Dipt. Culicidae		*maculipennis atroparvus* Van Thiel
Astacus fluviatilis Fabricius	Crust. Malacostraca	*Potamobius*	*astacus* Linnaeus
Corixa lateralis (Leach)	Ins. Heteropt. Corixidae	*Sigara*, *Vermicorixa*	*hieroglyphica* Dufour
C. limitata (Fieber)	Ins. Heteropt. Corixidae	*Sigara*, *Retrocorixa*	
C. nigrolineata (Fieber)	Ins. Heteropt. Corixidae	*Sigara*, *Vermicorixa*	*fabricii* Fieber
C. punctata Illiger		*Macrocorixa*	*geoffroyi* Leach
Erpobdella octoculata (Linnaeus)	Hirud.	*Herpobdella*	

Species	Group	Other generic names	Other species names
E. testacea (Savigny)	Hirud.	*Herpobdella*	
Glossiphonia complanata (Linnaeus)	,,	*Glossosiphonia*	
Heptagenia lateralis (Curtis)	Ins. Ephem.	*Ecdyonurus*	
Limnaea glabra (Müller)	Moll. Gast. Pulmonata	*Lymnaea, Leptolimnaea, Omphiscola*	
L. pereger (Müller)	Moll. Gast. Pulmonata	*Lymnaea, Radix*	*ovata* Draparnaud, *peregra*
L. stagnalis (Linnaeus)	Moll. Gast. Pulmonata	*Lymnaea*	
Margaritifera margaritifera (Linnaeus)	Moll. Bivalvia	*Margaritana*	
Planaria alpina Dana	Platyhelminthes	*Crenobia*	
P. vitta Duges	,,	*Phagocata*	
Planorbis carinatus Müller)	Moll. Gast. Pulmonata	*Tropidiscus*	
P. corneus (Linnaeus)	Moll. Gast. Pulmonata	*Planorbarius*	
P. vortex (Linnaeus)	Moll. Gast. Pulmonata	*Anisus, Spiralina*	
Procerodes ulvae (Oersted)	Platyhelminthes	*Gunda*	
Salmo trutta Linnaeus	Pisces	*Trutta*	*lacustris* Linnaeus *fario* Linnaeus

CHAPTER 12

Other Chemical Factors

Many workers comparing the fauna of two pieces of water, or studying the range of species, have analysed the water in the hope of finding correlations. They have generally been disappointed. I propose now to present a small selection of the many examples of this type of work and then to discuss why it has not been more fruitful.

The distribution of six species of sponge in the lakes of Wisconsin appears to be correlated with conductivity, the amount of SiO_2 and organic matter present, and the rate of flow (Jewell 1935). *Tubella pennsylvanica* and *Spongilla ingloviformis* were rarely found except in standing water with a high content of organic matter, and the second was nearly always in water with a low conductivity. In contrast *Ephydatia mülleri* and *Spongilla fragilis* were commonly taken in slowly flowing water; *E. mülleri* was restricted to water with a high SiO_2 content and high conductivity but *S. fragilis* was more tolerant and *S. lacustris*, often associated with them, more tolerant still. *E. everetti* occurred typically where there was little SiO_2 and little organic matter. Work of this kind is essential because it formulates the problem, but lacking in general interest because so many questions are still unanswered: which of these apparent correlations are direct, which indirect, and which fortuitous?

Edmondson (1944) collected rotifers from a large number of bodies of water, which he then divided into higher and lower categories according to the pH, and the concentration of calcium, magnesium, and bicarbonate. Each was treated separately and the division came at such a level that there were an equal number of places above and below it. The number of times each species had been found in the higher and lower category of each was set out in a table and the significance of the difference ascertained by means of a χ^2 calculation. Seven species were apparently limited by pH but not by bicarbonate, six by bicarbonate and not pH, and three by both. The range of one species was limited by calcium, that of another by magnesium and that of a third by conductivity. Edmondson dwells on the difficulty of judging the significance of results of this kind.

Lakes and ponds slowly fill up as the remains of plants that have grown in them accumulate on the bottom, and as material is washed in from the surrounding land. As the depth decreases, submerged plants are replaced by floating-leaved plants, which in their turn give way to emergent species. Which species take part in this succession depends on the fineness of the sediment falling to the bottom and on the rate at which it is accumulating. A copious supply of fine silt produces a substratum rich in the ions which plants need for good growth, and exigent species flourish; also, by causing good decomposition of plant remains, it prevents the development of an unfavourable substratum, which only a few hardy species can colonize. Where coarse silt accumulates fast, *Juncus bulbosus* is the common submerged plant, *Potamogeton natans*, which can tolerate highly organic soils, is the floating-leaved stage and the succession culminates in *Carex rostrata*. Bog develops from this. The large submerged species of *Potamogeton* flourish on soils where the silt is finest and the next stages in the succession are water-lilies and *Phragmites* or, where silt is particularly plentiful and fine, *Typha*. *Salix-Phragmites* swamp follows the reed-swamp. The various successions can be deflected at any time by a change in conditions and it is common to see *Phragmites* replaced by *Carex* at the apex of a bay because the reeds themselves, by hindering flow, have cut off the supply of silt which would make their continued existence possible. The development to *Salix-Phragmites* swamp, in other words to fen, is seen only in bays into which rivers or fairly large streams debouch (Pearsall 1920, 1921).

Corixa germari can live in deeper water than other species (Crisp 1962) and can probably thereby avoid the effects of large waves. This may be why it is sometimes found in large lakes with great areas a metre or two deep. Other species require some shelter and are found only in bays and reed-beds. If the site is fairly exposed, and the bottom in consequence sandy, *C. dorsalis* is an abundant species. In unproductive lakes it is likely to be accompanied by *C. scotti*, in productive ones by *C. falleni* (table 34) (Macan 1955b). There is a succession of species with increasing percentage of organic matter in the soil, which has been worked out by collecting in reed-beds at different stages of evolution. In Windermere, a moderately productive lake, *C. dorsalis* and the minute *Micronecta poweri* are generally the only water-bugs found where there is less than 10 per cent organic matter in the soil. With increasing amounts above this level *C. fossarum* and *C. distincta* become more and more important up to 40 per cent, at which *C. scotti* comes in. Between 51 and 55 per cent it is the most numerous species (Macan 1938), and experience elsewhere indicates that *C. castanea* is the final species in this series. *C. distincta* is characteristic of a succession leading to bog, *C. fossarum* of the richer conditions leading to fen. In the rich Danish lakes only the second of these two species was taken and the common species in the middle of the reed-beds was *C. linnei* (Macan 1954b). In the fen-pools *C. sahlbergi* was the main species, as it

Table 34

Occurrence in lakes of the indicator species of Corixidae (Macan 1955b)

	Oligotrophic → Eutrophic											
	Ennerdale	Crummock	Derwentwater	Bassenthwaite	Coniston	Ullswater	Windermere	Esthwaite	Blelham	Esrom	Fure	Neagh
C. dorsalis or striata	+	+	+	+	+	+	+	+	+	+	+	
C. scotti	+	+	+	+	+			+				
C. falleni							+	+	+	+	+	
C. fossarum							+	+	+	+	+	
C. germari												+

is around Windermere and neighbouring lakes (Macan 1954a). This succession can be summarized:

Development of vegetation and accumulation of organic matter

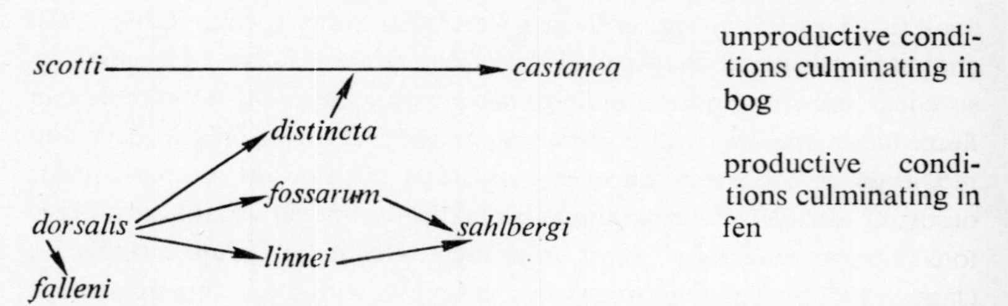

Why the range of each species is limited in this way is unknown.

Mackereth, Lund, and Macan (1957), respectively chemist, algologist and invertebrate zoologist on the staff of the Freshwater Biological Association, discussed chemical analysis in relation to ecology at a meeting of the Linnaean Society. Lund divided tarns, which are comparatively small bodies of water lying in side valleys, into low, medium, and high according to whether the concentration of total ions, expressed as milli-equivalents/l, fell below, between, or above 0·4 and 1·0. He points out that there is a general relation between production of algae and amount of ions available but that it is far from exact. If the Lake District lakes, which occupy the main valleys, be arranged in series, the most productive have two to five times as many ions as the least, but an algal production that may be over a thousand times as great.

The tarns fall very clearly into five groups according to the snails found in them (table 35). Group III, tarns with both *Limnaea pereger* and *Planorbis albus*, may be taken as a basis with which others are compared. All are in Lund's middle group except one, which is in his high group; its chemical advantages may be reduced by its small size, for smallness generally seems to be an unfavourable quality. Calcium ranges from 5·3 to 11·7 mg/l. Only one is heavily overgrown with emergent vegetation, and that was of very recent development at the time of the survey. All are surrounded by moorland or pasture. The tarns of group II lack *Planorbis albus*. Eight out of the twelve have less calcium than any tarn in group III. Of the remaining four, three are heavily overgrown with vegetation, and analysis of all the tarns indicates that, with increase in floating-leaved and emergent species, *P. albus* occurs less often and in smaller numbers. There remains Green Hows Reservoir, a large tarn, the absence of *P. albus* from which cannot be explained on the grounds of too little calcium or too much vegetation. However, it has an alkalinity that is unexpectedly low, which indicates perhaps that it has rather little bicarbonate; the amounts of chloride and sulphate are high. It is thus possible to find an explanation for all the absences of *P. albus* from the group II tarns, but how valuable analysis along these lines is time only will show. It certainly lays the analyst open to the charge that, given enough variables, time, and persistence, he could find something unique about each tarn to explain any faunistic peculiarity that needed explaining.

The tarns in group I are a selection from a larger number. No snails occurred in them. A correlation between this and a low concentration of ions is probable, because all these tarns and no others are in Lund's low class.

The four tarns in group IV harbour *L. palustris* in addition to the species of group III and at least one of three species found in no other group. The concentration of ions tends to be low compared with group III and the richer snail fauna cannot be linked with it. What does strike the eye is that the group contains the two largest tarns, Wise Een and Tarn Hows. Wray Mires tarn, not particularly large, is just below Wise Een, receives all its water, and cannot be considered apart from it. Wharton tarn also is not larger than many others. Its peculiarity is that it flows out through an artificial cut now filled with rubble that has fallen in from above. In consequence rain raises the tarn level considerably and the

Table 35

Lake District tarns showing the species of snail taken during a search lasting one hour and the concentration of certain ions in parts per million unless otherwise stated (Mackereth, Lund and Macan 1957)

Group	Tarn	*Limnaea stagnalis*	*Planorbis crista*	*P. complanatus*	*Limnaea palustris*	*Valvata piscinalis*	*Physa fontinalis*	*Planorbis contortus*	*P. albus*	*Limnaea pereger*	Total ions milliequiv./litre	Ca	Mg	Na	K	Cl	SO_4	NO_3	Area Dm^2 (approx)
	Rather Heath 6			X	X				X	X	1·5	20·0	2·6	4·7	0·9	9·8	12·2	0·4	64
	Rather Heath 1	X	X	X	X				X	X	1·4	19·8	2·1	5·1	0·7	9·1	12·2	0·4	210
	Lindeth			X	X				X	X	—	5·5	—	—	—	—	—	—	80
	Podnet			X					X	X	0·6	6·2	1·0	5·0	0·5	9·4	11·1	0·03	186
VI	New		X		X				X	X	0·8	6·0	1·7	6·2	0·7	12·5	11·4	0·03	83
	Cleabarrow			X	X				X	X	1·25	14·0	2·4	6·2	1·4	10·0	17·8	—	55
	Little Ludderburn				X				X	X	1·4	16·3	3·5	5·1	0·6	8·2	13·2	0·2	334
	Batemanfold				X					X	0·8	8·5	1·5	5·0	0·4	8·2	12·9	0·03	8
	Lost				X				X		0·9	10·5	1·4	5·0	0·8	9·1	10·3	0·1	16
	Cacer				X				X		0·7	8·8	0·9	4·2	0·4	8·8	8·9	—	48
V	Rose Castle								X		0·9	9·8	2·1	4·3	0·4	11·4	10·2	—	48
	Slew								X	X	0·6	6·5	0·6	3·5	0·5	8·0	9·6	0·3	52
	Wise Een				X			X	X	X	0·5	4·9	0·9	4·2	0·5	7·5	9·7	—	440
	Wray Mires				X	X		X	X	X	0·5	5·0	0·8	4·4	0·5	8·8	9·5	0·01	116

IV	Tarn Hows	X		X	X	X	X	X	X	0·6	5·3	1·3	4·0	0·2	8·9	5·3	—	430
	Wharton			X	X	X		X	X	0·7	7·4	1·3	3·8	0·5	7·3	8·7	0·01	128
	Blelham F.P.							X	X	0·8	9·6	1·5	4·1	0·6	12·3	9·8	0·5	28
	Clay Pond							X	X	1·1	11·7	2·4	7·8	0·8	11·4	15·0	—	6
III	Gill Head R.							X	X	0·7	5·9	1·6	5·5	0·6	11·4	12·0	—	308
	Arnside							X	X	0·6	5·3	1·1	4·5	0·6	8·1	7·8	0·08	83
	Knipe							X	X	0·9	9·7	1·9	5·3	0·3	10·6	9·6	—	213
	Black Beck Moss								X	0·7	7·7	1·3	5·1	0·5	9·2	10·7	0·2	30
	Wray Mires F.P. 2								X	0·6	4·7	1·3	6·2	0·6	12·5	10·3	0·03	8
	Robinson								X	0·6	6·2	1·1	4·7	0·5	8·7	10·8	0·02	32
	Moss Eccles								X	0·5	5·2	1·0	4·2	0·4	6·8	9·1	—	352
	Three Dubs								X	0·5	3·9	0·9	4·3	0·1	6·1	10·2	—	160
	Green Hows R.								X	0·7	6·6	1·2	6·0	0·6	11·2	13·6	0·07	104
II	Brigstone 2								X	0·5	5·1	0·6	3·3	0·3	6·1	6·1	0·1	25
	Gill Head F.P.								X	0·6	4·1	1·1	5·5	0·8	10·5	10·7	0·03	42
	Hodson								X	0·5	4·9	1·1	3·7	0·2	4·9	9·6	—	47
	Nor Moss								X	0·6	6·4	0·8	4·2	0·5	8·2	13·2	0·1	150
	Wray Mires F.P. 1								X	0·6	4·0	1·3	5·8	0·7	11·4	10·1	0·03	32
	Wray Mires F.P. 3								X	0·7	4·7	1·3	6·2	0·6	12·9	10·1	0·03	8
	Brigstone 1									0·4	2·8	0·5	3·2	0·5	5·8	7·1	0·1	62
	Blea									0·3	1·3	0·5	2·6	0·4	3·9	4·2	—	372
I	Lingmoor									0·3	2·2	0·7	3·3	0·3	5·9	6·4	—	52
	Harrop									0·2	1·5	0·4	2·1	0·1	3·4	4·8	—	150

level remains high for some time while the water percolates the debris in the blocked channel. Other tarns rise little after heavy rain, which passes through them quickly. Retention of water, characteristic of Wharton Tarn, is also a difference between a small tarn and a large one, and the richer snail fauna is likely to be connected with it. That the three species which, taken together, characterize the large tarns are characteristic also of lakes (Macan 1950a) lends support to the idea that large size favours their occurrence in the conditions prevailing in the Lake District.

Group V is one of the most striking because *L. pereger*, found in every tarn of every group except the first, is absent from three of the four tarns and scarce in the fourth. Again there is nothing different in the ions and again there is an obvious peculiarity; all of the tarns are surrounded by deciduous trees which shade them and drop dead leaves into them every autumn.

In group VI there tend to be more snails than in any other, though, as in group IV, not every species occurs in every tarn. All but one of the tarns that fall into Lund's high class are in this group, though some are not different from tarns in other groups. All lie to the east of Windermere not far from the limestone which marks the edge of the Lake District, and it is possible that the population of some is maintained by immigration from ponds that are calcareous. There can be little doubt, in the light of Boycott's findings, that the occurrence of *Limnaea stagnalis* is related to a calcium content unusually high for a Lake District tarn.

The outcome of this analysis is some correlation with calcium and total ionic content but generally failure to explain the occurrence of the various species in terms of the data supplied by the chemist. These failures could be due to three causes:

1. There is no correlation to find. This remains a possibility to be borne in mind, but not to be admitted until others have been explored.

2. The chemist has supplied the wrong data. As Mr Mackereth has remarked to me, there are (or were) ninety-two elements and he commonly analyses a mere half-dozen or so. Fox and Ramage (1931), using a spectrograph, found fourteen elements in the tissues of various animals and list twelve more which their method did not reveal but which other authors had found. Lund, in the symposium mentioned, stated that some algae need iron, manganese, boron, molybdenum, vanadium, and cobalt.

It may be that organic compounds rather than inorganic ones are what is important. Leaf infusion, such as is found in the tarns of group V, is evidently favourable to *P. albus* but the reverse to *L. pereger*, and whether the effect be direct or indirect, it would seem likely that organic compounds are involved. The same may be said of the different soils on which different species of plant and corixid are found, and organic matter was reckoned by Jewell a factor affecting the distribution of sponges. A similar conclusion, reached by several workers studying other groups, was noticed in the chapter on calcium. Unfortunately, as was seen when oviposition was being discussed, it has rarely been possible to record more than the total amount of organic matter and the nature of the components has remained unknown.

Many algae need accessory growth factors, such as vitamins. A productive lake in the Lake District has not only more algae than an unproductive lake but also different species, one of the most celebrated examples being *Asterionella formosa* which, never seen by Dr Lund in Ennerdale, produces a heavy crop in Windermere each spring. Something in the soil of the drainage basin of Windermere seems to be responsible (Lund, in Mackereth *et al.* 1957), but its nature remains obscure. It may be that the abundance of invertebrates that are not insects in lakes that are rich is due more to substances of this kind than to the greater supply of food. Similar mysterious substances are known in the sea. Up to 1930 *Sagitta elegans* was the common species of the genus in water in the English Channel off Plymouth but after that it was replaced by *S. setosa*. At the same time Wilson (1951) noticed that well-tried techniques for rearing polychaete worms were giving less satisfactory results than formerly. Successful culture was not re-established until 'Celtic' or '*elegans*' sea water was brought from further west. Evidently there are two masses of water, the boundary between which may shift to and fro, and the 'channel' or '*setosa*' water is less favourable to some organisms than the other.

Organic substances that are positively harmful are well known, and some produced by freshwater organisms were mentioned in the chapter on interrelationships. Others generally originate from land plants. Streams flowing through woods of *Thuya occidentalis* often contain very few animals, and fish are particularly sensitive to what is thought to be a toxin derived from the humus formed from the spruce needles (Ebeling 1930). Humic acids formed from acid peat where there is little inorganic matter are stated to be inimical to many species (e.g. Thienemann 1950) but there is little precise information, either on exactly what a 'humic acid' is, or on how it affects organisms.

3. The chemist has supplied the right data but the biologist does not know how to make use of it. In the first place he does not always know whether he should be looking for a deficiency of one substance or for an excess of something else. Secondly, and more important, there may be

complicated interactions. For example the concentration below which an animal is unable to take up an ion from the medium can limit its range. But it was also clear in chapter 10 that this critical concentration varied with pH (fig. 57), as well as with the concentration of other ions. Here may be an important reason why the search for correlations has not been as rewarding as many a field-worker had hoped. If there is a simple threshold for some substance above which a species does occur and below which it does not, a reasonable amount of collecting should reveal it. If it fluctuates according to the concentration of one or two other substances, the threshold will be much more difficult to detect from data of the kind presented in table 35, even with the latest mathematical techniques. The chances are that the collector will not visit enough pieces of water to obtain all the permutations necessary before its detection is possible.

Excess is certainly important as well as deficiency. Harnisch (1951) describes experiments which showed that solutions of KCl, $CaCl_2$ and $MgCl_2$ were toxic to *Daphnia magna* but harmless after NaCl had been added. He arranges the ions in a series according to the strength of their harmful effects:

$$Ca \rightarrow Na \rightarrow K \rightarrow NH_4 \text{ and}$$

acetate→phosphate→sulphite→nitrate→chloride→nitrite. Copper, zinc and lead are toxic. The first, however, was found in every sample of tissue from a number of animals which Fox and Ramage (1931) examined, and they believe it to be essential for the formation of protoplasm. Evidently the toxicity of an ion generally, perhaps always, depends entirely on what else is present and how much there is of it. The effect of various ions and combinations of ions on *Artemia* was noticed in chapter 10. R. Lloyd (1960) has shown that the toxicity of zinc to *Salmo gairdneri* is less in hard water than in soft.

It is probably exceptional to find in a water unaffected by human activity a concentration of any heavy metal sufficient to keep out any organism. Hutchinson (1932) cultured various species of Cladocera to find out the lethal concentration of magnesium, and concluded from his results that the high concentration of that element in Lake Tanganyika is not the cause of the absence of Cladocera. On the other hand the unusually high concentration of zinc in Bear Lake, Idaho, could account for the absence of Cladocera and other animals.

Toxic concentrations of heavy metals are unfortunately far from exceptional in places where man has been mining or carrying on certain industries. Lead mines reopened at various places in Wales during the period 1914-18 affected the streams for longer than that, because, for many years afterwards, heavy rainfall seeping through the spoil heaps would dissolve a harmful quantity of metal. Carpenter (1926) described four stages: 1, barren; 2, *Lemanea*, *Batrachospermum* and a few Bryophytes, and a small number of insect larvae none of which are Trichoptera; 3, flora enriched by Chlorophyceae, fauna by oligochaetes, turbellarians, Trichoptera, and further species of groups present in the previous stage; 4, flowering plants, molluscs and fishes. These four stages can be traced with passage downstream from a source of pollution and with the passage of time after mining has stopped. Laurie and Jones (1938), resurveying the River Rheidol, found 103 species in 1931/2 where Carpenter had found fourteen ten years before. There were forty-three mine workings along the river. All had been closed before 1914 but a few were opened again during the war years. Jones (1958) believes that zinc may have been more important than lead, which the earlier authors had regarded as the main cause of the reduced fauna. He suggests also that the Trichoptera were not more sensitive than other insects but were kept out by a poor food supply. Whether this was brought about by the zinc directly or whether by scouring by the grit washed out from the spoil heaps is not clear. Jones obviously makes an important point in drawing attention to biotic effects when part of a community is destroyed by something not concentrated enough to kill the rest.

In conclusion a word must be said about pH. In lagoons devoted to the culture of oysters, the pH may sometimes rise to 8·9, but, if nitrogen and phosphorus are added to increase the fertility, the pH may exceed 9·1. When it passes this point, the larvae die. Gaarder (1933) experimented with a range of pH brought about by adding CO_2 and $Ca(OH)_2$ to sea water and found that heavy mortality supervened rather suddenly at pH 9·1 under these conditions also.

Spirostomum ambiguum is most active at pH 7·4 and congregates there if placed in a gradient. Values below 6 and above 7·6 are lethal and the further they are beyond these limits, the quicker death ensues (Saunders 1924). The rate of movement of *Colpidium*, and the rate at which it forms food vacuoles in a suspension of Indian ink, rises as pH increases from about 5 and then dips to reach a minimum at pH 8 (fig. 65). Ciliates produce mucous to entangle food particles but in such small quantities that Mills (1931) was unable to make any tests on it. She did find, however, that snail mucous is most viscous at pH 8. She suggests that this is true of *Colpidium* mucous also, and that the changing degree of activity with changing pH depends on the extent to which the beating of the cilia is hindered by the viscosity of the mucous.

I doubt if it would be useful to mention even some of the many works in which range and pH have been correlated. Harnisch (1951) devotes a chapter to pH and ends it with the observation that it is very difficult to make out whether any of the limitations recorded are due to the concentration of the hydrogen or the hydroxyl ion or to whatever has caused one or the other to preponderate. Hutchinson (1941) writes: 'It is, however, exceedingly doubtful if more than a single case has ever been brought forward demonstrating unequivocally that the natural variation in numbers of

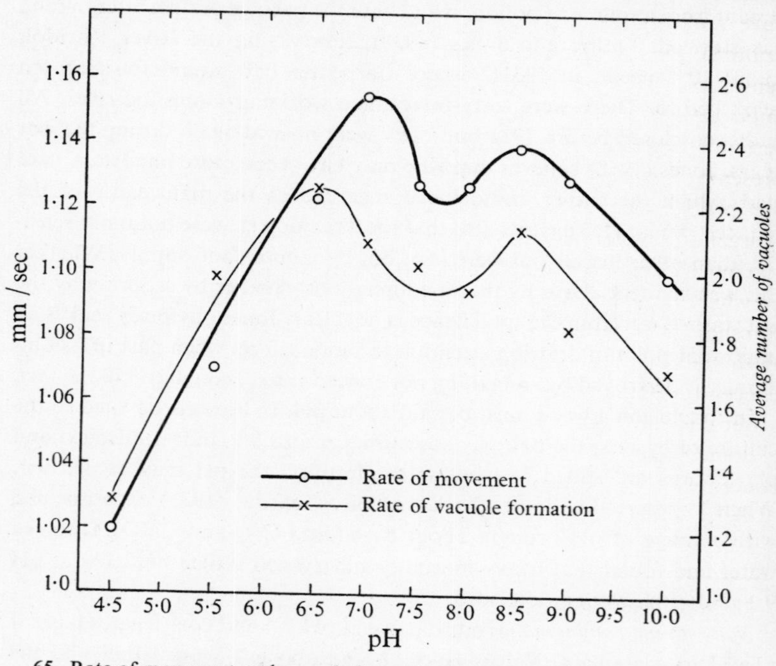

65. Rate of movement and vacuole formation in *Colpidium* at different hydrogen ion concentrations (Mills 1931) (*J. exp. Biol.* **8,** p. 24, fig. 4).

any species of animal is due to pH.' The single case is the work of Saunders on *Spirostomum.*

This type of work has tended to produce rather vague correlations or no correlation at all and has not come up to the expectations of thirty or forty years ago. Collaboration between field naturalist and chemist, or more often the results which one worker gets in those two fields, has not yielded as much as the former at least had hoped. That, however, is not a good reason to leap to the conclusion that few species are absent from certain waters because in them there is too much or too little of something.

The field may prove to be one where the inorganic chemist collaborates more fruitfully with the physiologist than with the field naturalist, and few, I think, will not expect advances when the organic chemist can isolate a greater range of compounds than at present.

SUMMARY

Attempts to explain distribution of species in terms of chemical differences have not had much success except where conditions are extreme. For example, the poor fauna of a lake in America is attributed to a high concentration of zinc, and pollution resulting from lead mining annihilates or impoverishes a fauna according to its degree. Field data suggest that certain species and certain types of organic matter are associated, but at present it is not possible to separate many of the organic compounds that are probably involved. It seems likely that there is correlation with inorganic compounds also. That more has not been found could be due to too much attention to common substances and too little to rare ones, some of which occur in many living tissues; or to the difficulty of detecting a threshold that varies according to the concentration of other substances. Both the level at which certain substances become toxic and the lowest concentration at which some can be taken up by an organism from the water depend on the amounts of other substances present.

Some Protozoa are probably the only animals affected directly by the concentration of hydrogen and hydroxyl ions.

SYNONYMS

Species	Group	Other generic names	Other species names
Carex rostrata Stokes	Angiospermae		*inflata* auct. non Hudson
Corixa castanea (Thomson)	Ins. Heteropt. Corixidae	*Sigara, Anticorixa, Hespero-corixa*	
C. distincta (Fieber)	Ins. Heteropt. Corixidae	*Sigara, Subsigara*	
C. dorsalis Leach	Ins. Heteropt. Corixidae	*Sigara*	*striata* (Linnaeus) *lacustris* Macan
C. falleni Fieber	Ins. Heteropt. Corixidae	*Sigara, Subsigara*	
C. fossarum (Leach)	Ins. Heteropt. Corixidae	*Sigara, Subsigara*	

Species	Group	Other generic names	Other species names
C. germari (Fieber)	Ins. Heteropt. Corixidae	*Sigara, Arcto-corisa*	
C. linnei (Fieber)	Ins. Heteropt. Corixidae	*Sigara, Anticorixa, Hespero-corixa*	
C. sahlbergi (Fieber)	Ins. Heteropt. Corixidae	*Sigara, Anticorixa, Hespero-corixa*	
C. scotti (Douglas and Scott)	Ins. Heteropt. Corixidae	*Sigara, Subsigara*	
Ephydatia everetti (Mills)	Parazoa	*Corvon-meyenia*	
E. mülleri Lieberkühn	Parazoa	*Meyenia*	
Limnaea palustris (Müller)	Moll. Gast. Pulmonata	*Lymnaea, Stagnicola*	
L. pereger (Müller)	Moll. Gast. Pulmonata	*Lymnaea, Radix*	*ovata* Draparnaud *peregra*
L. stagnalis (Linnaeus)	Moll. Gast. Pulmonata	*Lymnaea*	
Micronecta poweri (Douglas and Scott)	Inst. Heteropt. Corixidae	*Sigara*	*minutissima* Linnaeus
Planorbis albus Müller	Moll. Gast. Pulmonata	*Gyraulus*	

CHAPTER 13

Production

Various figures for standing crop, that is the number of animals per unit area at a given instant, and their weight or biomass, were quoted in chapter 3. These give a certain indication of production and can be used to compare lakes; such a comparison was one of the objectives which Berg (1938) had in view when he surveyed Esrom Lake. These gross figures, however, conceal certain information which is of theoretical interest, and of practical importance to those who are concerned with taking a crop from the water. Such people must know the production at various stages, commonly known as trophic levels. The primary producers are the plants, which create from inorganic substances living matter whose synthesis requires an input of energy. This comes from the sun. The primary consumers are the herbivorous animals. The secondary consumers are carnivores, which, in a European lake, will be fish and carnivorous invertebrates. However, as fish prey indifferently on carnivorous as well as on herbivorous invertebrates, to some extent they are tertiary consumers. In the last and final wholly aquatic stage are the piscivores such as pike. Any individual from any level whose death is not the result of being caught alive and eaten will be consumed by the saprophages.

The critical figure is the amount of food that each stage converts to flesh compared with the amount that it uses to maintain its tissues and to provide the energy for hunting and reproduction. This can vary considerably as an extreme example illustrates (in view of what Alm records about perch (chapter 6) it is possibly not as extreme as it may appear). There might be in a pond a stock of large fish that had to search so hard for its food that everything which it ate was used up by the effort of searching. If there was no mortality and any young produced were soon eaten, production would be very low. It would not be nil, for, as Le Cren (1962) points out, the weight of eggs may be a considerable fraction of the weight of the fish producing them. On the other hand, if the fish were never allowed to grow so large and some were removed each year, the survivors might find enough to eat to grow, and production could be good. It is important to stress at this point that production refers to all increase in weight up to a term fixed by the observer or up to the time of death, whenever and however that comes about, whereas the word crop is generally applied to the weight harvested by man. How many fish to remove, and at what size they should be removed, is the fish-producer's problem, and Allen (1952) set out to solve it when he undertook his survey of the Horokiwi Stream in New Zealand. For the moment I wish to consider this work more from the light that it throws on how to obtain the necessary information.

The Horokiwi is some ten miles long, and seven and a half miles were studied. Allen was able to estimate the number of eggs laid in two ways. He knew how many ripe females there were from his nettings, and the counting of the eggs in a sample of fish gave a figure for the average number per female. Secondly, he was able to count the redds, in which the eggs were laid. From various observations he obtained a figure for the number of eggs that would not hatch, and was thus able to start with a knowledge of the size of the population when it emerged from the eggs, which it did in October. The number was about 500,000 in 1940, but fewer in 1941 when floods washed away many eggs. Thereafter Allen sampled every three months with a seine net. On one occasion he made an effort to catch every fish in the stream but on the rest he sampled at certain selected stations. Every fish was weighed, measured and tagged and then returned to the water, except for some which were killed so that what they had been eating might be discovered. The proportion of unmarked specimens of any year-class in a catch indicated what proportion of the total population had not been caught in the previous sampling. The physical labour involved in this sampling programme, planned and carried out before the electric fish-shocker had been devised, must have been great but it left Allen with a far more accurate knowledge of how many fish there were in different parts of the stream than anybody had ever had for any piece of water inhabited by a wild population. Incidentally the marking showed that there was little movement upstream or down which simplified the subsequent calculations greatly.

Of 1,000 fry (table 36) there are only fifteen survivors after a lapse of six months, and in each subsequent period of the same duration numbers are roughly halved. What befell these fish was not discovered. The proportion removed by anglers was surprisingly small. Allen assumes that deaths are evenly spaced between two samplings and that increase in weight is steady, which enables him to calculate how much dead fish have contributed to production. For example the 985 fish which died during the first six months weighed $3\frac{3}{4}$ lb. Le Cren (1962) believes this to be an over-estimate because most of the 985 that failed to survive probably succumbed soon after hatching; the period between hatching and the first meal is a critical one. Otherwise there seems no reason not to accept Allen's assumptions and, therefore, his figures. It is noteworthy that, at the end of one year, the seven survivors weigh 2½ lb. and the fish that have died have contributed 5½ lb. making a total of 8 lb. of flesh produced. After a further year and a half only 2 lb. more have been added. Even though Allen overestimated the amount produced in the first six months, the contribution by the smallest specimens to total production is large.

Allen points out to anglers that, since the weight of living fish is greatest twelve months after hatching, and declines from then on, that is the time when the year-class should become takeable. If the fisherman waits for another year, other enemies of the trout will have taken a toll of the fish he might have caught, and, although the individual fish he catches will be larger, the total weight he removes from the water will be less.

Production in the Horokiwi is reckoned to be 500 lb/acre (560 kg/ha), which is high.

Table 36

Production of *Salmo trutta* in the Horokiwi Stream (Allen 1952)

The original units, ounces (oz.) and pounds (lb.), have been retained because the figures are approximations to the nearest whole number or major fraction. (16 oz. = 1 lb. = 0·453 kg)

Age	0	6	12	18	24	30	months
Number	1000	15	7	4	2	1	fish
Weight of a single fish	1/250	2	6	9	12	16	oz.
Weight of all living fish	¼	1¾	2½	2¼	1½	1	lb.
Weight of all fish that have died	0	3¾	5½	7	8¼	9	lb.
Total weight of fish produced	¼	5½	8	9¼	9¾	10	lb.

If data comparable to those in table 36 were available for every species in the stream, it would be possible to make a complete assessment of production. Unfortunately they are not. The trout possesses certain advantages which make it one of the easiest species to study in this way. In the first place it is large; even the egg is easily seen with the naked eye, and it is not necessary to search for any stage with a microscope. Secondly, it is unmistakable at all stages. Many invertebrates are not easy to distinguish even when fully grown and become increasingly similar at smaller sizes.

Few attempts have been made to distinguish the eggs, though such work as has been done holds out promise that they may prove distinct. Thirdly, all specimens hatch at about the same time. Fourthly, in the Horokiwi at least, the fish stayed in the same part of the stream all through their lives.

Ricker and Foerster (1948) calculated the production of sockeye salmon in Cultus Lake, British Columbia, but their field was not as comprehensive as that of Allen. In the first place they could not estimate the number of eggs, and in the second their work extended for the most part over a year only because few fish of each year-class remained longer in the lake. They based their calculations on three measurements:

1. The total number of fish and their weight. This was ascertained as the fish left the lake to set out for the sea.
2. The growth rate. This was determined by measuring freshly-eaten specimens obtained from the stomachs of predacious fish.
3. The death rate. This was discovered by releasing batches of marked fish at intervals through the year and recording the number in each batch that survived to migrate.

The number of fish at the end of their time in the lake, and growth and death rates during it, being known, numbers and biomass at any instant throughout the period of development could be calculated.

The passage of all the fish through the narrow outlet of the lake at the time of migration made this technique possible. It would be much less feasible for a non-migratory fish.

Hynes (1961b) has calculated the production in a stream by means of a method which is essentially the same as that used by Allen, except that he was not able to identify or count the eggs. He is dealing with species nearly every one of which completes development within one year. He takes the common species and shows how many specimens of each there were in each millimetre size-group. Numbers are highest in the smallest groups and fall rapidly, especially at first, just as did those of the Horokiwi trout. Subtraction of the numbers in one size group from the numbers in the one below it gives the number of specimens that died during the period when that particular millimetre of growth was being achieved. Hynes has no figures for weight, but he assumes that all the animals are of the same shape, calculates volume from this, and obtains weight by assuming that the specific gravity is 1·05. The errors from the assumptions about shape probably cancel each other. Various other sources of error are discussed, and Hynes stresses that his method is no more than very rough. He arrives at a figure of just under 30 kg/ha, which is poor production.

I have chosen to start with animals because they illustrate how the basic problem has been tackled. It would have been more logical to start with the plants, the primary producers. Most attention has been devoted to the phytoplankton, because, in any but a small lake, it contributes most to production. The individuals cannot usefully be counted, as trout or insects with their regular reproduction once a year were counted, because they reproduce frequently and rate of destruction is hard to assess; 100 diatoms to-day could represent 100 diatoms in a sample last week, or they could be survivors from a thousand that had originated from a hundred present a week ago. Production has, therefore, to be measured indirectly. The well-established method (e.g. Nygaard 1955) is to measure the amount of oxygen produced. Two similar bottles are filled with a sample of lake water and suspended at a given depth, one being covered with black paper. Oxygen concentration is measured at the beginning and at the end of the experiment. The fall in the dark bottle is the amount used in respiration; from the rise in the other the amount of assimilation, respiration having been allowed for, can be calculated. A newer method, more sensitive and consequently more suitable for unproductive waters, involves the use of C^{14}, a radioactive isotope. The experiment is set up in the same way, except that a small amount of $NaHC^{14}O_3$ is added to each bottle at the beginning of the experiment. At the end the algal contents are filtered off and the amount of C^{14} that they have built into their tissues measured by means of a Geiger or other suitable counter. Rodhe (1958) has given an account of some of the difficulties inherent in the method. He describes how, if a series of bottles is exposed for four hours each, production in them at the end of the day is greater than in a bottle exposed continuously during the same time. The nearer the bottle is to the surface the greater is the discrepancy between the two readings, which is attributed to the inhibitory effects of light. Photosynthesis is greatest at the surface only at the ends of the day and at some distance below it in between. It appears that, if an alga is held in light that is unfavourably strong for it, the inhibitory effect on photosynthesis may persist. The turbulence in a lake is continually carrying algae up to the surface and down below the zone of optimum light, and probably few individuals are exposed as long as a specimen in a bottle may be to light that is too bright. Consequently actual production is likely to be higher than the experimental results indicate.

Weekly measurements made by Rodhe (1958) in the Swedish Lake Erken showed a marked peak in production in spring, which is as expected because in temperate lakes algae usually increase rapidly in numbers at that time. Thereafter production was at a lower level and fluctuated violently from week to week. The irregularity was due in part to the usual wax and wane of species succeeding each other throughout the season, and in part to different water masses with different concentrations of algae brought to the sampling point by winds from different directions. The conclusion drawn from these results is that only the crudest comparison between lakes can be made unless production is measured at frequent intervals throughout the season. Production is expressed in two ways: the total production in the whole of the water column beneath a square metre of surface; or the production in that cubic metre in which most is produced. Figures from a number of lakes show a considerable range: 20-2,200 mg C/m^2 and 4-750 mg C/m^3. It is, however, too early yet to attempt to arrange the lakes in order according to their primary productivity, as on most there have not been weekly measurements. It will probably be possible in a year or two, as this is a field in which there is much activity at the moment. Jónasson and Mathiesen (1959) found in Esrom Lake and in Fure Lake no spring peak, except in years when ice persisted late. Greatest production was in August, and the fluctuations were not as great as in Erken Lake. Samples were taken monthly or fortnightly. Production was 215 g C/m^2/year in Fure Lake and 180 in Esrom compared with 104 in Erken. A figure of 660 is quoted for the shallow, polluted Lake Lyngby but it is not clear on how many measurements it was based. Gribsø appears to be about one third as productive as Esrom Lake.

Less attention has been paid to production by rooted plants, seemingly because limnologists like to work in the largest lake they can find, and in large lakes the contribution by rooted plants is small. Since they are not eaten much, harvesting and weighing the crop on a given area at the end of the season should give an accurate measure of their production. Odum (1957) has calculated it from oxygen concentration, but under peculiar conditions. At Silver Springs in Florida, water comes out of the ground at a constant temperature, which is very convenient for scientific work, and also with a concentration of oxygen well below saturation. It flows away down a channel in which rooted vegetation grows thickly. About three-quarters of a mile below the main spring the oxygen concentration was measured again and the increase, when allowance had been made for rate of flow, the water received from other springs on the way and exchange of gases between the water and the atmosphere, gave the production by the plants, both those rooted and the algae epiphytic on them. Measurements at night gave the respiration and, knowing how much oxygen certain species use in the laboratory and how many there are in the stretch of river, Odum goes so far as to work out the production at different trophic levels.

At one time great hopes were current that production in a lake could be assessed simply by measuring the amount of oxygen that disappears from the hypolimnion. This, the lower cold layer, becomes isolated by a warm upper layer early in the summer and remains cut off from any contact with the atmosphere until the surface layers cool down in the autumn. Organisms produced in the upper layers and dying there fall down into the hypolimnion and decompose, using oxygen in the process. The hypolimnion is generally well mixed and therefore, if its volume be known, an analysis to find the concentration of oxygen in it at the beginning of its existence and towards the end will show how much oxygen it has lost in the summer. This, it was once thought, should reflect productivity in the epilimnion. Gradually, however, it came to be realized that dead leaves washed in or blown in, and other organic matter of external origin, contributed far more than had been thought. The 'hypolimnetic areal deficit' is, therefore, no more than the roughest measure of production in a lake. Attention is now concentrated on measuring primary production.

It is difficult to find out how much a wild animal eats in the course of a day. There are plenty of data about what and how much was found inside the stomach of a given animal killed at a certain time, but how quickly it would have digested the contents and how rapidly any space in the stomach caused by digestion would have been filled remain largely unknown. Calculations of the efficiency of conversion of food by natural populations therefore rely heavily on laboratory findings. *Salmo trutta*, for example, kept in aquaria at a temperature between 4·4 and 10°C, increased 1 g in weight after eating 5 g of *Gammarus* (Pentelow 1939). This is an average figure and the range on which it is based is wide. M. E. Brown (1946) quotes a ratio of 6·3 :1 for weight of food eaten to weight of flesh produced. She was interested mainly in the factors that affect rate of growth and found them to be diverse. A hierarchy is soon established among a group of fish placed in an aquarium, and whichever establishes itself at the top of it gets more food than the rest and grows fastest. If the largest fish are removed, one that has been growing slowly will take over the vacant place and the rate of growth that goes with it. On the other hand, if food is plentiful, individuals in a group will grow faster than solitary specimens. There is an optimum density for best growth. External factors exert an influence, temperature being particularly important. Growth is most rapid at 8°C and 18°C. A trout weighing more than 100 g requires for maintenance 60 mg meat per week for every gram of its own weight. If it is lighter, its requirements will be more and if it is sexually mature, less. There are, therefore, adjustments to be made to any conversion figure, but the main difficulty confronting anyone attempting to base field calculations on laboratory data is that of knowing how much to allow for respiration. The published figures are based on fish living an idle life in a tank into which sufficient food is introduced at regular intervals. The wild fish is likely to require more, possibly much more, even all, of what it eats to provide the energy for finding and catching its next meal.

That of Lindeman (1942) is one of the most celebrated attempts to work out the production of successive trophic levels. He worked with published figures of the biomass in two American lakes, and on the physiological side he took results of the kind that have just been described. He finds the following levels of production expressed in g-cal/cm²/year. The first figure is that of the primary producers and it is followed by that of the primary, secondary and tertiary consumers.

Cedar Bog Lake

$$\overbrace{111\cdot3 : 14\cdot8}^{7\cdot5\,:\,1} : 3\cdot1 \qquad 111\cdot3 : \underbrace{14\cdot8 : 3\cdot1}_{4\cdot8\,:\,1}$$

Lake Mendota

$$\overbrace{480 : 41\cdot6}^{12\,:\,1} : \overbrace{2\cdot3 : 0\cdot3}^{7\cdot5\,:\,1} \qquad 480 : \underbrace{41\cdot6 : 2\cdot3}_{18\,:\,1} : 0\cdot3$$

Each level produces something of the order of one-sixth of the amount of what it feeds on in Cedar Bog Lake though less in Lake Mendota, a result in agreement with the laboratory findings. In the present state of knowledge inefficient production, as revealed by calculations of this kind, could be due to errors in the calculation or to real inefficiency in the system. The physiological basis rests on the assumption that all species are similar to the few that have been investigated in the laboratory. The biomass data are not entirely trustworthy because workers are still hampered by taxonomic difficulties, and also have not the knowledge they should have about life histories, food, number of eggs produced, and mortality, particularly in the early stages of the common species. As time goes on information on these subjects will accumulate and lead to calculations of ever-increasing accuracy. It will become possible to detect inefficiencies in the biological systems, which are almost certainly there. In a perfect system there would be in each level sufficient organisms to eat, throughout the year, just enough of those in the level below to ensure that they were never numerous enough to run short of food, and never so scarce that any of their food went to waste. Such a system operates on a farm, where an experienced farmer can judge exactly how much his grass can be grazed and will buy or sell stock accordingly. In a state of nature there are greater fluctuations. Whether, when these are fully understood, it will prove to be just a problem solved, or whether man will be able to step in and alter to make more efficient any production line in which he is interested, remains to be seen.

CHAPTER 14

Methods

A book by Welch (1948), ten Mitteilungen of the International Association of Limnology, and numerous papers in scientific journals have been devoted to methods, and within one chapter there is room for no more than generalities about them. I hope it will prove useful if I draw on my experience of long-term surveys of Ford Wood Beck and Hodson's Tarn and dwell particularly on how I came to leave undone those things that I ought to have done.

Anyone undertaking a survey should have a clear idea of the objective, which may seem too commonplace a remark to be worth making, but which I believe to be justified. I have in mind the unfortunate Ph.D. student who, expressing a taste for outdoor work, is advised by a supervisor who is an expert on an indoor branch of biology. Having delivered this heavy avuncular dictum, I must admit that it is rarely possible to plan work of this kind with the neatness and precision of a physiological experiment. So much being unknown, the investigator is bound to start with certain assumptions, some of which will certainly prove wrong and necessitate revision of the programme. Roughly a survey may have the following objectives:

1. To define communities and find out what determines their limits or the limits of component species. It is inevitable that valuable advance has come from studies of single groups in these days when the ecologist may still find that a taxonomic task lies between him and the research which he has planned. As, however, all organisms are part of the environment of every other species, these studies must ultimately be incorporated into a whole, and ecology of this kind must be based on community studies. Experimental work sometimes takes the form of altering one factor in the environment, by adding fertilizers to a fishpond for example, and observing the changes in the composition of the fauna. A reasonably complete species list is essential, and there must be investigation of the relative numbers of each as well. Knowledge of the absolute numbers is desirable. Those breaking new ground can only record these facts and such information about the environment as appears to them relevant. Those who follow are able to draw conclusions from comparison, and their programme will

be influenced to some extent by that of the worker with whose results they wish to make comparison.

2. To discover the structure of a community: where each species lives, what it eats, and what eats it.

3. To find how much is produced. Emphasis falls less on a complete list of species, as rare ones can be ignored, and more on accurate measurement of numbers per unit area. The student of production must be informed about the extent to which these numbers may be reduced by movement to another place, and he cannot make reliable calculations unless he knows the life histories of the main species. Attention is frequently concentrated on one species, generally a fish.

The next preliminary is to choose a suitable pond or stream, or suitable parts of a lake or river, the main consideration being time and manpower available. It is difficult even for an experienced worker to judge how far these will go, and how much to allow for the unforeseen.

The area having been chosen, stations where regular samples are to be taken must be selected. I have always found that the number I thought sufficient at first never has been. For example, it seemed reasonable to assume that the species inhabiting a sward of *Littorella* growing thickly and luxuriantly would be distributed fairly evenly in the bed. The error in this assumption is shown in table 37. When the collection was made in

Table 37

Numbers of *Enallagma* and *Pyrrhosoma* caught at two stations in *Littorella* between posts 6 and 7 in Hodson's Tarn

	1957			1958				1959			1960	
	Oct.	Nov.	Dec.	Mar.	Apr.	July	Sept.	Apr.	June	May	July	Sept.
Shallow Station												
yrrhosoma	141	69	41	45	75	76	83	35	67	18	8	45
nallagma	17	62	4	8	5	14	36	6	6	12	10	35
Deep Station												
yrrhosoma	26	10	17	17	27	101	52	41	89	14	9	25
nallagma	297	191	266	103	69	211	141	77	121	47	149	116

10-20 cm of water close to the bank, *Pyrrhosoma* was found to be predominant. At a depth of about 1 m *Enallagma* was the commoner. Oviposition habits accounts for this difference (chapter 5). As they grew older, nymphs of *Pyrrhosoma* wandered into deeper water, but the nymphs of *Enallagma* moved less.

A survey of the physiographical and botanical features is the first field operation when the preliminaries are complete. When I surveyed Hodson's Tarn, I mapped the outlines of the stands of the various plant species, but within each stand I did not measure numbers per unit area. This was an error, because *Carex* now grows more thinly in places and *Littorella* more thickly. The changes have been sufficiently great to be evident, but it would be more satisfactory if they could be expressed in numerical terms.

What non-biological variables in the environment should be recorded is always a difficult decision. The worker who starts with an excessive burden of routine sampling may find at the end that he has devoted much time to gathering data with which nothing correlates. The worker who waits till biological observations have given a clue to what may be important may miss records in a critical year. A right decision probably depends as much on good luck as good judgement. It is also difficult to know how often to sample. Chemical analyses after wet weather and after fine weather have prevailed for a long time, and when rain comes after a dry spell, may show extremes and may be all that is required. For other substances much more frequent sampling is essential; oxygen, for example, fluctuates more violently than any other substance and the view has already been expressed several times that continuous recording of its concentration will add considerably to knowledge about its importance as a limiting factor.

It is also difficult to obtain measurements of those temperature values that are critical to animals unless a device of some kind is in the water all the time. I have used a Cambridge Instrument Company recorder, which has a steel bulb full of mercury in the water, and a circular sheet of paper turned by clockwork on which the temperature is recorded. The contraction and expansion of the mercury in the bulb is transmitted to an arm bearing a pen through a capillary steel tube also filled with mercury. The sheet completes a circle in a week but the clock has on occasion, when an earlier visit was not possible, run for three weeks. The recording part is rather large and cannot be set up in places where the public is likely to interfere with it. The most modern method is the thermistor, the principle of which is to pass a current through a piece of metal whose resistance changes considerably with small changes in temperature. It is connected to a galvanometer whose arm, supplied with ink, hovers above a moving roll of paper. At intervals a bar descends and presses the arm against the paper, on which it makes a mark. It is an advantage to avoid in this way the resistance of a pen permanently pressed against the paper. This method also involves a large recorder and requires a source of electric current (Mortimer and Moore 1953).

Schmitz (1954) conceals in the water a glass tube containing sucrose, the amount of which that turns into glucose and fructose depends on the temperature. This, however, only sums the average temperature for the period of exposure, and gives no record of the maximum and minimum. Where a recorder is impracticable, I have found that the best substitute is a maximum and minimum thermometer protected by a metal tube and concealed in the water (Macan 1958c).

Flow has been measured by means of a small propeller, a paddle wheel, or a Pitot tube. However, these give a reading only of the current well above the stones and not in the layers inhabited by the organisms, and perhaps have no great advantage over the simple method of timing a float with a stopwatch. The traditional float is an orange which is conspicuous and, because it only just floats, offers insignificant wind resistance. Flow is another factor that fluctuates irregularly and within wide limits, and whose maximum value may be the significant one ecologically. Engineers calculate flow from level. They find or make a place where the cross-section is regular—an artificial V-shaped notch is preferred—measure the rate of flow at different surface levels and calculate the volume of water passing. Once a calibration table has been prepared, the height of the water surface is the only observation required, and in a river a reading of this once a day is enough. In a small stream considerable rise and fall can occur within a few hours. Professor Filteau has devised a simple method of measuring the greatest height reached. It consists of a vertical bar into which are inserted at intervals tubes, each sloping downwards sufficiently to retain water after the stream has sunk below its level.

Huet (*Verh. int. Ver. Limnol. 14*, 603), contributing to a discussion, asked that anyone reporting on studies of running water should record temperature, oxygen concentration, nature of the bottom, width, depth, and slope.

No complete list of animals is possible from collections in the water only; the immature stages of Trichoptera and Chironomidae are too ill-known for that. Some device for trapping emergent adults is essential if this objective is desired. On standing water any sort of transparent box will do. I have used one with a wooden frame and a top and some sides of celluloid. At least one side, preferably two, should be of gauze, because, if all are solid, there will be too much condensation on the inside. The most suitable trap in running water is, in my experience, what I have called the Mundie pyramid. It is a three-sided pyramid with a frame of angle-iron, two sides of strong perspex and one of gauze. The pyramid is, figuratively speaking, cut off below the point, and a collar is let into the resulting flat piece. Lengths of copper wire are attached to the inside of the collar in such a way that a screw-top honey-jar can be screwed into it. The inside of the jar is gained through an inverted cone of celluloid. Mundie (1956) has reviewed methods of trapping emerging insects.

Mud is the easiest substratum to sample quantitatively and the most satisfactory apparatus is the Birge-Ekman grab. Essentially it consists of a metal box open at top and bottom. The top is closed by two lids, which are pushed upwards by the pressure of water when the grab is descending and kept closed by the same means when it is being pulled up. The bottom is closed by two jaws held together by a strong spring. These open outwards and are held against the sides when the machine is set for action. It is lowered into the mud on a cable and then a messenger is sent down the cable to release the jaws. The sample is easy to obtain, but the subsequent sieving and washing to get the animals out of it are laborious.

An area of stony bottom reconstructed on a tray lowered into a suitable excavation in the bottom and left for a few weeks yields a good quantitative sample. The method is impracticable in running water, however, because the current fills with stones any excavation which the investigator tries to make. A popular method is the Surber sampler which consists essentially of two square frames hinged together. A net is attached to the one that is upright when the sampler is in operation, and the other, flat upon the substratum, serves to mark out the area that is to be sampled. The larger stones are then removed from this area and the rest stirred, so that all the animals are disturbed and washed into the net. The Surber sampler, cannot be used in swift water because the removal of one stone may cause others to shift, and the area from which animals are washed into the net is larger than that covered by the marker frame. Several workers have designed what I (Macan 1958a), in a review of the subject, have called a shovel sampler. My own consists of an upright square and a cutting edge which can be driven into the substratum by means of a handle. It is then pushed forward until a certain area has been sampled. The stones are retained in a coarse net attached to the upright frame, and any animals that have been washed off them are caught in a deeper fine net. All the stones are then tipped into a strong solution of calcium chloride or magnesium sulphate, in which they sink but animals float. Some stream-dwelling animals are powerful swimmers and probably flee from anything that gives warning of its approach. They could be caught by something that enclosed them when dropped from above so quickly that it was around them before they were aware of it. The difficulty is to obtain a seal on an uneven bottom. If the edge of the sampler comes to rest on the top of a stone, water may be deflected past the stone and animals may be washed away from the area that is to be sampled. Where a satisfactory seal can be made quickly, this is a good method in shallow water, and a square metal box without top or bottom was used in the surveys of Esrom Lake and Susaa.

Berg used a Birge-Ekman grab for obtaining samples from rooted vegetation and I have used a contraption that was similar except that the jaws were closed by handles, not by springs. The original apparatus took too large a sample (about 0·1 m^2), and I also decided that it was desirable that each stem or leaf should be severed as soon as the sampler touched it. The new apparatus, designed to do this, was made of two tubes, the inner 9·8 cm in diameter. The outer fitted closely round it and both were cut into sharp teeth at one end. A stud attached to the inner tube and passing through a slot in the outer kept the teeth level. The outer tube could accordingly be rotated through a segment of a circle, which caused the teeth to pass across each other like those of a hay-cutting machine. The operator lowers the sampler into the vegetation with one hand and moves the outer tube to and fro with the other. Then he drives the sampler into the soil to obtain a seal, closes the top of the inner tube with an air-tight cap, pulls the sampler out, and releases the sample into a large container.

These quantitative samplers cover a very small area and the sorting of the sample takes time. Moreover they cannot be used everywhere. I have, therefore, relied extensively on collections with a net either wielded for a given length of time or pushed along a given length of substratum. This is obviously a rough and ready quantitative method but it is also subject to certain errors of selection that might not be suspected at first. Table 38 shows a comparison of the catches in a coarse net and in a fine net.

Table 38

Numbers of *Baetis rhodani* of various sizes taken in sixty-five parallel collections with a coarse net (10 threads/cm) and a fine net (80 threads/cm) (Macan 1958a)

Size in mm	0–1	1–2	2–3	3–4	4–5	5–6	6–7	7–8	8–9	9–10	Over 10
Fine net	19986	21872	7730	2657	1244	787	320	174	107	10	17
Coarse net	25	226	252	424	724	761	409	311	152	76	46

Nearly equal numbers of *Baetis rhodani* nymphs 5-6 mm long were taken. The smaller the nymphs the smaller was the number caught in the coarse net, presumably because they were passing through the meshes, though the high proportion of the smallest ones doing this is a source of surprise to me at least. In contrast, the larger the nymphs the more were caught in the coarse net. Presumably, as the water took longer to pass through it, there was more opportunity to swim out of the fine net, and a greater proportion of larger nymphs availed themselves of it because they could swim faster than smaller ones.

Table 39

Comparison of twenty-five collections with a pond net and twenty-five collections with a quantitative sampler made at the same time and in approximately the same place

	Number of times taken		Total number taken		Ratio	
	Sampler	Net	Sampler	Net	Sampler	Net
Erpobdella octoculata	12	5	36	8	4·5	1
Triaenodes bicolor	8	14	78	120	1	1·5
Limnephilus marmoratus	6	7	8	17	1	2
Pyrrhosoma nymphula (over 10 mm)	18	20	69	169	1	2·4
Phryganea spp.	15	12	24	58	1	2·4
Deronectes spp.	6	12	7	25	1	3·6
Leptophlebia spp.	9*	9*	123	515	1	4
Limnaea pereger	11	15	55	229	1	4
Haliplus fulvus	7	13	8	34	1	4·2
Enallagma cyathigerum (over 10 mm)	14	19	27	220	1	8
Corixa castanea	3	15	4	109	1	27

* Only nine collections were compared, those made when the species was small enough to pass through the net having been omitted.

Table 39 contrasts catches with the saw-cylinder-sampler (SCS) and with the net in Hodson's Tarn. The last two columns show the ratio of numbers caught in one to numbers caught in the other. The purpose is to demonstrate how far one or the other is selective, and the first decision to be taken is which to treat as the standard. The SCS rapidly encloses a given area and nothing can escape it once it has descended. Odonata, Trichoptera, Mollusca, and Hirudinea do not move fast enough to avoid it, should they see it coming, and their numbers, it can be asserted with confidence, are those normally inhabiting that particular area. Their numbers in the SCS should, therefore, provide a standard with which other catches may be compared. About the swimmers, Ephemeroptera, Hemiptera, and Coleoptera, it is not possible to be so certain. Anyone passing over a *Littorella* sward in a boat and looking at it through a glass-bottomed box sees nothing alive, which indicates that any animal that has taken refuge in the *Littorella* is likely to be so deep that it cannot avoid the sampler. What cannot be ascertained is how many swimmers do not take cover but flee over the surface when they see a boat coming, keeping constantly ahead and out of sight.

The leech, *Erpobdella octoculata*, is the only species whose numbers in the sampler were larger than in the net (table 39). This is no doubt due to its ability to cling sufficiently tightly to avoid being dislodged by the net. Two Trichoptera, the next species in an order in which the ratio of numbers in the net to number in the sampler is increasing, possibly cling too. Then comes *Pyrrhosoma*, which, with a ratio of 1 : 2·4 must be contrasted with the similar and equally numerous *Enallagma*, which has a ratio of 1 : 8. It is difficult to doubt that so big a difference is significant of something, probably a difference in behaviour. I suggest that *Pyrrhosoma* nymphs tend to cling to the vegetation when disturbed, a reaction which saves a certain proportion from being caught, whereas *Enallagma* nymphs tend to swim, which, since they cannot swim fast, makes their capture more likely. So few *Corixa* were taken in the SCS that it seems likely that this is a species which does avoid it by swimming away when the boat first appears. The other active swimmers, the beetles *Deronectes* and *Haliplus*, would appear not to, since their ratios are much the same as that of *Limnaea pereger*. This is the most reliable animal because it cannot avoid the sampler by swimming away nor reduce the total caught in the net by clinging tightly to the vegetation. The sampler, then, gives an accurate measure of numbers per unit area of everything except Corixidae, but a pond-net, it must be concluded, is an unexpectedly selective instrument.

Simpson and Roe (1939) recommend that at least ten samples should be taken at each station, though in their wise discussion of the subject they point out that the ideal is rarely attainable. Time is always a factor that must determine how much a worker can do, and the size of the biotope may impose a limit also; a sampling must not leave a stand of vegetation so mutilated that it is not offering comparable conditions to the fauna on a later occasion.

From the statistical point of view, only a few specimens of any species per sample may be adequate. When the whole fauna is being studied, there must be a compromise which will catch more than is really necessary of the commoner species and fewer of the rarer.

When planning the sampling programme, the field worker who has not a good knowledge of statistics must consult either a book or a statistician in order to ensure that the sampling effort is used to the best advantage and that the fullest use is made of the results. For some statistical purposes it is essential to know whether the animals are randomly distributed, which is easy to discover by means of published tables. For example, in September 1958 the following numbers of *Pyrrhosoma* were taken in twenty SCS samples in *Littorella* in Hodson's Tarn:

6 14 9 6 5 5 9 7 7 16 4 8 5 10 8 6 9 8 12 5

The total number is 159 specimens and the mean per sample is 8. Table 40 in Pearson and Hartley (1954) shows 'the confidence limits for the expectation of a Poisson variable'. Against the figure 8 in the column headed 0·95 are the figures 3.45 and 15·76. These are the limits within which the catches would be expected to fall nineteen times out of twenty, or ninety-five times out of a hundred, if distribution were random. If the odds against an event are twenty to one or more, the statistical convention is that it is significant or not due to chance. In the present instance the actual limits are four and sixteen and the distribution may be treated as random. Then, since there were 159 specimens in 1,500 sq. cm, the area covered by twenty samples, which is 1/164 of the whole sward, there were $159 \times 164 = 26{,}076$ specimens in the whole sward. More accurately, the number lay somewhere between 22,055 and 30,153, the confidence limits for the figure quoted. The object of the sampling was to discover this.

In September 1960, in another *Littorella* sward, numbers in five samples were:

2 2 2 1 2

When the figures are as close as this, the organism is obviously distributed very evenly.

In September 1957 numbers of *Leptophlebia* in *Littorella* were:

32 2 18 18 101 29 27 29 45 78 59

The total is 438, the mean 39, and the confidence limits from the table 27·7 and 52·2. The actual limits, 2 and 101, are so far beyond these that distribution is obviously not random, and any calculation of number of specimens in the sward from mean per sample is subject to an error of unknown dimensions. This is an extreme example chosen from many samples; possibly the clumping was due to recent hatching of the eggs, and the nymphs had not had time to distribute themselves more evenly, as other samples suggest that they do later. Tadpoles show this phenomenon very obviously.

In the next example, number of *Leptophlebia* in *Carex*, the samples are in an arranged order, not, as previously, in the order in which they were taken, the station on the left being nearest the edge of the bed and the others progressively further away from it.

110 137 85 98 82 51 80 31 71 90 69 33 67 35 27

The total is 1,066 and the mean 71. The extremes, 137 and 27, exceed those in the tables, which are 89·2 and 55·0, and distribution is evidently not random. The highest figures lie to the left, the lowest to the right and there is some tendency towards a regular decrease between the two. Possibly this indicates that the nymphs are entering the *Carex* from the vegetation that adjoins it, or it may be related to the more vigorous growth of the sedge at the periphery of the bed.

Incidentally the tables of Pearson and Hartley (1954) give confidence limits for numbers from 0 to 50. The tables of Crow and Gardner (1959) run up to 300.

Further than this it would be imprudent to go. There are many tests for significance that the faunistic worker may need to carry out. Each differs slightly from the rest according to the exact circumstances and any selection of examples might mislead. In addition to the book by Simpson and Roe, already mentioned, those by David (1953) and Bailey (1959) are recommended.

Another method of estimating population is to mark a number of specimens and return them to the water. The total number of marked specimens to the total population stands in the same ratio as the number of marked specimens to total in a subsequent catch, provided that the marked specimens distribute themselves evenly throughout the population. Fish are marked with small metal tags attached to the gill covers or elsewhere, with plastic capsules inserted under the skin, or by mutilation of a fin. Cellulose paint has been used to mark insects.

REFERENCES

ALLEE, W. C., Emerson, A. E., Park, O., Park, T., and Schmidt, K. P. (1949). *Principles of Animal Ecology*. Philadelphia: Saunders. pp. xii + 837.

ALLEN, K. R. (1940). 'Studies on the biology of the early stages of the salmon (*Salmo salar*). 1. Growth in the River Eden.' *J. Anim. Ecol.* **9,** 1-23.

ALLEN, K. R. (1952). *A New Zealand trout stream some facts and figures.* New Zealand Marine Dept. Fish. Bull. 10A. pp. 70.

ALM, G. (1952). 'Year class fluctuations and span of life of perch.' *Rep. Inst. Freshw. Res. Drottning*. **33,** 17-38.

ALSTERBERG, G. (1930). *Wichtige Züge in der Biologie der Süsswasser-gastropoden.* Lund: Gleerupska Univ.-Bokhandeln. pp. 130.

AMBÜHL, H. (1959). 'Die Bedeutung der Strömung als ökologischer Faktor.' *Schweiz. Z. Hydrol.* **21**, 133-264.

ANDERSON, G. C. (1958). 'Some limnological features of a shallow saline meromictic lake.' *Limnol. Oceanogr*. **3**, 259-270.

ANDERSON, W. L. (1948). 'Results of an analysis of farm ponds in the Midwest.' *Progr. Fish Cult*. **10,** 111-116.

ANDREWARTHA, H. G. (1952). 'Diapause in relation to the ecology of insects.' *Biol. Rev*. **27,** 50-107.

ANDREWARTHA, H. G. and Birch, L. C. (1954). *The distribution and abundance of animals.* Chicago U.P. pp. xvi + 782.

BAGENAL, T. B. (1951). 'A note on the papers of Elton and Williams on the generic relations of species in small ecological communities.' *J. Anim. Ecol.* **20,** 242-245.

BAGGERMAN, B. (1957). 'An experimental study of the timing of breeding and migration in the three-spined stickle-back (*Gasterosteus aculeatus* L.).' *Arch. néerl Zool.* **7,** 105-318.

BAILEY, N. T. J. (1959). *Statistical methods in biology.* London: English Univ. P. pp. x + 200.

BALDWIN, N. S. (1957). 'Food consumption and growth of Brook Trout at different temperatures.' *Trans. Amer. Fish. Soc.* **86,** 323-328.

BALFOUR-BROWNE, F. (1940). *British water beetles*, 1. London: Ray Soc. pp. xx + 375.

BALFOUR-BROWNE, F. (1953). 'The aquatic Coleoptera of the western Scottish islands with a discussion on their sources of origin and means of arrival.' *Ent. Gaz.* **4,** 79-127.

BARNES, H. (1958a). 'Regarding the southern limits of *Balanus balanoides* (L.).' *Oikos* **9,** 139-157.

BARNES, H. (1958b). 'Process of restoration and synchronization in marine ecology. The spring diatom increase and the "spawning" of the common barnacle, *Balanus balanoides* (L.).' *Un. int. Sci. biol.* **24,** 67-85.

BAR-ZEEV, M. (1958). 'The effect of temperature on the growth rate and survival of the immature stages of *Aëdes aegypti* (L.).' *Bull. ent. Res.* **49,** 157-163.

BATES, M. (1939). 'The use of salt solutions for the demonstration of physiological differences between the larvae of certain European anopheline mosquitoes.' *Amer. J. trop. Med.* **19,** 357-384.

BATES, M. (1940). 'Oviposition experiments with anopheline mosquitoes.' *Amer. J. trop. Med.* **20,** 569-583.

BATES, M. (1949). *The natural history of mosquitoes.* New York: Macmillan. pp. xvi + 379.

BEADLE, L. C. (1932). 'The waters of some East African Lakes in relation to their fauna and flora.' *J. Linn. Soc.* (*Zool.*) **38,** 157-211.

BEADLE, L. C. (1934). 'Osmotic regulation in *Gunda ulvae.*' *J. exp. Biol.* **11,** 382-396.

BEADLE, L. C. (1939). 'Regulation of the haemolymph in the saline water mosquito larva *Aëdes detritus* Edw.' *J. exp. Biol.* **16,** 346-362.

BEADLE, L. C. (1943). 'An ecological survey of some inland saline waters of Algeria.' *J. Linn. Soc.* (*Zool.*) **41,** 218-242.

BEADLE, L. C. (1957). 'Comparative physiology: osmotic and ionic regulation in aquatic animals.' *Annu. Rev. Physiol.* **19,** 329-358.

BEADLE, L. C. (1959). 'Osmotic and ionic regulation in relation to the classification of brackish and inland saline waters.' *Arch. Oceanogr. Limnol. Roma* **11,** 143-151.

BEADLE, L. C. and Shaw, J. (1950). 'The retention of salt and the regulation of the non-protein nitrogen fraction in the blood of the aquatic larva, *Sialis lutaria.*' *J. exp. Biol.* **27,** 96-109.

BEATTIE, M. V. F. (1930). 'Physico-chemical factors in relation to mosquito prevalence in ponds.' *J. Ecol.* **18,** 67-80.

BEAUCHAMP, R. S. A. (1932). 'Some ecological factors and their influence on competition between stream and lake-living triclads.' *J. Anim. Ecol.* **1,** 175-190.

BEAUCHAMP, R. S. A. (1933). 'Rheotaxis in *Planaria alpina*.' *J. exp. Biol.* **10,** 113-129.

BEAUCHAMP, R. S. A. (1935). 'The rate of movement of *Planaria alpina*.' *J. exp. Biol.* **12,** 271-285.

BEAUCHAMP, R. S. A. (1937). 'Rate of movement and rheotaxis in *Planaria alpina*.' *J. exp. Biol.* **14,** 104-116.

BEAUCHAMP, R. S. A. and Ullyott, P. (1932). 'Competitive relationships between certain species of fresh-water triclads.' *J. Ecol.* **20,** 200-208.

BEAUFORT, L. F. de (1954) ed. *Veranderingen in de flora en fauna van de Zuiderzee (thans Ijsselmeer) na de afsluiting in 1932.* Nederlandse Dierkundige Vereniging. pp. vii + 359.

BERG, K. (1938). 'Studies on the bottom animals of Esrom Lake.' *K. danske vidensk. Selsk. Skr.* **7,** pp. 255.

BERG, K. (1941). 'Contributions to the biology of the aquatic moth *Acentropus niveus* (Oliv.).' *Vidensk. Medd. dansk. naturh. Foren. Kbh.* **105,** 59-139.

BERG, K. (1943). 'Physiographical studies on the River Susaa.' *Folia limnol. scand.* **1,** pp. 174.

BERG, K. (1948). 'Biological studies on the River Susaa.' *Folia. limnol. scand.* **4,** pp. 318.

BERG, K. (1951). 'On the respiration of some molluscs from running and stagnant water.' *Ann. Biol.* **27,** 329-335.

BERG, K. (1952). 'On the oxygen consumption of Ancylidae (Gastropoda) from an ecological point of view.' *Hydrobiologia* **4,** 225-267.

BERG, K. (1953). 'The problem of respiratory acclimatization.' *Hydrobiologia* **5,** 331-350.

BERG, K., Jónasson, P. M., and Ockelmann, K. W. (1962). 'The respiration of some animals from the profundal zone of a lake.' *Hydrobiologia* **19,** 1-39.

BERG, K., Lumbye, J., and Ockelmann, K. W. (1958). 'Seasonal and experimental variations of the oxygen consumption of the limpet *Ancylus fluviatilis* (O. F. Müller).' *J. exp. Biol.* **35,** 43-73.

BERG, K. and Ockelmann, K. W. (1959). 'The respiration of fresh-water snails.' *J. exp. Biol.* **36,** 690-708.

BERG, K. and Petersen, I. C. (1956). 'Studies on the humic, acid Lake Gribsø.' *Folia limnol. scand.* **8,** pp. 273.

BLÄSING, I. (1953). 'Experimentelle Untersuchungen über den Umfang der ökologischen und physiologischen Toleranz von *Planaria alpina* Dana und *Planaria gonocephala* Dugès.' *Zool. Jb.* (Physiol.) **64,** 112-152.

BLUNCK, H. (1914). 'Die Entwicklung des *Dytiscus marginalis* L. vom Ei bis zur Imago. 1. Teil. Das Embryonalleben.' *Z. wiss. Zool.* **111,** 76-151.

BLUNCK, H. (1924). 'Die Entwicklung des *Dytiscus marginalis* L. vom Ei bis zur Imago. 2. Teil. Die Metamorphose (B. Das Larven und das Puppenleben)'. *Z. wiss. Zool.* **121,** 171-391.

BOISEN BENNIKE, S. A. (1943). 'Contributions to the ecology and biology of the Danish fresh-water leeches (Hirudinea).' *Folia limnol. scand.* **2,** 1-109.

BOND, R. M. (1935). 'Investigation of some Hispaniolan Lakes.' *Arch. Hydrobiol.* **28,** 137-161.

BONDESEN, P. and Kaiser, E. W. (1949). '*Hydrobia* (*Potamopyrgus*) *jenkinsi* Smith in Denmark illustrated by its ecology.' *Oikos* **1,** 252-281.

BOONE, E. and Baas-Becking, L. G. M. (1931). 'Salt effects on eggs and nauplii of *Artemia salina* L.' *J. gen. Physiol.* **14,** 753-763.

BOURRELLY, P. (1958). 'Algues microscopiques de quelques cuvettes supralittorales de la région de Dinard.' *Verh. int. Ver. Limnol.* **13,** 683-686.

BOVBJERG, R. V. (1952). 'Comparative ecology and physiology of the crayfish *Orconectes propinquus* and *Cambarus fodiens*.' *Physiol. Zool.* **25,** 34-56.

BOYCOTT, A. E. (1927). 'Oecological notes. X. Transplantation experiments on the habitats of *Planorbis corneus* and *Bithinia tentaculata*.' *Proc. malac. Soc. Lond.* **17,** 156-158.

BOYCOTT, A. E. (1936). 'The habitats of fresh-water Mollusca in Britain.' *J. Anim. Ecol.* **5,** 116-186.

BOYD, M. F. (1949), ed. *Malariology. A comprehensive survey of all aspects of this group of diseases from a global standpoint* (sixty-five contributors). Philadelphia and London: W. B. Saunders. 2 vols. pp. 1,643.

BRAND, T. VON (1944). 'Occurrence of anaerobiosis among invertebrates.' *Biodynamica* **4,** 185-328.

BRAND, T. VON (1945). 'The anaerobic metabolism of invertebrates.' *Biodynamica* **5,** 165-295.

BRETT, J. R. (1944). 'Some lethal temperature relations of Algonquin Park fishes.' *Publ. Ont. Fish. Res. Lab.* **63,** 1-49.

BRETT, J. R. (1956). 'Some principles in the thermal requirements of fishes.' *Quart. Rev. Biol.* **31,** 75-87.

BRIGGS, R., Dyke, G. V., and Knowles, G. (1958). 'Use of the wide-bore dropping-mercury electrode for long-period recording of concentration of dissolved oxygen.' *The Analyst,* **83,** 304-311.

BRINKHURST, R. O. (1959). 'The habitats and distribution of British *Gerris* and *Velia* species.' *J. Soc. Brit. Ent.* **6,** 37-44.

BRINKHURST, R. O. (1960). 'The British distribution of the water-bug *Velia saulii* Tamanini with some notes on alary polymorphism.' *Proc. R. ent. Soc. Lond. A.* **35,** 91-92.

BROWN, E. S. (1951). 'The relation between migration-rate and type of habitat in aquatic insects, with special reference to certain species of Corixidae.' *Proc. zool. Soc. Lond.* **121,** 539-545.

BROWN, E. S. (1954). 'Report on Corixidae (Hemiptera) taken in light-traps at Rothamsted Experimental Station.' *Proc. R. ent. Soc. Lond. A.* **29,** 17–22.

BROWN, L. A. (1927). 'Temperature coefficients for development in cladocerans.' *Proc. Soc. exp. Biol. N.Y.* **25,** 164-165.

BROWN, L. A. (1929). 'Natural history of cladocerans in relation to temperature.' *Amer. Nat.* **63,** 248-264, 346-352, 443-454.

BROWN, M. E. (1946). 'The growth of Brown Trout (*Salmo trutta* Linn.).' *J. exp. Biol.* **22,** 118-155.

BRUNDIN, L. (1951). 'The relation of O_2-microstratification at the mud surface to the ecology of the profundal bottom fauna.' *Rep. Inst. Freshw. Res. Drottning.* **32,** 32-42.

BRYAN, G. W. (1960). 'Sodium regulation in the crayfish *Astacus fluviatilis.* I. The normal animal.' *J. exp. Biol.* **37,** 83-99.

Bryan, G. W. (1960). 'Sodium regulation in the crayfish, *Astacus fluviatilis*. II. Experiments with sodium-depleted animals.' *J. exp. Biol.* **37,** 100-112.

Bullough, W. S. (1939). 'A study of the reproductive cycle of the minnow in relation to the environment.' *Proc. zool. Soc. Lond.* A.**109,** 79-102.

Buscemi, P. A. (1958). 'Littoral oxygen depletion produced by a cover of *Elodea canadensis*.' *Oikos* **9,** 239-245.

Butcher, R. W., Longwell, J., and Pentelow, F. T. K. (1937). 'Survey of the River Tees. III—The non-tidal reaches—chemical and biological.' *Tech. Pap. Wat. Pollut. Res. Lond.* **6,** pp. xiii + 189.

Butcher, R. W., Pentelow, F. T. K., and Woodley, J. W. A. (1930). 'Variations in composition of river waters.' *Int. Rev. Hydrobiol.* **24,** 47-80.

Buxton, P. A. and Hopkins, G. H. E. (1927). 'Researches in Polynesia and Melanesia.' *Mem. London. Sch. trop. Med. Hyg.* **1,** pp. 260.

Buxton, P. A. and Leeson, H. S. (1949). 'Anopheline mosquitoes: life history.' Chap. 12 in *Malariology*, ed. M. F. Boyd. London: W. B. Saunders. pp. 257-283.

Carpenter, K. E. (1926). 'The lead mine as an active agent in river pollution.' *Ann. appl. Biol.* **13,** 395-401.

Chimits, P. (1956). 'Le Brochet.' *Bull. franç. Piscic.* **28,** 81-96.

Christophers, S. R. (1960). *Aëdes aegypti (L.). The yellow fever mosquito.* Cambridge U.P. pp.xii + 739.

Claridge, M. F. and Staddon, B. W. (1961). '*Stenelmis canaliculata* Gyll. (Col., Elmidae): a species new to the British list.' *Ent. mon. Mag.* **96,** 141-144.

Claus, A. (1937). 'Vergleichend-physiologische Untersuchungen zur Ökologie der Wasserwanzen.' *Zool. Jb.* **58,** 365-432.

Clemens, H. P. (1950). 'Life cycle and ecology of *Gammarus fasciatus* Say.' *Contr. Stone Lab. Ohio Univ.* **12,** 1-63.

Coker, R. E. (1954). *Streams, lakes, ponds.* Univ. N. Carolina P. pp. xviii + 327.

Colless, D. H. (1956). 'The *Anopheles leucosphyrus* group.' *Trans. R. ent. Soc. Lond.* **108,** 37-116.

Corbet, P. S., Longfield, C., and Moore, N. W. (1960). *Dragonflies* (New Naturalist 41). London: Collins. pp. xii + 260.

COUÉGNAS, J. (1920). 'L'aire de distribution géographique des écrivisses de la région de Sussac (Haute-Vienne) et ses rapports avec les données géologiques.' *Arch. zool. exp. gén.* **59,** 11-13.

CREASER, C. W. (1930). 'Relative importance of hydrogen-ion concentration, temperature, dissolved oxygen, and carbon dioxide tensions on habitat selection by brook trout.' *Ecology* **10,** 246-262.

CRISP, D. T. (1962). 'Estimates of the annual production of *Corixa germari* (Fieb.) in an upland reservoir.' *Arch. Hydrobiol.* **58,** 210-223.

CRISP, D. T. and Heal, O. W. (1958). 'The Corixidae (O. Hemiptera), Gyrinidae (O. Coleoptera) and Cladocera (Subphylum Crustacea) of a bog in Western Ireland.' *Irish. Nat. J.* **12,** 1-14.

CROGHAN, P. C. (1958a). 'The survival of *Artemia salina* (L.) in various media.' *J. exp. Biol.* **35,** 213-218.

CROGHAN, P. C. (1958b). 'The osmotic and ionic regulation of *Artemia salina* (L.)'. *J. exp. Biol.* **35,** 219-233.

CROGHAN, P. C. (1958c). 'The mechanism of osmotic regulation in *Artemia salina* (L.): the physiology of the branchiae.' *J. exp. Biol.* **35,** 234-242.

CROGHAN, P. C. (1958d). 'The mechanism of osmotic regulation in *Artemia salina* (L.): the physiology of the gut.' *J. exp. Biol.* **35,** 243-249.

CROGHAN, P. C. (1958e). 'Ionic fluxes in *Artemia salina* (L.).' *J. exp. Biol.* **35,** 425-436.

CROW, E. L. and Gardner, R. S. (1959). 'Table of confidence limits for the expectation of a Poisson variable.' *Biometrika* **46,** 441-453.

DAHM, A. G. (1958). *Taxonomy and ecology of five species groups in the family Planariidae.* Malmo: Nya Litografen. pp. 241.

DALLINGER, W. H. (1887). 'The President's address.' *J. R. Micros. Soc.* 185-199.

DAVID, F. N. (1953). *A statistical primer.* London: Griffin. pp. xii + 226.

DAVIES, L. and Smith, C. D. (1958). 'The distribution and growth of *Prosimulium* larvae (Diptera: Simuliidae) in hill streams in northern England.' *J. Anim. Ecol.* **27,** 335-348.

DAVIS, N. C. (1932). 'The effects of heat and cold upon *Aëdes* (*Stegomyia*) *aegypti*.' *Amer. J. Hyg.* **16,** 177-191.

DECKSBACH, N. K. (1924). 'Seen und Flüsse des Turgai-Gebietes (Kirgisen-Steppen).' *Verh. int. Ver. Limnol.* **2,** 252-288.

DEGERBØL, M. and Krog, H. (1951). 'Den europaeiske Sumpskild-padde (*Emys orbicularis* L.) i Danmark.' *Danm. geol. Unders.* 2 (78), 5-130 (English summary 91-112).

DIVER, C. (1944). In 'Symposium on the interrelations of plants and insects; the place of both in the eco-system.' *Proc. R. ent. Soc. Lond. C.* **8,** 44-48.

DODDS, G. S. and Hisaw, F. L. (1924). 'Ecological studies of aquatic insects. 2. Size of respiratory organs in relation to environmental conditions.' *Ecology* **5,** 262-271.

DONALDSON, L. R. and Foster, F. J. (1941). 'Experimental study of the effect of various water temperatures on the growth, food utilization, and mortality rates of fingerling sockeye salmon.' *Trans. Amer. Fish. Soc.* **70,** 339-346.

DORIER, A. and Vaillant, F. (1954). 'Observations et expériences relatives à la resistance au courant de divers invertébrés aquatiques.' *Trav. Lab. Hydrobiol. Grenoble* **45** and **46,** 9-31.

DUFFIELD, J. E. (1933). 'Fluctuations in numbers among fresh-water crayfish, *Potamobius pallipes* Lereboullet.' *J. Anim. Ecol.* **2,** 184-195.

DUFFIELD, J. E. (1936). 'Fluctuations in numbers of Crayfish.' *J. Anim. Ecol.* **5,** 396.

DUNN, L. H. (1927a). 'Mosquito breeding in "test" water-containers.' *Bull. ent. Res.* **18,** 17-22.

DUNN, L. H. (1927b). 'Observations on the oviposition of *Aëdes aegypti* Linn., in relation to distance from habitations.' *Bull. ent. Res.* **18,** 145-148.

DUSSART, B. (1955). 'La température des lacs et ses causes de variations.' *Verh. int. Ver. Limnol.* **12,** 78-96.

DYMOND, J. R. (1955). 'The introduction of foreign fishes in Canada.' *Verh. int. Ver. Limnol.* **12,** 543-553.

EBELING, G. (1930). 'Fischereischädigungen durch *Thuya occidentalis.*' *Z. Fisch.* **28,** 433-453.

ECKEL, O. (1953). 'Zur Thermik der Fliesgewässer: Über die Aenderung der Wassertemperatur entlang des Flusslaufs.' *Wett. u. Leben.* **2,** 41-47.

ECKEL, O. and Reuter, H. (1950). 'Zur Berechnung des sommerlichen Wärmeumsatzes in Flussläufen.' *Geogr. Ann. Stockh.* **32,** 188-209.

EDMONDSON, W. T. (1944). 'Ecological studies of sessile Rotatoria. 1. Factors affecting distribution.' *Ecol. Monogr.* **14,** 31-66.

EHRENBERG, H. (1957). 'Die Steinfauna der Brandungsufer ostholsteinischer Seen.' *Arch. Hydrobiol.* **53,** 87-159.

EINSELE, W. (1961). 'Fischereiwissenschaft.' *Verh. int. Ver. Limnol.* **14,** 806-819.

ELLIS, A. E. (1951). 'Census of the distribution of British non-marine Mollusca.' *J. Conch.* **23,** 171-243.

ELLIS, W. G. (1933). 'Calcium and the resistance of *Nereis* to brackish water.' *Nature, Lond.* **132,** 748.

ELSON, P. F. (1942). 'Effect of temperature on activity of *Salvelinus fontinalis*.' *J. Fish. Res. Bd Can.* **5,** 461-470.

ELTON, C. (1927). *Animal ecology.* London: Sidgwick and Jackson. pp. xxi + 207.

ELTON, C. (1946). 'Competition and the structure of ecological communities.' *J. Anim. Ecol.* **15,** 54-68.

ELTON, C. S. and Miller, R. S. (1954). 'The ecological survey of animal communities: with a practical system of classifying habitats by structural characters.' *J. Ecol.* **42,** 460-496.

ENGELHARDT, W. (1951). 'Faunistisch—ökologische Untersuchungen über Wasserinsekten an den südlichen Zuflüssen des Ammersees.' *Mitt. Münchner. ent. Ges.* **41,** 1-135.

EWER, R. F. (1942). 'On the function of haemoglobin in *Chironomus*.' *J. exp. Biol.* **18,** 197-205.

EYLES, D. E. (1944). 'A critical review of the literature relating to the flight and dispersion habits of anopheline mosquitoes.' *Publ. Hlth. Bull., Wash.* **287,** 1-39.

FABRICIUS, E. (1950). 'Heterogeneous stimulus summation in the release of spawning activities in fish.' *Rep. Inst. Freshw. Res. Drottning.* **31,** 57-99.

FAURÉ-FREMIET, E. (1950). 'Ecology of ciliate infusoria.' *Endeavour* **9,** 183-187.

FERGUSON, R. G. (1958). 'The preferred temperature of fish and their midsummer distribution in temperate lakes and streams.' *J. Fish. Res. Bd Can.* **15,** 607-624.

FERNANDO, C. H. (1958). 'The colonization of small freshwater habitats by aquatic insects. 1. General discussion, methods and colonization in the aquatic Coleoptera.' *Ceylon J. Sci.* **1,** 117-154.

FERNANDO, C. H. (1959). 'The colonization of small freshwater habitats by aquatic insects. 2. Hemiptera (the water-bugs).' *Ceylon J. Sci.* **2,** 5-32.

FISH, F. F. (1948). 'The return of blueback salmon to the Columbia River.' *Scientific Monthly* **66,** 283-292.

FISHER, K. C. and Elson, P. F. (1950). 'The selected temperature of Atlantic salmon and speckled trout and the effect of temperature on the response to an electrical stimulus.' *Physiol. Zool.* **23,** 27-34.

FORSMAN, B. (1952). 'Studies on *Gammarus duebeni* Lillj. with notes on some rock pool organisms in Sweden.' *Zool. Bidrag. Uppsala* **29,** 215-237.

FOX, H. M. (1921). 'An investigation into the cause of the spontaneous aggregation of flagellates and into the reactions of flagellates to dissolved oxygen.' *J. gen. Physiol.* **3,** 483-511.

FOX, H. M. (1948). 'The haemoglobin of *Daphnia.*' *Proc. roy. Soc. B.* **135,** 195-212.

FOX, H. M. (1955). 'The effect of oxygen on the concentration of haem in invertebrates.' *Proc. roy. Soc. B.* **143,** 203-214.

FOX, H. M., Gilchrist, B. M., and Phear, E. A. (1951). 'Functions of haemoglobin in *Daphnia.*' *Proc. roy. Soc. B.* **138,** 514-528.

FOX, H. M. and Phear, E. A. (1953). 'Factors influencing haemoglobin synthesis by *Daphnia.*' *Proc. roy. Soc. B.* **141,** 179-189.

FOX, H. M. and Ramage, H. (1931). 'A spectographic analysis of animal tissues.' *Proc. roy. Soc. B.* **108,** 157-173.

FOX, H. M. and Simmonds, B. G. (1933). 'Metabolic rates of aquatic arthropods from different habitats.' *J. exp. Biol.* **10,** 67-74.

FOX, H. M., Simmonds, B. G., and Washbourn, R. (1935). 'Metabolic rates of ephemerid nymphs from swiftly flowing and from still waters.' *J. exp. Biol.* **12,** 179-184.

FOX, H. M. and Taylor, A. E. R. (1955). 'The tolerance of oxygen by aquatic invertebrates.' *Proc. roy. Soc. B.* **143,** 214-225.

FOX, H. M., Wingfield, C. A., and Simmonds, B. G. (1937). 'The oxygen consumption of ephemerid nymphs from flowing and from still waters in relation to the concentration of oxygen in the water.' *J. exp. Biol.* **14,** 210-218.

FRASER, F. C. (1952). 'Methods of exophytic oviposition in Odonata.' *Ent. mon. Mag.* **88,** 261-262.

FREY, D. G. and Fry, F. E. J. (1953). *Fundamentals of limnology* (translation of Ruttner, F., *Grundriss der Limnologie*). Toronto U.P. pp. xii + 242.

FRÖMMING, E. (1936). 'Über den Einfluss der Wasserstoffionenkonzentration auf unsere Süsswasserschnecken.' *Int. Rev. Hydrobiol.* **33,** 25-37.

FRÖMMING, E. (1938). 'Untersuchungen über den Einfluss der Härte des Wohngewässers auf das Vorkommen unserer Süsswassermollusken.' *Int. Rev. Hydrobiol.* **36,** 531-561.

FROST, W. E. and Macan, T. T. (1948). 'Corixidae (Hemiptera) as food for fish.' *J. Anim. Ecol.* **17,** 174-179.

FRY, F. E. J. (1947). 'Effects of the environment on animal activity.' *Publ. Ont. Fish. Res. Lab.* **68,** 1-62.

FRY, F. E. J. (1951). 'Some environmental relations of the speckled trout (*Salvelinus fontinalis*).' *Proc. N.E. Atlantic Fisheries Conference* (cyclostyled).

FRY, F. E. J. (1957). 'The aquatic respiration of fish.' Chap. 1. 1. in *The physiology of fishes.* Vol. 1, ed. M. E. Brown. New York: Academic Press.

FRY, F. E. J., Brett, J. R., and Clawson, G. H. (1942). 'Lethal limits of temperature for young goldfish.' *Rev. canad. Biol.* **1,** 50-56.

FRY, F. E. J. and Hart, J. S. (1948). 'The relation of temperature to oxygen consumption in the goldfish.' *Biol. Bull., Wood's Hole* **94,** 66-77.

FRY, F. E. J., Hart, J. S., and Walker, K. F. (1946). 'Lethal temperature relations for a sample of young speckled trout, *Salvelinus fontinalis*.' *Publ. Ont. Fish. Res. Lab.* **66,** 1-35.

FRYER, G. (1959a). 'Some aspects of evolution in Lake Nyasa.' *Evolution* **13,** 440-451.

FRYER, G. (1959b). 'The trophic interrelationships and ecology of some littoral communities of Lake Nyasa with especial reference to the fishes, and a discussion of the evolution of a group of rock-frequenting Cichlidae.' *Proc. zool. Soc. Lond.* **132,** 153-281.

GAARDER, T. (1933). 'Untersuchungen über Produktions-und Lebensbedingungen in norwegischen Austern-Pollen.' *Bergens Mus. Aarb.* 1932 **3,** 1-64.

GAJEVSKAJA, N. S. (1958). 'Le rôle de groupes principaux de la flore aquatique dans les cycles trophiques des différents bassins d'eau douce.' *Verh. int. Ver. Limnol.* **13,** 350-362.

GAMESON, A. L. H. (1957). 'Weirs and the aeration of rivers.' *J. Inst. Water Engineers* **11,** 477-490.

GAMESON, A. L. H. and Griffith, S. D. (1959). 'Six months' oxygen records for a polluted stream.' *Water and Waste Treatment J.* Jan./Feb. 1959.

GARDNER, A. E. (1950a). 'The life-history of *Sympetrum sanguineum* Müller (Odonata).' *Ent. Gaz.* **1,** 21-26.

GARDNER, A. E. (1950b). 'The life-history of *Aeshna mixta* Latreille (Odonata).' *Ent. Gaz.* **1,** 128-138.

GARDNER, A. E. (1953). 'Further notes on exophytic oviposition in Odonata.' *Ent. mon. Mag.* **89,** 98-99.

GARSIDE, E. T. and Tait, J. S. (1958). 'Preferred temperature of rainbow trout (*Salmo gairdneri* Richardson) and its unusual relationship to acclimation temperature.' *Canad. J. Zool.* **36,** 563-567.

GEIJSKES, D. C. (1935). 'Faunistisch-ökologische Untersuchungen am Röserenbach bei Liestal im Basler Tafeljura.' *Tijdschr. Ent.* **78,** 249-382.

GELDIAY, R. (1956). 'Studies on local populations of the freshwater limpet *Ancylus fluviatilis* Müller.' *J. Anim. Ecol.* **25,** 389-402.

GESSNER, F. (1955). 'Die limnologischen Verhältnisse in den Seen und Flüssen von Venezuela.' *Verh. int. Ver. Limnol.* **12,** 284-295.

GESSNER, F. (1957). 'Van Gölü. Zur Limnologie des grossen Soda-Sees in Ostanatolien (Türkei).' *Arch. Hydrobiol.* **53,** 1-22.

GIBSON, M. B. (1954). 'Upper lethal temperature relations of the guppy, *Lebistes reticulatus*.' *Canad. J. Zool.* **32,** 393-407.

GIBSON, E. S. and Fry, F. E. J. (1954). 'The performance of the lake trout, *Salvelinus namaycush*, at various levels of temperature and oxygen pressure.' *Canad. J. Zool.* **32,** 252-260.

GIBSON, M. B. and Hirst, B. (1955). 'The effect of salinity and temperature on the pre-adult growth of guppies.' *Copeia* **1955,** 241-243.

GILBERT, O., Reynoldson, T. B., and Hobart, J. (1952). 'Gause's hypothesis: an examination.' *J. Anim. Ecol.* **21,** 310-312.

GILCHRIST, B. M. (1954). 'Haemoglobin in *Artemia.*' *Proc. roy. Soc. B.* **143,** 136-146.

GILLETT, J. D. (1955a). 'Variation in the hatching-response of *Aëdes* eggs (Diptera: Culicidae).' *Bull. ent. Res.* **46,** 241-254.

GILLETT, J. D. (1955b). 'The inherited basis of variation in the hatching-response of *Aëdes* eggs (Diptera: Culicidae).' *Bull. ent. Res.* **46,** 255-265.

GILLIES, M. T. (1950). 'Egg-laying of olives.' *Salm. Trout Mag.* **129,** 106-108.

GISLÉN, T. (1948). 'Aerial plankton and its conditions of life.' *Biol. Rev.* **23,** 109-126.

GLEDHILL, T. (1960). 'The Ephemeroptera, Plecoptera and Trichoptera caught by emergence traps in two streams during 1958.' *Hydrobiologia* **15,** 179-188.

GODWIN, H. (1923). 'Dispersal of pond floras.' *J. Ecol.* **11,** 160-164.

GORHAM, E. (1958). 'The physical limnology of northern Britain: an epitome of The Bathymetrical Survey of the Scottish Freshwater Lochs, 1897-1909.' *Limnol. Oceanogr.* **3,** 40-50.

GRAHAM, J. M. (1949). 'Some effects of temperature and oxygen pressure on the metabolism and activity of the speckled trout, *Salvelinus fontinalis.*' *Canad. J. Res. D.* **27,** 270-288.

GREENBANK, J. (1945). 'Limnological conditions in ice-covered lakes, especially as related to winter-kill of fish.' *Ecol. Monogr.* **15,** 343-392.

GRENIER, P. (1949). 'Contribution à l'étude biologique des simuliides de France.' *Physiol. comp.* **1,** 165-330.

HALL, R. E. (1953). 'Observations on the hatching of eggs of *Chirocephalus diaphanus* Prévost.' *Proc. zool. Soc. Lond.* **123,** 95-109.

HALL, R. E. (1959). 'The development of eggs of *Chirocephalus diaphanus* Prévost in relation to depth of water.' *Hydrobiologia* **14,** 79-84.

HALSBAND, E. (1953). 'Untersuchungen über das Verhalten von Forelle (*Trutta iridea* W. Gibb.) und Döbel (*Squalius cephalus* Heck.) bei Einwirkung verschiedener Aussenfaktoren.' *Z. Fisch.* **2,** 227-270.

HAMILTON, J. D. (1958). 'On the biology of *Holopedium gibberum* Zaddach (Crustacea: Cladocera).' *Verh. int. Ver. Limnol.* **13,** 785-788.

HARDIN, G. (1960). 'The competitive exclusion principle.' *Science* **131,** 1292-1297.

HARDY, A. C. and Milne, P. S. (1937). 'Insect drift over the North Sea.' *Nature, Lond.* **139,** 510.

HARDY, A. C. and Milne, P. S. (1938). 'Studies in the distribution of insects by aerial currents.' *J. Anim. Ecol.* **7,** 199-229.

HARKER, J. E. (1953). 'An investigation of the distribution of the mayfly fauna of a Lancashire stream.' *J. Anim. Ecol.* **22,** 1-13.

HARNISCH, O. (1951). 'Hydrophysiologie der Tiere.' *Die Binnengewässer* **19,** pp. viii + 299.

HARNISCH, O. (1958). 'Leben ohne Sauerstoff namentlich bei euryoxybionten Süsswassertieren.' *Gewässer und Abwässer* **20,** 7-12.

HARRIS, J. R. (1952). *An angler's entomology.* London: Collins. pp. xvi + 268.

HART, J. S. (1947). 'Lethal temperature relations of certain fish of the Toronto region.' *Trans. roy. Soc. Can.* **41,** 57-71.

HECHT, O. (1931). 'Ueber den Wärmesinn der Stechmücken bei der Eiablage.' *Riv. Malariol.* **9,** 706-724.

HECHT, O. (1934). 'Experimentelle Beiträge zur Biologie der Stechmücken V. Über den Wärmesinn der *Anopheles maculipennis* Rassen bei der Eiablage.' *Arch. Schiffs- u. Tropenhyg.* **38,** 124-131.

HEDGPETH, J. W. (1957). 'Estuaries and Lagoons II. Biological aspects.' *Mem. geol. Soc. Amer.* **67,** (1), 693-729.

HEDGPETH, J. W. (1959). 'Some preliminary considerations of the biology of inland mineral waters.' *Arch. Oceanogr. Limnol. Roma* suppl. **11,** 111-141.

HENDELBERG, J. (1960). 'The fresh-water pearl mussel, *Margaritifera margaritifera* (L).' *Rep. Inst. Freshw. Res. Drottning.* **41,** 149-171.

HERBERT, M. R. (1954). 'The tolerance of oxygen deficiency in the water by certain Cladocera.' *Mem. Ist. Ital. Idrobiol.* **8,** 99-107.

HERS, M. J. (1942). 'Anaerobiose et régulation minérale chez les larves de *Chironomus*.' *Ann. Soc. zool. Belg.* **73**, 173-179.

HETHERINGTON, A. (1932). 'The constant culture of *Stentor coeruleus*.' *Arch. Protistenk.* **76**, 118-129.

HINTON, H. E. (1954). 'Resistance of the dry eggs of *Artemia salina* L. to high temperatures.' *Ann. Mag. nat. Hist.* **7**, 158-160.

HOESTLANDT, H. (1959). 'Répartition actuelle du crabe chinois.' *Bull. franç. Piscic.* **194**, 1-13.

HOLMQUIST, C. (1959). *Problems on marine-glacial relicts on account of investigations of the genus* Mysis. Lund: Berling. pp. 270.

HOWLAND, L. J. (1930). 'Bionomical investigation of English mosquito larvae with special reference to their algal food.' *J. Ecol.* **18**, 81-125.

HUBAULT, E. (1927). 'Contribution a l'étude des invertébrés torrenticoles.' *Bull. biol.* suppl. **9**, 1-388.

HUBAULT, E. (1955). 'Introduction of species of fish into lakes in the eastern part of France.' *Verh. int. Ver. Limnol.* **12**, 515-519.

HUBENDICK, B. (1947). 'Die Verbreitungsverhältnisse der limnischen Gastropoden in Südschweden.' *Zool. Bidr. Uppsala* **24**, 419-559.

HUDSON, B. N. A. (1956). 'The behaviour of the female mosquito in selecting water for oviposition.' *J. exp. Biol.* **33**, 478-492.

HUET, M. (1942). 'Esquisse hydrobiologique des eaux piscicoles de la Haute-Belgique.' *Trav. Sta. Rech. Groenendael.* D **2**, 1-40.

HUFFAKER, C. B. (1944). 'The temperature relations of the immature stages of the malarial mosquito, *Anopheles quadrimaculatus* Say, with a comparison of the development power of constant and variable temperatures in insect metabolism.' *Ann. Soc. Ent. Amer.* **37**, 1-27.

HUNTER, W. R. (1953a). 'The condition of the mantle cavity in two pulmonate snails living in Loch Lomond.' *Proc. roy. Soc. Edinb.* B **65**, 143-165.

HUNTER, W. R. (1953b). 'On the migrations of *Lymnaea peregra* (Müller) on the shores of Loch Lomond.' *Proc. roy. Soc. Edinb.* B **65**, 84-105.

HUNTER, W. R. 1957. 'Studies on freshwater snails at Loch Lomond.' *Glasg. Univ. Publ., Stud. Loch Lomond* **1**, 56-95.

HUNTER, W. R. and Warwick, T. (1957). 'Records of *Potamopyrgus jenkinsi* (Smith) in Scottish fresh waters over fifty years (1906-56).' *Proc. roy. Soc. Edinb.* B.**66,** 360-373.

HUNTSMAN, A. G. (1946). 'Heat stroke in Canadian maritime stream fishes.' *J. Fish. Res. Bd Can.* **6,** 476-482.

HURLBUT, H. S. (1943). 'The rate of growth of *Anopheles quadrimaculatus* in relation to temperature.' *J. Parasit.* **29,** 107-113.

HUTCHINSON, G. E. (1932). 'Experimental studies in ecology. 1. The magnesium tolerance of Daphniidae and its ecological significance.' *Int. Rev. Hydrob.* **28,** 90-108.

HUTCHINSON, G. E. (1937). 'A contribution to the limnology of arid regions.' *Trans. Conn. Acad. Arts. Sci.* **33,** 49-132.

HUTCHINSON, G. E. (1937). 'Limnological studies in Indian Tibet.' *Int. Rev. Hydrob.* **35,** 134-176.

HUTCHINSON, G. E. (1941). 'Ecological aspects of succession in natural populations.' *Amer. Nat.* **75,** 406-418.

HUTCHINSON, G. E. (1957a). *A treatise on limnology.* 1. *Geography, physics, and chemistry.* New York: John Wiley. pp xiv + 1015.

HUTCHINSON, G. E. (1957b). 'Concluding remarks.' *Cold Spring Harbour Symposium on Quantitative Biology* **22,** 415-427.

HUTCHINSON, G. E. (1959). 'Homage to Santa Rosalia or why are there so many kinds of animals?' *Amer. Nat.* **93,** 145-159.

HYNES, H. B. N. (1941). 'The taxonomy and ecology of the nymphs of British Plecoptera with notes on the adults and eggs.' *Trans. R. ent. Soc. Lond.* **91,** 459-557.

HYNES, H. B. N. (1952). 'The Plecoptera of the Isle of Man.' *Proc. R. ent. Soc. Lond.* A. **27,** 71-76.

HYNES, H. B. N. (1954). 'The ecology of *Gammarus duebeni* Lilljeborg and its occurrence in fresh water in western Britain.' *J. Anim. Ecol.* **23,** 38-84.

HYNES, H. B. N. (1955a). 'Distribution of some freshwater Amphipoda in Britain.' *Verh. int. Ver. Limnol.* **12,** 620-628.

HYNES, H. B. N. (1955b). 'The reproductive cycle of some British freshwater Gammaridae.' *J. Anim. Ecol.* **24,** 352-387.

HYNES, H. B. N. (1958a). 'A key to the adults and nymphs of British stoneflies (Plecoptera).' *Sci. Pub. freshwat. biol. Assoc.* **17.** p. 87.

HYNES, H. B. N. (1958b). 'The effect of drought on the fauna of a small mountain stream in Wales.' *Verh. int. Ver. Limnol.* **13,** 826-833.

HYNES, H. B. N. (1960). *The biology of polluted waters.* Liverpool U.P. pp. xiv + 202.

HYNES, H. B. N. (1961a). 'The effect of water-level fluctuations on littoral fauna.' *Verh. int. Ver. Limnol.* **14,** 652-656.

HYNES, H. B. N. (1961b). 'The invertebrate fauna of a Welsh mountain stream.' *Arch. Hydrobiol.* **57,** 344-388.

ILLIES, J. (1952a). 'Die Mölle. Faunistisch-ökologische Untersuchungen an einem Forellenbach im Lipper Bergland.' *Arch. Hydrobiol.* **46,** 424-612.

ILLIES, J. (1952b). 'Die Plecopteren und das Monardsche Prinzip.' *Ber. Limnol. Flussst. Freudenthal.* **3,** 53-69.

ILLIES, J. (1953). 'Beitrag zur Verbreitungsgeschichte der europäischen Plecopteren.' *Arch. Hydrobiol.* **48,** 35-74.

ILLIES, J. (1955a). 'Steinfleigen oder Plecoptera' in *Die Tierwelt Deutschlands.* Jena: Fischer. 43 Teil, pp. 150

ILLIES, J. (1955b). 'Die Bedeutung der Plecopteren für die Verbreitungsgeschichte der Süsswasserorganismen.' *Verh. int. Ver. Limnol.* **12,** 643-653.

ILLIES, J. (1956). 'Seeausfluss—Biozönosen lappländischer Waldbäche.' *Ent. Tidskr.* **77,** 138-153.

ILLIES, J. (1959). 'Retardierte Schlupfzeit von *Baetis*-Gelegen (Ins., Ephem.).' *Naturwissenchaften* **46,** 119-120.

JACKSON, D. J. (1952). 'Observations on the capacity for flight of water beetles.' *Proc. R. ent. Soc. Lond.* A. **27,** 57-70.

JACKSON, D. J. (1956a). 'Observations on flying and flightless water beetles.' *J. Linn. Soc.* (*Zool.*) **43,** 18-42.

JACKSON, D. J. (1956b). 'The capacity for flight of certain water beetles and its bearing on their origin in the western Scottish Isles.' *Proc. Linn. Soc. Lond.* **167,** 76-96.

JACOBI, E. F. and Baas-Becking, L. G. M. (1933). 'Salt antagonism and effect of concentration in nauplii of *Artemia salina* L.' *Tijdschr. ned. dierk. Ver.* **3,** 145-153.

JENKIN, P. M. (1932 and 1936). 'Reports on the Percy Sladen Expedition to some Rift Valley Lakes in Kenya in 1929. I and VII.' *Ann. Mag. nat. Hist.* **9,** 533-553, **18,** 133-181.

JENKIN, P. M. (1942). 'Seasonal changes in the temperature of Windermere (English Lake District).' *J. Anim. Ecol.* **11,** 248-269.

JEWELL, M. E. (1935). 'An ecological study of the fresh-water sponges of north-eastern Wisconsin.' *Ecol. Monogr.* **5,** 461-504.

JEWELL, M. E. (1939). 'An ecological study of the fresh-water sponges of Wisconsin, II. The influence of calcium.' *Ecology* **20,** 11-28.

JOBLING, B. (1935). 'The effect of light and darkness on oviposition in mosquitoes.' *Trans. R. Soc. trop. Med. Hyg.* **29,** 157-166.

JÔNASSON, P. M. and Mathiesen, H. (1959). 'Measurements of primary production in two Danish eutrophic lakes, Esrom Sø and Furesø.' *Oikos* **10,** 137-167.

JONES, J. R. E. (1950). 'A further ecological study of the River Rheidol: the food of the common insects of the main stream.' *J. Anim. Ecol.* **19,** 159-174.

JONES, J. R. E. (1952). 'The reactions of fish to water of low oxygen concentration'. *J. exp. Biol.* **29,** 403-415.

JONES, J. R. E. (1958). 'A further study of the zinc-polluted River Ystwyth.' *J. Anim. Ecol.* **27,** 1-14.

KALLEBERG, H. (1958). 'Observations in a stream tank of territoriality and competition in juvenile salmon and trout (*Salmo salar* L. and *S. trutta* L.).' *Rep. Inst. Freshw. Res. Drottning.* **39,** 55-98.

KIMMINS, D. E. (1943). 'A list of the Trichoptera (caddis-flies) of the Lake District with distributional and seasonal data.' *J. Soc. Brit. Ent.* **2,** 136-157.

KIMMINS, D. E. (1944). 'Supplementary notes on the Trichoptera of the English Lake District.' *Entomologist* **77,** 81-84.

KINNE, O. (1953). 'Zur Biologie und Physiologie von *Gammarus duebeni* Lillj., II. Über die Häutungsfrequenz, ihre Abhängigkeit von Temperatur und Salzgehalt, sowie über ihr Verhalten bei isoliert gehaltenen und amputieren Versuchstieren.' *Zool. Jb.* (*Abt. Zool. Phys.*) **64,** 183-206.

KINNE, O. (1954). 'Zur Biologie und Physiologie von *Gammarus duebeni* Lillj.' *Z. wiss, Zool.* **157,** 427-491.

KITCHING, J. A. (1934). 'The physiology of contractile vacuoles. I. Osmotic relations.' *J. exp. Biol.* **11,** 364-381.

KITCHING, J. A. (1938a). 'The physiology of contractile vacuoles III. The water balance of fresh-water Peritricha.' *J. exp. Biol.* **15,** 143-151.

KITCHING, J. A. (1938b). 'Contractile vacuoles.' *Biol. Rev.* **13,** 403-444.

KITCHING, J. A. (1939). 'The physiology of contractile vacuoles IV. A note on the sources of the water evacuated, and on the function of contractile vacuoles in marine Protozoa.' *J. exp. Biol.* **16,** 34-37.

KITCHING, J. A. (1948). 'The physiology of contractile vacuoles V, VI.' *J. exp. Biol.* **25,** 406-436.

KITCHING, J. A. (1951). 'The physiology of contractile vacuoles. VII. The osmotic relations in a suctorian, with special reference to the mechanism of control of vacuolar output.' *J. exp. Biol.* **28,** 203-214.

KITCHING, J. A., Sloane, J. F., and Ebling, F. J. (1959). 'The ecology of Lough Ine VIII. Mussels and their predators.' *J. Anim. Ecol.* **28,** 331-341.

KLIGLER, I. J. and Theodor, O. (1925). 'Effect of salt concentration and reaction on the development of *Anopheles* larvae.' *Bull. Ent. Res.* **16,** 45-49.

KLIGLER, I. J. and Mer, G. (1930). 'Studies on Malaria VI. Long-range dispersion of *Anopheles* during the prehibernating period.' *Rapp. Cong. int. Paludisme. Alger.* **2,** 1-14.

KORRINGA, P. (1957). 'Water temperature and breeding throughout the geographical range of *Ostrea edulis.*' *Un. int. Sci. biol.* **24,** 1-17.

KRAMER, S. D. (1915). 'The effect of temperature on the life cycle of *Musca domestica* and *Culex pipiens.*' *Science,* **41,** 874-877.

KREUZER, R. (1940). 'Limnologisch-ökologische Untersuchungen an holsteinischen Kleingewässern.' *Arch. Hydrobiol.* **10.** suppl., 359-572.

KROGH, A. (1939). *Osmotic regulation in aquatic animals.* Cambridge U.P. pp. viii + 242.

LACK, D. (1933). 'Habitat selection in birds with special reference to the effects of afforestation on the Breckland avifauna.' *J. Anim. Ecol.* **2,** 239-262.

LACK, D. (1954). *The natural regulation of animal numbers.* Oxford U.P. pp. viii + 343.

LAMBORN, W. A. (1922). 'Some problems of the breeding-places of the anophelines of Malaya; a contribution towards their solution.' *Bull. ent. Res.* **13,** 1-23.

LANSBURY, I. (1955). 'Some notes on invertebrates other than Insecta found attached to water-bugs (Hemipt.-Heteroptera).' *Entomologist* **88,** 139-140.

LANSBURY, I. (1961). '*Gerris rufoscutellatus* (Latreille), (Hem. Het. Gerridae) new to Co. Clare, Eire.' *Entomologist* **94,** 149-150.

LAURIE, E. M. O. (1942). 'The dissolved oxygen of an upland pond and its inflowing stream, at Ystumtuen, North Cardiganshire, Wales.' *J. Ecol.* **30,** 357-382.

LAURIE, R. D. and Jones, J. R. E. (1938). 'The faunistic recovery of a lead-polluted river in North Cardiganshire, Wales.' *J. Anim. Ecol.* **7,** 272-289.

LE CREN, E. D. (1955). 'Year to year variation in the year-class strength of *Perca fluviatilis*.' *Verh. int. Ver. Limnol.* **12,** 187-192.

LE CREN, E. D. (1962). 'The efficiency of reproduction and recruitment in freshwater fish' in *The Exploitation of natural animal populations*, ed. E. D. Le Cren and M. W. Holdgate. Oxford: Blackwell, pp. 283-296.

LEDEBUR, VON, J. F. (1939). 'Der Sauerstoff als ökologischer Faktor.' *Ergebn. Biol.* **16.**

LEFÉVRE, M., Hedwig, J., and Nisbet, M. (1951). 'Compatibilités et antagonismes entre algues d'eau douce dans les collections d'eau naturelles.' *Verh. int. Ver. Limnol.* **11,** 224-229.

LEFÉVRE, M., Jakob, H., and Nisbet, M. (1952). 'Auto. et hétéro-antagonisme chez les algues d'eau douce.' *Ann. Sta. cent. Hydrobiol. appl.* **4,** 5-198.

LEITCH, I. (1916). 'The function of haemoglobin in invertebrates with special reference to *Planorbis* and *Chironomus* larvae.' *J. Physiol.* **50,** 370-379.

LELOUP, E. (1944). 'Recherches sur les triclades dulcicoles épigés de la forêt de Soignes.' *Mém. Mus. roy. Hist. nat. Belg.* **102,** 1-112.

LESTON, D. (1956). 'The status of the pondskater *Limnoporus rufoscutellatus* (Latr.) (Hem. Gerridae) in Britain.' *Ent. mon. Mag.* **92,** 189-193.

LINDBERG, H. (1937). 'Ökologische Studien über die Coleopteren-und Hemipterenfauna im Meere in der Pojo—Wiek und im Schärenarchipel von Ekenäs in Südfinnland.' *Acta. Soc. Fauna Flor. fenn.* **60,** 516-572.

LINDBERG, H. (1944). 'Ökologisch-geographische Untersuchungen zur Insektenfauna der Felsentümpel an den Küsten Finnlands.' *Acta. zool. fenn.* **41,** 1-178.

LINDBERG, H. (1948). 'Zur Kenntnis der Insektenfauna im Brackwasser des Baltischen Meeres.' *Comment. biol. Helsingf.* **10,** 1-206.

LINDEMAN, R. L. (1942). 'Trophic-dynamic aspect of ecology.' *Ecology* **23,** 399-418.

LINDROTH, A. (1941). 'Sauerstoffverbrauch der Fische bei verscheidenem Sauerstoffdruck und verschiedenem Sauerstoffbedarf.' *Z. vergl. Physiol.* **28,** 142-152.

LINDROTH, A. (1942). 'Sauerstoffverbrauch der Fische. II. Verschiedene Entwicklungs- und Altersstadien vom Lachs und Hecht.' *Z. vergl. Physiol.* **29,** 583-594.

LINDUSKA, J. P. (1942). 'Bottom type as a factor influencing the local distribution of mayfly nymphs.' *Canad. Ent.* **74,** 26-30.

LLOYD, LL. (1937). 'Observations on sewage flies: their seasonal incidence and abundance.' *J. and Proc. Inst. Sewage Purification,* pp. 2-16.

LLOYD, LL., Graham, J. F., and Reynoldson, T. B. (1940). 'Materials for a study in animal competition. The fauna of sewage bacteria beds.' *Ann. appl. Biol.* **27,** 122-150.

LLOYD, R. (1960). 'The toxicity of zinc sulphate to rainbow trout.' *Ann. appl. Biol.* **48,** 84-94.

LOCKWOOD, A. P. M. (1959a). 'The osmotic and ionic regulation of *Asellus aquaticus* (L.).' *J. exp. Biol.* **36,** 546-555.

LOCKWOOD, A. P. M. (1959b). 'The regulation of the internal sodium concentration of *Asellus aquaticus* in the absence of sodium chloride in the medium.' *J. exp. Biol.* **36,** 556-561.

LOCKWOOD, A. P. M. (1959c). 'The extra-haemolymph sodium of *Asellus aquaticus* (L.).' *J. exp. Biol.* **36,** 562-565.

LONGFIELD, C. (1949). *The dragonflies of the British Isles.* London: Warne. pp. 256.

Lumbye, J. (1958). 'The oxygen consumption of *Theodoxus fluviatilis* (L.) and *Potamopyrgus jenkinsi* (Smith) in brackish and fresh water.' *Hydrobiologia* **10,** 245-262.

Lund, J. W. G. (1950). 'Studies on *Asterionella formosa* Hass. II. Nutrient depletion and the spring maximum.' *J. Ecol.* **38,** 1-35.

Lundblad, O. (1921). 'Vergleichende Studien über die Nahrungsaufnahme einiger schwedischer Phyllopoden, nebst synonymischen, morphologischen und biologischen Bemerkungen.' *Ark. Zool.* **13,** 1-114.

Macan, T. T. (1938). 'Evolution of aquatic habitats with special reference to the distribution of Corixidae.' *J. Anim. Ecol.* **7,** 1-19.

Macan, T. T. (1939). 'The Culicidae of the Cambridge District.' *Parasitology* **31,** 263-269.

Macan, T. T. (1940). 'Dytiscidae and Haliplidae (Col.) in the Lake District.' *Trans. Soc. Brit. Ent.* **7,** 1-20.

Macan, T. T. (1949). 'Survey of a moorland fishpond.' *J. Anim. Ecol.* **18,** 160-186.

Macan, T. T. (1950a). 'Ecology of fresh-water Mollusca in the English Lake District.' *J. Anim. Ecol.* **19,** 124-146.

Macan, T. T. (1950b). 'The anopheline mosquitoes of Iraq and North Persia.' *Mem. London Sch. Hyg. Trop. Med.* **7,** 109-223.

Macan, T. T. (1951). 'Mosquito records from the southern part of the Lake District.' *Ent. Gaz.* **2,** 141-147.

Macan, T. T. (1954a). 'A contribution to the study of the ecology of Corixidae (Hemipt.). *J. Anim. Ecol.* **23,** 115-141.

Macan, T. T. (1954b). 'The Corixidae (Hemipt.) of some Danish lakes.' *Hydrobiologia* **6,** 44-69.

Macan, T. T. (1955a). 'A plea for restraint in the adoption of new generic names.' *Ent. mon. Mag.* **91,** 279-282.

Macan, T. T. (1955b). 'Littoral fauna and lake types.' *Verh. int. Ver. Limnol.* **12,** 608-612.

Macan, T. T. (1957). 'The Ephemeroptera of a stony stream.' *J. Anim. Ecol.* **26,** 317-342.

Macan, T. T. (1958a). 'Methods of sampling the bottom fauna in stony streams.' *Mitt. int. Ver. Limnol.* **8,** pp. 21.

MACAN, T. T. (1958b). 'Causes and effects of short emergence periods in insects.' *Verh. int. Ver. Limnol.* **13,** 845-849.

MACAN, T. T. (1958c). 'The temperature of a small stony stream.' *Hydrobiologia* **12,** 89-106.

MACAN, T. T. (1960a). 'The effect of temperature on *Rhithrogena semicolorata* (Ephem.)'. *Int. Rev. Hydrobiol.* **45,** 197-201.

MACAN, T. T. (1960b). 'The occurrence of *Heptagenia lateralis* (Ephem.) in streams in the English Lake District.' *Wett. u. Leben, Jg.* **12,** 231-234.

MACAN, T. T. (1961). 'A review of running water studies.' *Verh. int. Ver. Limnol.* **14,** 587-602.

MACAN, T. T. (1962a). 'Why do some pieces of water hold more species of Corixidae than others?' *Arch. Hydrobiol.* **58,** 224-232.

MACAN, T. T. (1962b). 'Biotic factors in running waters.' *Schweiz. z. Hydrol.*

MACAN, T. T. and Macfadyen, A. (1941). 'The water bugs of dewponds.' *J. Anim. Ecol.* **10,** 175-183.

MACAN, T. T. and Mackereth, J. C. (1957). 'Notes on *Gammarus pulex* in the English Lake District.' *Hydrobiologia* **9,** 1-12.

MACFADYEN, A. (1957). *Animal ecology aims and methods.* London: Pitman, pp. xx + 264.

MACKERETH, J. C. (1957). 'Notes on the Plecoptera from a stony stream.' *J. Anim. Ecol.* **26,** 343-351.

MACKERETH, J. C. (1960). 'Notes on the Trichoptera of a stony stream.' *Proc. R. ent. Soc. Lond.* A. **33,** 17-23.

MACKERETH, F. J. H., Lund, J. W. G., and Macan, T. T. (1957). 'Chemical analysis in ecology illustrated from Lake District tarns and lakes.' *Proc. Linn. Soc. Lond.* **167,** 159-175.

MCLEESE, D. W. (1956). 'Effects of temperature, salinity and oxygen on the survival of the American lobster.' *J. Fish. Res. Bd Can.* **13,** 247-272.

MANN, H. (1952). 'Die Giftwirkung von erhöhtem Sauerstoffgehalt im Wasser.' *Z. Aquarien- u. TerrarVer.* **5,** 210-213.

MANN, K. H. (1955a). 'The ecology of the British freshwater leeches.' *J. Anim. Ecol.* **24,** 98-119.

MANN, K. H. (1955b). 'Some factors influencing the distribution of freshwater leeches in Britain.' *Verh. int. Ver. Limnol.* **12,** 582-587.

MANN, K. H. (1956). 'A study of the oxygen consumption of five species of leech.' *J. exp. Biol.* **33,** 615-626.

MANN, K. H. (1959). 'On *Trocheta bykowskii* Gedroyć, 1913, a leech new to the British fauna, with notes on the taxonomy and ecology of other Erpobdellidae.' *Proc. zool. Soc. Lond.* **132,** 369-379.

MANN, K. H. (1961a). 'The life history of the leech *Erpobdella testacea* Sav. and its adaptive significance.' *Oikos* **12,** 164-169.

MANN, K. H. (1961b). 'The oxygen requirements of leeches considered in relation to their habitats.' *Verh. int. Ver. Limnol.* **14,** 1009-1013.

MARGALEF, R. (1947). 'Observaciones sobre el desarrollo de la vida en pequeños volúmenes de agua dulce y sobre la ecologia de las larvas de *Aëdes aegypti.*' *Publ. Inst. Biol. apl. Barcelona* **3,** 79-112.

MARSHALL, J. F. (1938). *The British Mosquitoes.* London: Brit. Mus. (Nat. Hist.), pp. xii + 341.

MASON, I. L. (1939). 'Studies on the fauna of an Algerian hot spring.' *J. exp. Biol.* **16,** 487-498.

MATONICKIN, I. (1957). 'La faune des eaux des thermes absolument chauds en Croatie.' *Bioloski Glasnik* **10,** 5-12. (In Serbo-Croatian: French summary.)

MILLS, S. M. (1931). 'The effect of the H-ion concentration on Protozoa, as demonstrated by the rate of food vacuole formation in *Colpidium.*' *J. exp. Biol.* **8,** 17-29.

MONARD, A. (1920). 'La faune profonde du lac de Neuchâtel.' *Bull. Soc. neuchâteloise Sci. Nat.* **44,** 65-236.

MOON, H. P. (1934). 'An investigation of the littoral region of Windermere.' *J. Anim. Ecol.* **3,** 8-28.

MOON, H. P. (1935). 'Flood movements of the littoral fauna of Windermere.' *J. Anim. Ecol.,* **4,** 216-228.

MOON, H. P. (1936). 'The shallow littoral region of a bay at the north west end of Windermere.' *Proc. zool. Soc. Lond.* 490-515.

MOON, H. P. (1957a). 'The distribution of *Asellus* in Windermere.' *J. Anim. Ecol.* **26,** 113-123.

MOON, H. P. (1957b). 'The distribution of *Asellus* in the English Lake District and adjoining areas.' *J. Anim. Ecol.* **26,** 403-409.

MOORE, J. A. (1939). 'Temperature tolerance and rates of development in the eggs of Amphibia.' *Ecology* **20**, 459-478.

MOORE, J. A. (1949). 'Patterns of evolution in the genus *Rana*.' Chap. 17 in Jepsen, G. L., Mayr, E., and Simpson, G. G., *Genetics, palaeontology and evolution*. Princeton U.P.

MOORE, N. W. (1952). 'Notes on the oviposition behaviour of the dragonfly *Sympetrum striolatum* Charpentier.' *Behaviour* **4**, 101-103.

MORAWA, F. W. F. (1958). 'Einige Beobachtungen über die Schwankungen des Fett- und Wassergehaltes von Fischen aus verschiedenen Umweltverhältnissen.' *Verh. int. Ver. Limnol.* **13**, 770-775.

MORTIMER, C. H. (1956). 'The oxygen content of air-saturated fresh waters, and aids in calculating percentage saturation.' *Mitt. int. Ver. Limnol.* **6**, pp. 20.

MORTIMER, C. H. and Moore, W. H. (1953). 'The use of thermistors for the measurement of lake temperatures.' *Mitt. int. Verh. Limnol.* **2**, pp. 42.

MUIRHEAD THOMSON, R. C. (1940). 'Studies on the behaviour of *Anopheles minimus* 1, 2, 3.' *J. Malar. Inst. India* **3**, 265-348.

MUIRHEAD THOMSON, R. C. (1941). 'Studies on the behaviour of *Anopheles minimus* 4.' *J. Malar. Inst. India* **4**, 63-102.

MUIRHEAD-THOMSON, R. C. (1951). *Mosquito behaviour in relation to malaria transmission and control in the tropics*. London: Arnold. pp. viii + 219.

MÜLLER-LIEBENAU, I. (1956). 'Die Besiedlung der *Potamogeton*-Zone ostholsteinischer Seen.' *Arch. Hydrobiol.* **52**, 470-606.

MUNDIE, J. H. (1956). 'Emergence traps for aquatic insects.' *Mitt. int. Ver. Limnol.* **7**, pp. 13.

MUNDIE, J. H. (1959). 'The diurnal activity of the larger invertebrates at the surface of Lac la Ronge, Saskatchewan.' *Canad. J. Zool.* **37**, 945-956.

MUNRO, W. R. and Balmain, K. H. (1956). 'Observations on the spawning runs of brown trout in the South Queich, Loch Leven.' *Freshw. Salmon Fish. Res. Edinb.* **13**, 1-17.

NAYLOR, E. and Slinn, D. J. (1958). 'Observations on the ecology of some brackish water organisms in pools at Scarlett Point, Isle of Man.' *J. Anim. Ecol.* **27**, 15-25.

NIELSEN, A. (1950a). 'On the zoogeography of springs.' *Hydrobiologia*, **2**, 313-321.

NIELSEN, A. (1950b). 'The torrential invertebrate fauna.' *Oikos* **2**, 176-196.

NIELSEN, A. (1951). 'Spring fauna and speciation.' *Verh. int. Ver. Limnol.* **11**, 261-263.

NILSSON, N.-A. (1955). 'Studies on the feeding habits of trout and char in north Swedish lakes.' *Rep. Inst. Freshw. Res. Drottning.* **36**, 163-225.

NOLAND, L. E. (1925). 'Factors influencing the distribution of freshwater ciliates.' *Ecology* **6**, 437-452.

NYGAARD, G. (1955). 'On the productivity of five Danish Waters.' *Verh. int. Ver. Limnol.* **12**, 123-133.

ODUM, H. T. (1957). 'Trophic structure and productivity of Silver Springs, Florida.' *Ecol. Monogr.* **27**, 55-112.

ODUM, E. P. and H. T. (1959). *Fundamentals of ecology.* Philadelphia: Saunders. pp. xvii + 546.

O'GOWER, A. K. (1955). 'The influence of the physical properties of a water container surface upon its selection by the gravid females of *Aëdes scutellaris scutellaris* (Walker) for oviposition (Diptera, Culicidae).' *Proc. Linn. Soc. N.S.W.* **79**, 211-218.

O'GOWER, A. K. (1958). 'The oviposition behaviour of *Aëdes australis* (Erickson) (Diptera, Culicidae).' *Proc. Linn. Soc. N.S.W.* **83**, 245-250.

ORTON, J. H. (1919). 'Sea-temperature, breeding and distribution in marine animals.' *J. Mar. biol. Ass. U.K.* **12**, 339-366.

ORTON, J. H. and Lewis, H. M. (1930). 'On the effect of the severe winter of 1928-1929 on the oyster drills of the Blackwater Estuary.' *J. Mar. biol. Ass. U.K.* **17**, 301-313.

PACAUD, A. (1939). 'Contribution à l'écologie des cladocères.' *Bull. biol. suppl.* **25**, 1-260.

PACAUD, A. (1944). 'Sur les amphipodes gammariens des eaux superficielles dans la region centrale du Bassin parisien.' *Bull. Soc. Zool. Fr.* **69**, 33-46.

PACAUD, A. (1948). 'Température et relations de nutrition dans les biocoenoses limniques.' *Bull. Soc. cent. Aquic. Pêche.* **4**, 1-5.

PACAUD, A. (1949). 'Elevages combinés mollusques-cladocères. Introduction à l'étude d'une biocoenose limnique.' *J. Rech.* **9**, 1-16.

PALMÉN, E. (1953). 'Hatching of *Acentropus niveus* (Oliv.) (Lep., Pyralidae) in the brackish waters of Tvärminne, S. Finland.' *Ann. ent. fenn.* **19**, 181-186.

PANIKKAR, N. K. (1940). 'Influence of temperature on osmotic behaviour of some Crustacea and its bearing on problems of animal distribution.' *Nature, Lond.* **146**, 366.

PANTIN, C. F. A. (1931). 'The adaptation of *Gunda ulvae* to salinity.' *J. exp. Biol.* **8**, 63-94.

PARK, T. (1954). 'Experimental studies of interspecies competition. II. Temperature, humidity, and competition in two species of *Tribolium*.' *Physiol. Zoöl.* **27**, 177-238.

PARK, T. (1957). 'Experimental studies of interspecies competition. III. Relation of initial species proportion to competitive outcome in populations of *Tribolium*.' *Physiol. Zoöl.* **30**, 22-40.

PAUSE, J. (1919). 'Beiträge zur Biologie und Physiologie der Larve von *Chironomus gregarius*.' *Zool. Jb.* **36**, 339-452.

PEARSALL, W. H. (1920). 'The aquatic vegetation of the English Lakes.' *J. Ecol.* **8**, 163-201.

PEARSALL, W. H. (1921). 'The development of vegetation in the English Lakes, considered in relation to the general evolution of glacial lakes and rock basins.' *Proc. roy. Soc. B.* **92**, 259-284.

PEARSON, E. S. and Hartley, H. O. (1954). *Biometrika tables for statisticians*, Vol. 1. Cambridge U.P. pp. xiv + 238.

PENNINGTON, W. (1941). 'The control of the numbers of fresh-water phytoplankton by small invertebrate animals.' *J. Ecol.* **29**, 204-211.

PENTELOW, F. T. K. (1939). 'The relation between growth and food consumption in the brown trout (*Salmo trutta*).' *J. exp. Biol.* **16**, 446-473.

PERCIVAL, E. and Whitehead, H. (1928). 'Observations on the ova and oviposition of certain Ephemeroptera and Plecoptera.' *Proc. Leeds phil. lit. Soc.* **1**, 271-288.

PERCIVAL, E. and Whitehead, H. (1929). 'A quantitative study of the fauna of some types of stream-bed.' *J. Ecol.* **17**, 282-314.

PERCIVAL, E. and Whitehead, H. (1930). 'Biological survey of the River Wharfe.' *J. Ecol.* **18**, 286-302.

PESTA, O. (1929). 'Der Hochgebirgssee der Alpen.' *Binnengewässer* **8,** pp. xii + 156.

PESTA, O. (1933). 'Beiträge zur Kenntnis der limnologischen Beschaffenheit ostalpiner Tümpelgewässer.' *Arch. Hydrobiol.* **25,** 68-80.

PHILIPSON, G. N. (1954). 'The effect of water flow and oxygen concentration on six species of caddis fly (Trichoptera) larvae.' *Proc. zool. Soc. Lond.* **124,** 547-564.

PHILLIPSON, J. (1956). 'A study of factors determining the distribution of the larvae of the black-fly, *Simulium ornatum* Mg.' *Bull. ent. Res.* **47,** 227-238.

PICKEN, L. E. R. (1937). 'The structure of some protozoan communities.' *J. Ecol.* **25,** 368-384.

PITT, T. K., Garside, E. T., and Hepburn, R. L. (1956). 'Temperature selection of the carp (*Cyprinus carpio* Linn.)' *Canad. J. Zool.* **34,** 555-557.

PLESKOT, G. (1951). 'Wassertemperatur und Leben im Bach.' *Wett u. Leben* **3,** 129-143.

PLESKOT, G. (1953). 'Zur Ökologie der Leptophlebiiden (Ins., Ephemeroptera).' *Öst. zool. Z.* **4,** 45-107.

PLESKOT, G. (1958). 'Die Periodizität einiger Ephemeropteren der Schwechat.' *Wasser u. Abwasser* 1958, **1-32.**

PLESKOT, G. (1961). 'Die Periodizität der Ephemeropteren-Fauna einiger österreichischer Fliessgewässer.' *Verh. int. Ver. Limnol.* **14,** 410-416.

POPHAM, E. J. (1941). 'The variation in the colour of certain species of *Arctocorisa* (Hemiptera, Corixidae) and its significance.' *Proc. zool. soc. Lond.* A. **111,** 135-172.

POPHAM, E. J. (1943). 'Further experimental studies of the selective action of predators.' *Proc. zool. soc. Lond.* A. **112,** 105-117.

POTTS, W. T. W. (1954). 'The energetics of osmotic regulation in brackish and freshwater animals.' *J. exp. Biol.* **31,** 618-630.

PRATT, R. *et al.* (1944). 'Chlorellin, an antibacterial substance from *Chlorella*.' *Science* **99,** 351-352.

PRATT, D. M. (1943). 'Analysis of population development in *Daphnia* at different temperatures.' *Biol. Bull. Wood's Hole* **85,** 116-140.

PROVOST, M. W. (1952). 'The dispersal of *Aëdes taeniorhynchus*. 1. Preliminary studies.' *Mosquito News* **12,** 174-190.

QUENNERSTEDT, N. (1958). 'Effect of water level fluctuation on lake vegetation.' *Verh. int. Ver. Limnol.* **13,** 901-906.

RAMSAY, J. A. (1950). 'Osmotic regulation in mosquito larvae.' *J. exp. Biol.* **27,** 145-157.

RAMSAY, J. A. (1951). 'Osmotic regulation in mosquito larvae: the role of the Malpighian tubules.' *J. exp. Biol.* **28,** 62-73.

RAMSAY, J. A. (1953). 'Exchanges of sodium and potassium in mosquito larvae.' *J. exp. Biol.* **30,** 79-89.

RAWSON, D. S. (1945). 'The experimental introduction of small mouth black bass into lakes of the Prince Albert National Park, Saskatchewan.' *Trans. Amer. Fish. Soc.* **73,** 19-31.

RAWSON, D. S. and Moore, J. E. (1944). 'The saline lakes of Saskatchewan.' *Canad. J. Res.* D. **22,** 141-201.

RAWSON, D. S. and Ruttan, R. A. (1952). 'Pond fish studies in Saskatchewan.' *J. Wildlife Mgmt.* **16,** 283-288.

REICH, K. and Aschner, M. (1947). 'Mass development and control of the phytoflagellate *Prymnesium parvum* in fish ponds in Palestine.' *Palestine J. Bot. Jerusalem* **4,** 14-23.

REMANE, A. (1958). 'Ökologie des Brackwassers. (I. Teil der *Biologie des Brackwassers* von A. Remane and C. Schlieper).' *Binnengewässer* **22,** 1-216.

REYNOLDSON, T. B. (1947a). 'An ecological study of the enchytraeid worm population of sewage bacteria beds. 1. Field investigations.' *J. Anim. Ecol.* **16,** 26-37.

REYNOLDSON, T. B. (1947b). 'An ecological study of the enchytraeid worm population of sewage bacteria beds. 2. Laboratory experiments.' *Ann. appl. Biol.* **34,** 331-345.

REYNOLDSON, T. B. (1948). 'An ecological study of the enchytraeid worm population of sewage bacteria beds. 3. Synthesis of field and laboratory data.' *J. Anim. Ecol.* **17,** 27-38.

REYNOLDSON, T. B. (1957). 'Population fluctuations in *Urceolaria mitra* (Peritricha) and *Enchytraeus albidus* (Oligochaeta) and their bearing on regulation.' *Cold Spring Harbor Symposia on Quantitative Biology* **22,** 313-327.

REYNOLDSON, T. B. (1958a). 'Observations on the comparative ecology of lake-dwelling triclads in southern Sweden, Finland and northern Britain.' *Hydrobiologia* **12,** 129-141.

REYNOLDSON, T. B. (1958b). 'The quantitative ecology of lake-dwelling triclads in northern Britain.' *Oikos* **9**, 94-138.

REYNOLDSON, T. B. (1958c). 'Triclads and lake typology in northern Britain—qualitative aspects.' *Verh. int. Ver. Limnol.* **13**, 320-330.

REYNOLDSON, T. B. (1960). 'A quantitative study of the population biology of *Polycelis tenuis* (Ijima) (Turbellaria, Tricladida).' *Oikos* **11**, 125-141.

REYNOLDSON, T. B. (1961a). 'Observations on the occurrence of *Asellus* (Isopoda, Crustacea) in some lakes of northern Britain.' *Verh. int. Ver. Limnol.* **14**, 988-994.

REYNOLDSON, T. B. (1961b). 'A quantitative study of the population biology of *Dugesia lugubris* (O. Schmidt) (Turbellaria, Tricladida).' *Oikos* **12**, 111-125.

RICHARD, G. (1961). 'Observations nouvelles sur les migrations d'insectes hétéroptères Corixidae.' *Verh. int. Ver. Limnol.* **14**, 995-998.

RICHARDS, C. M. (1958). 'The inhibition of growth in crowded *Rana pipiens* tadpoles.' *Physiol. Zoöl.* **31**, 138-151.

RICKER, W. E. and Foerster, R. E. (1948). 'Computation of fish production.' *Bull. Bing. Ocean. Col.* **11**, 173-211.

RODHE, W. (1958). 'The primary production in lakes: some results and restrictions of the ^{14}C method.' *Rapp. Cons. Explor. Mer.* **144**, 122-128.

ROLLINAT, R. (1934). *La vie des reptiles de la France centrale.* Chaps. 2 and 3. 'La cistude d'Europe (*Emys orbicularis* (L.))'. Paris: Delagrave.

ROOS, T. (1957). 'Studies on upstream migration in adult stream-dwelling insects. 1.' *Rep. Inst. Freshw. Res. Drottning.* **38**, 167-193.

ROSE, S. M. (1959). 'Failure of survival of slowly growing members of a population.' *Science* **129**, 1026.

RUSSELL, P. F. and Rao, T. R. (1942a). 'On the relation of mechanical obstruction and shade to ovipositing of *Anopheles culicifacies*.' *J. exp. Zool.* **91**, 303-329.

RUSSELL, P. F. and Rao, T. R. (1942b). 'On the ecology of larvae of *Anopheles culicifacies* Giles in borrow-pits.' *Bull. ent. Res.* **32**, 341-361.

RUTTNER, F. (1953). *Fundamentals of Limnology* (trans. D. G. Frey and F. E. J. Fry). Univ. Toronto Press. pp. xii + 242.

RUTTNER-KOLISKO, A. (1961). 'Biotop and Biozonöse des Sandufers einiger österreichischer Flüsse.' *Verh. int. Ver. Limnol.* **14**, 362-368.

RYTHER, J. H. (1954). 'Inhibitory effects of phytoplankton upon the feeding of *Daphnia magna* with reference to growth, reproduction, and survival.' *Ecology* **35**, 522-533.

SAUNDERS, J. T. (1924). 'The effect of the hydrogen ion concentration on the behaviour, growth and occurrence of *Spirostomum*.' *Proc. Camb. phil. Soc.* **1**, 189-203.

SAVAGE, R. M. (1939). 'The distribution of the spawn-ponds of the common frog, *Rana temporaria temporaria* Linn., over a portion of the London Clay and associated drift.' *Proc. zool. Soc. Lond.* **109**, 1-19.

SAWYER, F. E. (1950). 'B.W.O. Studies of the Sherry Spinner.' *Salm. Trout Mag.* **129**, 110-114.

SAWYER, F. (1952). *Keeper of the stream.* London: Black. pp. x + 214.

SCHACHTER, D. (1958). 'Contribution à l'étude écologique d'un étang mediterranéen.' *Verh. int. Ver. Limnol.* **13**, 676-682.

SCHÄPERCLAUS, W. (1954). *Fischkrankheiten.* 3 Aufl. Berlin: Akademie-Verlag. pp xii + 708.

SCHIEMENZ, H. (1954). 'Über die angebliche Bindung der Libelle *Leucorrhinia dubia* v.d.L. an das Hochmoor.' *Zool. Jb.* (*Syst.*) **82**, 473-480.

SCHLIEPER, C. (1952a). 'Versuch einer physiologischen Analyse der besonderen Eigenschaften einiger eurythermer Wassertiere.' *Biol. Zbl.* **71**, 449-461.

SCHLIEPER, C. (1952b). 'Über die Temperatur-Stoffwechsel-Relation einiger eurythermer Wassertiere.' *Zool. Anz.* suppl. **16**, 267-272.

SCHLIEPER, C. (1958a). 'Physiologie des Brackwassers' (2. Teil der *Biologie des Brackwassers* von A. Remane und C. Schlieper). *Binnengewässer* **22**, 217-348.

SCHLIEPER, C. (1958b). 'Über die Physiologie der Brackwassertiere.' *Verh. int. Ver. Limnol.* **13**, 710-717.

SCHLIEPER, C. and Bläsing, J. (1952). 'Über Unterschiede in dem individuellen und ökologischen Temperaturbereich von *Planaria alpina* Dana.' *Arch. Hydrobiol.* **47**, 288-294.

SCHLIEPER, C., Bläsing, J., and Halsband, E. (1952). 'Experimentelle Veränderungen der Temperaturtoleranz bei stenothermen und eurythermen Wassertieren.' *Zool. Anz.* **149**, 163-169.

SCHMEING-ENGBERDING, F. (1953). 'Die Vorzugstemperaturen einiger Knochenfische und ihre physiologische Bedeutung.' *Z. Fisch.* **2**, 125-155.

SCHMITZ, W. (1954). 'Grundlagen der Untersuchung der Temperaturverhältnisse in den Fliessgewässern.' *Ber. Limnol. Flussstation Freudenthal.* **6**, 29-50.

SCHMITZ, W. (1959). 'Zur Frage der Klassifikation der binnenländischen Brackwässer.' *Arch. Oceanogr. Limnol. Roma* suppl. **11**, 179-226.

SCHULTZ, V. (1952). 'A limnological study of an Ohio farm pond.' *Ohio J. Sci.* **52**, 267-285.

SCOTT, D. (1958). 'Ecological studies on the Trichoptera of the River Dean, Cheshire.' *Arch. Hydrobiol.* **54**, 340-392.

SEGERSTRÅLE, S. G. (1946). 'On the occurrence of the Amphipod, *Gammarus duebeni* Lillj. in Finland, with notes on the ecology of the species.' *Comment. biol., Helsingf.* **9**, 1-22.

SEGERSTRÅLE, S. G. (1948). 'New observations on the distribution and morphology of the amphipod, *Gammarus zaddachi* Sexton, with notes on related species.' *J. Mar. biol. Ass. U.K.* **27**, 219-244.

SEGERSTRÅLE, S. G. (1949). 'The brackish-water fauna of Finland.' *Oikos* **1**, 127-141.

SEGERSTRÅLE, S. G. (1957). 'Baltic Sea.' *Mem. Geol. Soc. Amer.* **67**, 751-800.

SEGERSTRÅLE, S. G. (1958). 'A quarter century of brackishwater research.' *Verh. int. Ver. Limnol.* **13**, 646-671.

SENIOR-WHITE, R. (1926). 'Physical factors in mosquito ecology.' *Bull. ent. Res.* **16**, 187-248.

SHAW, J. (1959a). 'Salt and water balance in the East African fresh-water crab, *Potamon niloticus* (M. Edw.).' *J. exp. Biol.* **36**, 157-176.

SHAW, J. (1959b). 'The absorption of sodium ions by the crayfish, *Astacus pallipes* Lereboullet. I. The effect of external and internal sodium concentrations.' *J. exp. Biol.* **36**, 126-144.

SHAW, J. (1960). 'The absorption of sodium ions by the crayfish, *Astacus pallipes* Lereboullet. II. The effect of the external anion.' *J. exp. Biol.* **37**, 534-547.

SHAW, J. (1960). 'The absorption of sodium ions by the crayfish, *Astacus pallipes* Lereboullet. III. The effect of other cations in the external solution.' *J. exp. Biol.* **37,** 548-556.

SHAW, J. (1960). 'The absorption of chloride ions by the crayfish, *Astacus pallipes* Lereboullet.' *J. exp. Biol.* **37,** 557-572.

SHELUBSKY, M. (1951). 'Observations on the properties of a toxin produced by *Microcystis*,' *Verh. int. Ver. Limnol.* **11,** 362-366.

SHEPARD, M. P. (1955). 'Resistance and tolerance of young speckled trout (*Salvelinus fontinalis*) to oxygen lack, with special reference to low oxygen acclimation.' *J. Fish. Res. Bd Can.* **12,** 387-446.

SIMPSON, G. G. and Roe, A. (1939). *Quantitative zoology*. New York: McGraw pp. xviii + 414.

SMITH, M. (1951). *The British Amphibians and Reptiles* (New Naturalist 20). London: Collins. pp. xiv + 318.

SMYLY, W. J. P. (1952). 'The Entomostraca of the weeds of a moorland fishpond.' *J. Anim. Ecol.* **21,** 1-11.

SMYLY, W. J. P. (1955). 'Comparison of the Entomostraca of two artificial moorland ponds near Windermere.' *Verh. int. Ver. Limnol.* **12,** 421-424.

SMYLY, W. J. P. (1958a). 'The Cladocera and Copepoda (Crustacea) of the tarns of the English Lake District.' *J. Anim. Ecol.* **27,** 87-103.

SMYLY, W. J. P. (1958b). 'Distribution and seasonal abundance of Entomostraca in moorland ponds near Windermere.' *Hydrobiologia* **11,** 59-72.

SONNEBORN, M. T. (1939). '*Paramecium aurelia:* mating types and groups; lethal interactions; determination and inheritance.' *Amer. Nat.* **73,** 390-413.

SOLOMON, M. E. (1949). 'The natural control of animal populations.' *J. Anim. Ecol.* **18,** 1-35.

SOUTHWARD, A. J. and Crisp, D. J. (1956). 'Fluctuations in the distribution and abundance of intertidal barnacles.' *J. Mar. biol. Ass. U.K.* **35,** 211-229.

SPOONER, G. M. (1948). 'The distribution of *Gammarus* species in estuaries, pt. 1.' *J. Mar. biol. Ass. U.K.* **27,** 1-52.

SPOONER, G. M. (1951). 'On *Gammarus zaddachi oceanicus* Segerstråle.' *J. Mar. biol. Ass. U.K.* **30,** 129-147.

STARMÜHLNER, F. (1961). 'Biologische Untersuchungen in isländischen, mitteleuropäischen und madegassischen Warmbächen.' *Verh. int. Ver. Limnol.* **14,** 404-409.

STEINBÖCK, O. (1942). 'Das Verhalten von "*Planaria alpina*" Dana in der Natur und im Laboratoriumsversuch.' *Mem. Ist. ital. Idrobiol. de Marchi* **21,** 63-75.

STEINER, H. (1948). 'Die Bindung der Hochmoorlibelle *Leucorrhinia dubia* Vand. an ihren Biotop.' *Zool. Jb.* (Syst.) **78,** 65-96.

STRENGER, A. (1953). 'Zur Kopfmorphologie der Ephemeridenlarven Erster Teil. *Ecdyonurus* und *Rhithrogena*.' *Österreichische zool. Zeitschrift*. **4,** 191-228.

STUART, T. A. (1941). 'Chironomid larvae of the Millport shore pools.' *Trans. roy. Soc. Edinb.* **60,** 475-502.

STUART, T. A. (1953). 'Spawning migration, reproduction and young stages of loch trout (*Salmo trutta* L.).' *Freshw. Salm. Fish. Res. Edinb.* **5,** 1-39.

SULLIVAN, C. M. and Fisher, K. C. (1953). 'Seasonal fluctuations in the selected temperature of speckled trout, *Salvelinus fontinalis* (Mitchill).' *J. Fish. Res. Bd Can.* **10,** 187-195.

SVERDRUP, H. U., Johnson, M. W. and Fleming, R. H. (1942). *The Oceans.* New York: Prentice-Hall, pp. x + 1087.

TANSLEY, A. G. (1939). *The British Islands and their vegetation.* Cambridge U.P. pp. xxxviii + 930.

TAYLOR, E. W. (1958). *The examination of waters and water supplies* (Thresh, Beale, and Suckling). London: Churchill. 7th edn. pp. viii + 841.

THIENEMANN, A. (1948). 'Die Tierwelt eines astatischen Gartenbeckens in vier aufeinanderfolgenden Jahren.' *Schweiz. Z. Hydrol. Festgabe* 15-41.

THIENEMANN, A. (1950). 'Verbreitungsgeschichte der Süsswassertierwelt Europas.' *Die Binnengewässer* **18.** pp. xvi + 809.

THORPE, W. H. (1930). 'The biology of the petroleum fly (*Psilopa petrolii* Coq.).' *Trans. R. ent. Soc. Lond.* **78,** 331-344.

TREHERNE, J. E. (1954). 'The exchange of labelled sodium in the larva of *Aëdes aegypti* L.' *J. exp. Biol.* **31,** 386-401.

TUCKER, D. S. (1958). 'The distribution of some fresh-water invertebrates in ponds in relation to annual fluctuations in the chemical composition of the water.' *J. Anim. Ecol.* **27**, 105-123.

TUCKER, D. W. (1959). 'A new solution to the Atlantic eel problem.' *Nature, Lond.* **183**, 495-501.

VAAS, K. F. and Sachlan, M. (1955). 'Limnological studies on diurnal fluctuations in shallow ponds in Indonesia.' *Verh. int. Ver. Limnol.* **12**, 309-319.

VERRIER, M.-L. (1949). 'Les tropismes et la répartition des Éphémères.' *Int. Congr. Zool.* **13**, 184-185.

VERRIER, M.-L. (1954). 'Rassemblements et migrations chez les Ephémères.' *Bull. biol.* **88**, 68-89.

VERRIER, M.-L. (1956). *Biologie des Ephémères.* Paris: Colin. pp. 216.

VIVIER, P. (1955). 'Sur l'introduction des salmonidés exotiques en France.' *Verh. int. Ver. Limnol.* **12**, 527-535.

VOLLENWEIDER, R. A. (1948). 'Zum Gesellschaftsproblem in der Limnobiocoenologie.' *Schweiz. Z. Hydrol.* **10**, 53-64.

VOLLENWEIDER, R. A. and Ravera, O. (1958). 'Preliminary observations on the oxygen uptake by some freshwater zooplankters.' *Verh. int. Ver. Limnol.* **13**, 369-380.

VORSTMAN, A. G. (1951). 'A year's investigations on the life cycle of *Neomysis vulgaris* Thompson.' *Verh. int. Ver. Limnol.* **11**, 437-445.

WALSHE, B. M. (1947a). 'On the function of haemoglobin in *Chironomus* after oxygen lack.' *J. exp. Biol.* **24**, 329-342.

WALSHE, B. M. (1947b). 'The function of haemoglobin in *Tanytarsus* (Chironomidae).' *J. exp. Biol.* **24**, 343-351.

WALSHE, B. M. (1948). 'The oxygen requirements and thermal resistance of chironomid larvae from flowing and from still waters.' *J. exp. Biol.* **25**, 35-44.

WALSHE, B. M. (1950). 'The function of haemoglobin in *Chironomus plumosus* under natural conditions.' *J. exp. Biol.* **27**, 73-95.

WALTON, G. A. (1936). 'Oviposition in the British species of *Notonecta* (Hemipt.).' *Trans. soc. Brit. Ent.* **3**, 49-57.

WASHBOURN, R. (1936). 'Metabolic rates of trout fry from swift and slow-running waters.' *J. exp. Biol.* **13**, 145-147.

WAUTIER, J. and Pattée, E. (1955). 'Expérience physiologique et expérience écologique. L'influence du substrat sur la consommation d'oxygène chez les larves d'Ephéméroptères.' *Bull. mens. Soc. linn. Lyon* **24,** 178-183.

WEBER, H. H. (1955). 'Zum Wohngewässerwechsel der Corixiden (Hem. Het. Corixidae).' *Verh. Ver. naturwiss. Heimatforsch. Hamburg* **32,** 5-10.

WEEREKOON, A. C. J. (1956). 'Studies on the biology of Loch Lomond. 2. The repopulation of McDougall Bank.' *Ceylon J. Sci.* **7,** 95-133.

WELCH, P. S. (1948). *Limnological Methods.* Philadelphia: Blakiston. pp. xviii + 381.

WESENBERG-LUND, C. (1913). 'Odonaten-Studien.' *Int. Rev. Hydrobiol.* **6,** 155-422.

WESENBERG-LUND, C. (1943). *Biologie der Süsswasserinsekten.* Berlin: Springer. pp. 682.

WHITNEY, R. J. (1939). 'The thermal resistance of mayfly nymphs from ponds and streams.' *J. exp. Biol.* **16,** 374-385.

WHITNEY, R. J. (1942). 'Diurnal fluctuations of oxygen and pH in two small ponds and a stream.' *J. exp. Biol.* **19,** 92-99.

WIEBE, A. H. and McGavock, A. M. (1932). 'The ability of several species of fish to survive on prolonged exposure to abnormally high concentration of dissolved oxygen.' *Trans. Amer. Fish. Soc.* **62,** 267-274.

WIGGLESWORTH, V. B. (1933a). 'The function of the anal gills of the mosquito larva.' *J. exp. Biol.* **10,** 16-26.

WIGGLESWORTH, V. B. (1933b). 'The adaptation of mosquito larvae to salt water.' *J. exp. Biol.* **10,** 27-37.

WIGGLESWORTH, V. B. (1938). 'The regulation of osmotic pressure and chloride concentration in the haemolymph of mosquito larvae.' *J. exp. Biol.* **15,** 235-247.

WIKGREN, B. (1953). 'Osmotic regulation in some aquatic animals with special reference to the influence of temperature.' *Acta. zool. fenn.* **71,** 1-102.

WILLIAMS, C. B. (1947). 'The generic relations of species in small ecological communities.' *J. Anim. Ecol.* **16,** 11-18.

WILLIAMS, C. B. (1958). *Insect Migration* (New Naturalist 36). London: Collins. pp. xiv + 235.

WILSON, D. P. (1951). 'A biological difference between natural sea waters.' *J. Mar. biol. Ass. U.K.* **30,** 1-20.

WINGFIELD, C. A. (1939). 'The function of the gills of mayfly nymphs from different habitats.' *J. exp. Biol.* **16,** 363-373.

WOODBURY, L. A. (1942). 'A sudden mortality of fishes accompanying a supersaturation of oxygen in Lake Waubesa, Wisconsin.' *Trans. Amer. Fish. Soc.* **71,** 112-117.

WUHRMANN, K. and Woker, H. (1955). 'Influence of temperature and oxygen tension on the toxicity of poisons to fish.' *Verh. int. Ver. Limnol.* **12,** 795-801.

WUNDSCH, H. H. (1930). 'Ausscheidungen der Wasserschnecke *Limnaea peregra* (Müll.) als raschwirkendes Fischgift.' *Z. Fisch.* **28,** 1-12.

ZAHNER, R. (1959), (1960), 'Über die Bindung der mitteleuropaischen *Calopteryx*-Arten (Odonata, Zygoptera) an den Lebensraum des strömenden Wassers.' *Int. Rev. Hydrobiol.* **44,** 51-130; **45,** 101-123.

ZWICKY, K. and Wigglesworth, V. B. (1956). 'The course of oxygen consumption during the moulting cycle of *Rhodnius prolixus* Stål (Hemiptera)'. *Proc. R. ent. Soc. Lond.* A. **31**, 153-160.

Index to Species Names

(*Species that appear only in the tables in chapter 3 are not included*)

Ablabesmyia nemorum, 133
Abramis brama, 214
  vimba, 94
Acantholebris curvirostris, 57, 250
Acerina cernua, 32, 214
Acentropus niveus, 41, 158
Acroloxus lacustris, 34, 48, 49, 63, 128, 137, 189, 205
Aëdes, 78, 133, 215
  aegypti, 83, 85-7, 89, 134, 137, 158, 159, 164, 165, 174, 235, 236, 243
  africanus, 134
  albopicta (*see* Stegomyia)
  argenteus (*see* aegypti)
  australis, 85, 86
  cantans, 249, 252
  detritus, 216, 235-7, 243
  punctor, 249, 252
  rusticus, 249, 252
  scutellaris, 85, 86
  taeniorhynchus, 71
  variegatus, 83
Aeshna mixta, 80
Agapetus fuscipes, 21, 29, 144
Agrion, 124
  splendens, 32, 35, 81, 82, 89, 157, 197, 198
  virgo, 81, 82, 89, 197, 198
Alanopsis elongata, 57, 250
Alburnus tarihi, 241
Alnus, 15
Ameiurus nebulosus, 151, 153
Ameletus inopinatus, 21
Amphinemura standfussi, 23
Amphora commutata, 238
Anabaenopsis circularis var. javanica, 240
Anabolia, 122, 196
  nervosa, 196, 197
Anacystis thermalis, 240, 243
Anatopynia varia, 204
Ancylastrum fluviatile, (*see* Ancylus)
Ancylus fluviatilis, 17, 30, 35, 44, 47-9, 63, 120, 124, 127-9, 133, 137, 181, 182, 189, 205
  lacustris (*see* Acroloxus)
Anguilla anguilla, 214, 216, 223, 225, 226, 243
Anisus vortex (*see* Planorbis)
Anodonta cygnaea, 223, 224
Anomoeoneis sphaerophora (*see* Navicula)
Anopheles, 10, 78, 92, 100
  barbirostris, 156
  bifurcatus (*see* claviger)
  claviger, 85, 89, 156, 174
  culicifacies, 86, 87
  elutus (*see* sacharovi)
  gambiae, 71
  hyrcanus, 156
  leucosphyrus, 86
  maculipennis, 85, 251
  maculipennis atroparvus, 85, 251, 252
  minimus, 84, 85, 156-8
  pharoensis, 71
  pulcherrimus, 71, 145, 146
  quadrimaculatus, 158
  sacharovi, 71, 72, 77, 145, 174
  stephensi, 145, 146
  superpictus, 146, 251
  vagus, 156
Anticorisa (*see* Corixa)
Apatidea, 165
  muliebris, 144
Aphanizomenon flos-aquae f. gracile (*see* gracile)
  gracile, 94, 114
Aphelocheirus aestivalis, 31, 35, 63
  montandoni (*see* aestivalis)
Apium inundatum, 54
Aquarius najas (*see* Gerris)
Arctocorisa (*see* Corixa)
Arctodiaptomus salinus (*see* Diaptomus)
Arenicola, 217, 218
  marina, 216
Artemia salina, 92, 163, 202, 204, 237, 238, 240-3, 262
Arthrospira platensis, 240, 243
Asellus, 55, 75, 76, 249, 250
  aquaticus, 33, 39, 44, 48, 58, 60, 192, 213, 228
  meridianus, 48, 213
Astacus, 226, 230
  astacus (*see* fluviatilis)
  fluviatilis, 200, 205, 223, 225, 227-9, 244, 248, 252
  pallipes, 113, 228, 229
Asterias rubens, 216, 219, 220
Asterionella formosa, 261

Atyaephyra desmaresti, 214

Baetis, 22, 34, 79, 120, 121, 124, 193, 195, 196
  pumilus, 18, 27, 102
  rhodani, 18, 22, 27, 102, 156, 192, 193, 281, 282
  scambus, 30, 32, 125, 126, 192, 193
  tenax, 23, 30, 32
  vernus, 121, 125, 126
Balanus balanoides, 109, 163, 164
Bathyomphalus contortus (*see* Planorbis)
Batrachobdella paludosa, 48
Batrachospermum, 263
Bithynia leachii, 34, 43, 48, 63
  tentaculata, 32, 43, 48, 63, 74, 77
Bodo sulcatus, 180
Bosmina coregoni var. lilljeborgi, 57, 250
Brachionus angularis, 240, 241
  calyciflorus var. pala (*see* pala)
  mülleri (*see* plicatilis)
  pala, 110, 114, 240
  plicatilis, 237, 238, 240, 241, 244
  plicatilis spatuosus, 238
  satanicus, 238
Bulimus (*see* Bithynia)
Bythotrephes, 188

Caenis, 73
Caenis moesta, 33, 40, 45
Callicorixa (*see* Corixa)
Calopteryx (*see* Agrion)
Calothrix, 104
Cambarus fodiens, 130
Capnia bifrons, 47
Carassius auratus, 148, 149, 153, 154, 184-7, 192, 225, 227
Carassius vulgaris (*see* Cyprinus carassius)
Carcinus, 199, 220, 226
  maenas, 214, 216, 219, 220, 225
Carex, 51, 54, 55, 61, 80, 81, 278, 285
  inflata (*see* rostrata)
  rostrata, 50, 255, 265
Centroptilum luteolum, 40, 45, 79
Ceratium hirundinella, 238
Ceriodaphnia pulchella, 55, 56
  rotunda, 93, 188
Chaetoceras orientalis, 240
Chara, 240
Chara Hornemannii, 241
Cheirocephalus, 164
  diaphannus, 134
  grubii, 133
Chironomus anthracinus bathophilus, 201, 205
  cingulatus, 202, 205
  gregarius, 202, 205
  plumosus, 201-4, 206
  riparius (*see* thummi)
  thummi, 201, 202, 206
Chlamydomonas, 204
Chlorella, 95, 182
Chloroperla torrentium, 47
Chroococcus, 238
  thermalis (*see* Anacystis)
Chthamalus stellatus, 163
Chydorus sphaericus, 55, 63, 189
Cladophora, 129, 238
  fracta, 238
  glomerata, 177, 238
Cletocamptus albuquerquensis, 238, 241
  retrogressus, 237, 244
Cloeon, 45, 61
  dipterum, 33, 51, 54, 156, 192, 193
  simile, 33, 54
Coleps hirtus, 93
Colpidium, 263, 264
  campylum, 93
  colpoda, 93
Colurella adriatica, 241
Conochilus unicornis, 238
Corambe batava, 214
Cordylophora caspia, 213
  lacustris, 213
Coregonus, 94, 111
  albula, 181
Coretus corneus (*see* Planorbis)
Corixa, 49, 283
  carinata, 147, 174, 213, 244
  castanea, 52, 54, 255, 256, 265, 282
  dentipes, 52, 105
  distincta, 52, 90, 212, 230-2, 244, 255, 256, 265
  dorsalis, 69, 77, 255, 256, 265
  fabricii (*see* nigrolineata)
  falleni, 34, 69, 77, 212, 244, 255, 256, 265
  fossarum, 70, 77, 212, 230-2, 244, 255, 256, 265
  geoffroyi (*see* punctata)
  germari, 255, 256, 266
  hieroglyphica (*see* lateralis)
  kilimandjaronis (*see* lateralis)
  lacustris (*see* dorsalis)
  lateralis, 212, 240, 241, 244, 249, 252
  limitata, 249, 252
  linnei, 52, 212, 256, 266
  lugubris (*see* stagnalis)
  nigrolineata, 249, 252
  praeusta, 52, 69, 77, 212
  punctata, 70, 77, 249, 252
  sahlbergi, 52, 212, 244, 256, 266
  scotti, 52, 54, 255, 256, 266
  selecta, 212, 216, 244
  semistriata, 212, 244
  stagnalis, 198-200, 206 212, 216, 230-2, 244
  striata, 34, 45, 77, 212, 213, 244, 256
  wollastoni, 147, 174
Corophium lacustre, 213
  volutator, 213

Cothurnia curvula, 221
Cottus gobio, 132
Covonmeyenia (*see* Ephydatia)
Crangon crangon, 214
Crangonyx pseudogracilis, 48, 67, 77
Craspedacusta sowerbii, 213
Crenobia alpina (*see* Planaria)
Culex, 85
Culex fatigans, 86
  molestus, 86
  nebulosa, 86
  pipiens, 78, 158
Culicella, 78
Cyathura carinata, 213
Cyclops viridis, 39, 56, 63, 237, 240, 241
Cyclotella Meneghiniana, 238, 240
Cymatia bondsdorffi, 212
  coleoptrata, 212
Cymatopleura solea, 238
Cymbella tumida, 238
Cyprinodon fasciatus, 238
Cyprinus carassius, 227, 244
  carpio, 154, 155, 184
Cyrnus, 57, 59
  trimaculatus, 54

Daphnia, 95, 202
  hyalina, 188
  longispina, 55, 188, 238
  magna, 162, 188, 189, 223, 262
  obtusa, 93, 182, 188
  pulex, 92, 110, 159, 188, 189
  thomsoni, 188
Daphnella brandtianum (*see* Diaphanosoma brachyurum)
  wingii (*see* Diaphanosoma brachyurum)
Dendrocoelum lacteum, 37, 58, 124, 248
Deronectes spp., 282, 283
Deronectes assimilis, 52, 54
  depressus, 48, 49, 54
Diaphanosoma brachyurum, 56, 63
Diaptomus salinus, 237, 244
  silicoides, 238
  spinosus, 241
Dictyopterygella recta (*see* Diura bicaudata)
Dina lineata, 247
Dinocras cephalotes, 129, 137
Diogenes rotundus, 110
Diura bicaudata, 21-23, 26, 47, 97, 114
Dreissena polymorpha, 32, 35, 43-5, 72, 77, 214, 244
Dreissensia (*see* Dreissena)
Dugesia lugubris, 30, 37, 44, 54, 64, 111, 114, 248
  subtentaculata (*see* Planaria gonocephala)
Dunaliella, 241
Dunaliella viridis, 237

Dytiscus marginalis, 152-4
  punctulatus (*see* semisulcatus)
  semisulcatus, 52, 152-4, 174

Ecclisopteryx guttulata, 130
Ecdyonurus, 120, 121, 144
  austriacus, 142, 143
  dispar, 47, 50
  torrentis, 18, 22, 23, 27, 50, 122
  venosus, 18, 23, 27, 50, 125, 142, 143, 195, 196
  zelleri, 142, 143
Echinogammarus berilloni, 189
Eleocharis, 80
Elmis, 49
Elodea canadensis, 176
Emys orbicularis, 147
Enallagma cyathigerum, 49, 51, 54, 57, 80, 81, 282, 283
Enchytraeus albidus, 108, 114
Enochrus diffusus, 240
Enteromorpha prolifera tubulosa, 238
Epeorus alpicola, 124
Ephemera danica, 27, 29, 79, 183
  vulgata, 192, 193, 223
Ephemerella ignita, 18, 27, 80, 120, 121, 125, 129, 195, 196
  notota, 27
Ephydatia everetti, 254, 266
  mülleri, 254, 266
Ephydra glauca, 238
  hians, 240
  macellaria, 237
  subopaca, 240
Equisetum, 88
Eriocheir sinensis, 214, 217, 218, 225-7, 229
Eristalis, 183
Erpobdella octoculata, 29-33, 39, 48, 51, 54, 63, 190, 191, 247, 252, 282, 283
  testacea, 39, 48, 58, 60, 190-2, 247, 253
Erythromma, 157
  najas, 33, 45, 58
Esox lucius, 60, 88
Eucrangonyx gracilis (*see* Crangonyx pseudogracilis)
Eucricotopus atritarsis, 241
  brevipalpis, 92
Euplanaria gonocephala (*see* Planaria)
Eurycercus lamellatus, 39, 56, 188

Filinia major, 241
Fontinalis dalecarlica, 57, 61

Gammarus, 44, 120, 132, 273
  duebeni, 7, 97, 98, 160, 162, 213, 215, 232-4
  fasciatus, 159, 174
  locusta, 214, 215
  pulex, 17, 23, 32, 39, 45, 48, 51, 60, 78, 97, 98, 112, 121, 124, 127, 129, 133, 157, 189, 213-15, 249

Gammarus—*cont.*
tigrinus (*see* fasciatus)
zaddachi, 213, 235
zaddachi oceanicus, 214, 215
zaddachi salinus, 214, 215
zaddachi zaddachi, 214, 215
Gasterosteus aculeatus, 166, 188
Gerris najas, 136, 137
rufoscutellatus, 144
Girella nigricans, 154
Glaenocorisa cavifrons (*see* propinqua)
propinqua, 147, 148, 174
Gloetrichia, 130
Glossiphonia complanata, 29-33, 39, 51, 63, 124, 137, 190, 191, 206, 247, 253
heteroclita, 39, 48, 63
Glossosiphonia (*see* Glossiphonia)
Glossosoma boltoni, 19, 130
Glyphotaelius, 57, 59
Gobio fluviatilis (*see* gobio)
gobio, 168, 169, 174
Gomphosphaeria aponina, 240
Gonium pectorale, 93
Gunda ulvae (*see* Procerodes)

Habrophlebia fusca, 18, 27
lauta, 27
Habroleptoides modesta, 79, 168
Haemopis sanguisuga, 48
Halicorixa (*see* Corixa)
Haliplus, 45, 282
flavicollis, 49
fulvus, 40, 49, 52, 282
Halmopota hutchinsoni, 238
Helobdella stagnalis, 33, 39, 48, 190, 191, 247
Helophorus brevipalpis, 70, 72, 105
Hemiclepsis marginata, 38, 48, 247
Heptagenia dalecarlica, 22
fuscogrisea, 249
lateralis, 8, 18, 27, 34, 47, 122, 124, 141-4, 162, 249, 253
sulphurea, 27, 34, 249
Herpobdella atomaria (*see* Erpobdella octoculata)
Heterocypris incongruens, 204
Hesperocorixa (*see* Corixa)
Hirudo medicinalis, 223
Holocentropus, 57
dubius, 52, 54
Holopedium gibberum, 250
Hucho hucho, 94, 115
Hyas arenea, 225, 226
Hydrobia jenkinsi (*see* Potamopyrgus)
stagnalis (*see* ventrosa)
ventrosa, 214, 215, 244
Hydrodictyon, 176
Hydropsyche, 122, 196
angustipennis, 30-32, 35, 120, 121, 125, 197
fulvipes, 119, 122, 123

Isogenus nubecula, 27

Juncus bulbosus, 255

Keratella divergens (*see* quadrata)
quadrata, 240, 245

Lampetra fluviatilis (*see* Petromyzon)
Lebistes reticulatus, 154
Lemanea, 263
Lemna, 93, 177
Lepomis macrochirus (*see* microchirus)
microchirus, 88, 90
Leptodora, 188
Leptolimnaea glabra (*see* Limnaea)
Leptophlebia, 57, 282, 284, 285
marginata, 49, 51, 54, 193
vespertina, 49, 51, 54, 58, 192, 193
Lestes sponsa, 51, 54, 58
Leucioperca leucioperca (*see* sandra)
sandra, 94, 115, 214, 245
Leuciscus erythrophthalmus, 223, 245
rutilus, 187, 214, 227
Leucorrhinia dubia, 81
rubicunda, 81
Leuctra, 102
fusca, 18, 21, 27, 47, 64
fusciventris (*see* fusca)
geniculata, 27, 127
hippopus, 18, 29, 47
Leydigia leydigii, 93
Limnaea, 181
glabra, 246, 253
ovata (*see* pereger)
palustris, 42, 48, 64, 257, 258, 266
pereger, 42, 48, 49, 51, 54, 60, 64, 95, 115, 124, 137, 147, 168, 174, 223, 224, 245, 249, 253, 257, 258, 260, 266, 282, 283
stagnalis, 204, 206, 227, 247, 249-50, 253, 258, 260, 266
Limnephilus marmoratus, 52, 282
Limnesia undulata, 41, 45
Limnius tuberculatus, 27, 30-2, 40
Limnocalanus macrurus, 73
Limnoporus rufoscutellatus (*see* Gerris)
Liponeura cinerascens, 124
Lithotanytarsus, 252
Littorella uniflora, 50, 54, 81, 277, 278, 284
Lumbricillus lineatus, 107, 108, 115
variegatus, 27
Lymnaea (*see* Limnaea)
Lyngbya aestuarii, 238
Birgei, 238

Macrocorixa (*see* Corixa)
Macrothix laticornis, 93
Maja, 220
Margaritifera margaritifera, 246, 253
Margaritana (*see* Margaritifera)

Megacyclops vulgaris (*see* Cyclops viridis)
Melosira nyassensis, 240, 245
deuriesii (*see* nyassensis)
minor (*see* nyassensis)
Merismopedia, 238
Metriocnemus, 108
hirticollis, 107
hygropetricus (*see* longitarsus)
longitarsus, 107, 115
Meyenia (*see* Ephydatia)
Microchironomus, 241
Micronecta borealis (*see* poweri)
jenkinae, 240
poweri, 41, 45, 64, 212, 245, 255, 266
scutellaris, 240
Micropterus, 165
dolomieu, 150, 152
Mochlonyx culiciformis, 133
Moina, 241
brachiata, 92, 188
hutchinsoni, 241
macrocopa, 158
Molanna, 57
Myriophyllum, 54, 55, 80, 81
alternaeflorum, 50
Mysis oculata (*see* relicta)
relicta, 73, 77, 164, 180, 181, 198, 206, 235
Mytilus, 251
edulis, 216, 219

Navicula sphaerophora, 240, 245
Nemoura avicularis, 40, 47
cinerea, 23, 29, 31, 51, 64
variegata (*see* cinerea)
Neomysis integer, 162, 174, 214, 245
vulgaris (*see* integer)
Nereis diversicolor, 199, 200, 213, 216, 218, 251
Neritina fluviatilis, 30, 43, 44, 48, 64, 124, 137, 200, 206
Neumania callosa, 41, 45
Niphargus, 27
Nitella, 54, 55
Nitocra lacustris, 237
Notonecta furcata (*see* obliqua)
glauca, 61, 88, 157
maculata, 88
obliqua, 61, 64

Ochlerotatus (*see* Aëdes)
maculatus (*see* A. cantans)
nemorosus (*see* A. punctor)
waterhousei (*see* A. cantans)
Odontocerum albicorne, 19, 130, 144
Oedogonium sp., 238
Omphiscola glabra (*see* Limnaea)
Oncorhynchus nerka, 153, 154, 270
tshawytscha, 154
Oocystis Borgei, 241
Orchestia bottae, 213
Orconectes propinquus, 130
Oreodytes halensis, 48
Orthetrum, 81
septentrionalis, 49
Oscillatoria chlorina, 240
Osmerus eperlanus, 214
Ostraea, 165, 263
madrasensis, 7
virginica, 218

Pachydrilus lineatus (*see* Lumbricillus)
Palaemon longirostris, 214
varians, 214
Pallasea quadrispinosa, 39, 45
Paludestrina jenkinsi (*see* Potamopyrgus)
stagnalis (*see* Hydrobia ventrosa)
Paramecium, 131
aurelia, 95
caudatum, 223
Pedalia, 238
fennica, 237, 241
fennica var. polydonta, 241
jenkinae, 241
Pediastrum Boryanum, 238
duplex, 238
Pedicia, 19, 120
Pelmatohydra oligactis, 32, 33, 37, 213
Perca flavescens, 154, 187
fluviatilis, 60, 112, 113, 214, 223
Pericoma, 252
Perla bipunctata, 18, 20, 22, 64, 129, 132, 137
carlukiana (*see* bipunctata)
cephalotes (*see* Dinocras)
marginata (*see* bipunctata)
Perlodes microcephala, 20, 26, 27, 64, 97, 115
mortoni (*see* microcephala)
Petromyzon fluviatilis, 217, 223, 225, 227, 245
Phagocata vitta (*see* Planaria)
Phalacrocera diversa (*see* replicata)
replicata, 59, 60, 64
Phormidium, 107
Phoxinus laevis, 6
Phragmites, 15, 49, 57, 61, 255
Phryganea, 57, 73, 282
Physa acuta, 147
fontinalis, 42, 48, 258
Piscicola geometra, 32, 33, 38, 190, 191, 247
Pisidium, 45, 49, 68
Planaria alpina, 9, 17, 21, 23, 24, 27, 34, 64, 98, 111, 115, 124, 127, 137, 156, 157, 163, 164, 168-74, 183, 206, 251, 253
gonocephala, 27, 98, 115, 168-74, 183, 206
lugubris (*see* Dugesia)
montenegrina, 98, 115
teratophila (*see* montenegrina)

Planaria—*cont.*
torva, 37, 48
vitta, 247, 248, 253
Planorbarius corneus (*see* Planorbis)
Planorbis albus, 30, 32, 35, 42, 48, 60, 64, 257, 258, 260, 266
carinatus, 42, 48, 64, 246, 251, 253
complanatus, 258
contortus, 34, 42, 48, 65, 258
corneus, 34, 65, 74, 77, 93, 115, 202, 204, 206, 249, 253
crista, 34, 42, 65, 258
planorbis, 34, 42, 48, 65
vortex, 249, 253
Plasmodium falciparum, 100
malariae, 100
vivax, 100
Platambus maculatus, 49
Platichthys flesus (*see* Pleuronectes)
Plectonema nostocorum, 240
Plectrocnemia conspersa, 19, 29, 122
Pleuronectes flesus, 214, 216, 227, 245
platessa, 216, 222-4
Polycelis cornuta (*see* felina)
felina, 17, 22, 23, 27, 34, 44, 65, 92, 98, 99, 115, 124
hepta, 248
nigra, 30, 37, 58, 111, 247, 248, 250
tenuis, 37, 111, 248
Polymitarcis virgo, 69
Potamanthus luteus, 69
Potamobius (*see* Astacus)
Potamogeton, 45, 241, 255
crispus, 197
perfoliatus, 129
natans, 50, 54, 80, 92, 177, 255
Potamon niloticus, 229, 230
Potamophylax latipennis (*see* Stenophylax)
stellatus (*see* Stenophylax)
Potamopyrgus crystallinus carinatus (*see* jenkinsi)
jenkinsi, 9, 67, 68, 77, 200, 206, 214
Procerodes ulvae, 215, 219, 245, 251, 253
Protodrilus flavocapitatus, 215
Protonemura montana, 21
Prymnesium parvum, 95
Pseudosida bidentata, 159
Psychoda alternata, 108
severini, 108
Ptilocolepus granulatus, 109
Ptygora brevis, 130
Pungitius pungitius, 240
Pyrrhosoma nymphula, 51, 54, 58, 80, 277, 282-4.

Radix pereger (*see* Limnaea)
Rana, 160, 161
Rana pipiens, 160, 161, 235
Ranunculus fluitans, 197
Retrocorixa (*see* Corixa)
Rhithrogena haarupi, 27
semicolorata, 8, 18, 22, 23, 27, 34, 50, 122, 125, 129, 133, 135, 142, 143, 156, 162, 194-7
Rhodnius prolixus, 182
Rhyacophila, 124, 196, 197
dorsalis, 19, 123
nubila, 125, 127, 196
Riola, 252

Sagitta elegans, 261
setosa, 261
Salix, 255
Salmo gairdneri, 100, 155, 170-4, 262
hucho (*see* Hucho hucho)
iridaeus (*see* gairdneri)
salar, 99, 115, 162, 174, 214, 216, 245
trutta 20, 54, 65, 78, 88, 90, 94, 99, 100, 115, 148, 153, 164, 168, 169, 174, 184, 206, 251, 253, 269, 273
Salvelinus, 153, 181
alpinus, 99
fontinalis, 100, 148-54, 161, 163, 184-7, 192
namaycush, 153, 187
Scapholebris mucronata, 93, 189
Scardinius erythrophthalmus (*see* Leuciscus)
Schizothrix lardacea, 238
Sialis flavilatera (*see* lutaria)
Sialis lutaria, 59, 65, 228
Sida crystallina, 56, 188
Sigara (*see* Corixa)
poweri (*see* Micronecta)
Simocephalus exspinosus, 56, 65, 93, 188
vetulus, 57, 93, 188, 189, 250
Simosa exspinosus (*see* Simocephalus)
Simulium, 19, 22, 23, 35, 50, 120, 132, 133
austeni (*see* venustum)
hirtipes, 19, 142, 174
ornatum, 19, 122, 124
venustum, 30, 35, 65
Siphlonurus lacustris, 23, 47
Sphagnum, 57, 60, 80, 81, 247
Spiralina vortex (*see* Planorbis)
Spirostomum ambiguum, 263, 264
Spirulina jenneri (*see* Arthrospira platensis)
laxissima, 240
subsalsa, 238
subtilissima, 240
Spongilla fragilis, 254
ingloviformis, 254
lacustris, 31-3, 58, 254
Squalius cephalus, 168-74
Stagnicola palustris (*see* Limnaea)
Stegomyia albopicta, 87, 90
argenteus (*see* Aëdes aegypti)
calopus (*see* Aëdes aegypti)

Stegomyia—*cont.*
 fasciata (*see* Aëdes aegypti)
 sugens, 87, 90
 vittata (*see* sugens)
Stenelmis canaliculatus, 49
Stenophylax latipennis, 19, 65, 130, 137
 stellatus, 19, 29, 65, 123, 130, 137
Stentor, 93
Streblocerus serricaudatus, 250
Stylodrilus heringianus, 27
Subsigara (*see* Corixa)
Surirella armenica, 241
 ovalis, 238
Sympetrum striolatum, 80

Tanytarsus brunnipes, 202
 gregarius (*see* Chironomus)
Tendipes (*see* Chironomus)
Theobaldia annulata, 78
Theodoxus fluviatilis (*see* Neritina)
Theromyzon tessulatum, 48, 65
Thuya occidentalis, 261
Tinodes, 44
Torleya major, 132
Triaenodes bicolor, 282
Tribolium castaneum, 95-7
 confusum, 95, 96
Trichocerca taurocephala, 241
Trichocorixa verticalis, 240, 241
Trocheta bykowskii, 48
Tropidiscus carinatus (*see* Planorbis)
 complanatus (*see* Planorbis)
 marginatus (*see* Planorbis)
 umbilicatus (*see* Planorbis)
Trutta fario (*see* Salmo)
 lacustris (*see* Salmo)
 salar (*see* Salmo)
Tubella pennsylvanica, 254
Tubifera, 183
Tubifex, 204
Typha, 255

Urceolaria mitra, 109, 110
Utricularia vulgaris americana, 130

Valvata piscinalis, 34, 43, 48, 258
Velia caprai, 19, 65, 71, 77
 currens (*see* caprai)
Velletia (*see* Acroloxus)
Vermicorixa (*see* Corixa)

Wolterstorffia blanchardi (*see* Cletocamptus retrogressus)
Wormaldia, 122
 occipitalis, 19, 144

Index to English Names

Buffalo fish, 88
Bream (see *Abramis brama*)

Carp (see *Cyprinus carpio*)
Char (see *Salvelinus alpinus*)
Crucian carp (see *Cyprinus carassius*)

Fairy shrimp (see *Cheirocephalus* spp.)
Flounder (see *Pleuronectes flesus*)
Flour beetle (see *Tribolium* spp.)

Goldfish (see *Carassius auratus*)
Guppy (see *Lebistes reticulatus*)

Lamprey (see *Petromyzon fluviatilis*)
Leeches (see Hirudinea)
Lobster, American, 5, 7, 155

Minnow (see *Phoxinus laevis*)
Mitten crab (see *Eriocheir senensis*)
Mussel (see *Mytilus edulis*)

Oyster (see *Ostrea* spp.)

Perch (see *Perca fluviatilis*)
Pikeperch (see *Leucioperca sandra*)
Plaice (see *Pleuronectes platessa*)
Pope (see *Acerina cernua*)

Roach (see *Leuciscus rutilus*)

Salmon, Atlantic (see *Salmo salar*)
  blueback, 155
  sockeye (see *Oncorhynchus nerka*)
Smelt (see *Osmerus eperlanus*)
Starfish (see *Asterias rubens*)
Stickleback, three-spined (see *Gasterosteus aculeatus*)
  nine-spined (see *Pungitius pungitius*)

Tadpoles, 54, 92, 113
Trout, Brown (see *Salmo trutta*)
  Rainbow (see *Salmo gairdneri*)

Water-mites (see Hydrachnellae)

Zebra Mussel (see *Dreissena polymorpha*)

Index to Authors' Names

Allee, 1, 141
Allen, 162, 268-70
Alm, 113, 267
Alsterberg, 181
Alexander, 210
Ambühl, 118-22, 125-7, 194-6
Anderson, G. C., 139, 239
Anderson, W. L., 89
Andrewartha, 3, 166
Aschner, 95

Baas-Becking, 92, 242
Băcescu, 211
Bagenal, 103
Baggerman, 166
Bailey, 285
Baldwin, 150
Balfour-Browne, 47, 70, 71
Balmain, 88
Barnes, 109, 164
Bar–Zeev, 158, 159
Bassindale, 210
Bates, 2, 84, 85, 162, 251
Beadle, 199, 209, 216, 217, 219, 220, 228, 236-40
Beattie, 78
Beauchamp, 98, 111, 127, 156, 157
Berg, 28, 36, 44, 45, 48, 57, 58, 60-2, 131, 138, 158, 181-3, 189, 192, 198, 200, 267, 281
Birch, 3
Bläsing, 157, 168, 172, 173, 251
Blunck, 152, 153, 154, 162
Boisen Bennike, 44, 131, 247
Bond, 237, 241
Bondesen, 67, 68
Boone, 92, 242
Bourrelly, 74
Bovbjerg, 130
Boycott, 48, 68, 74, 246, 247, 260
Boyd, 100
Brand, 201
Brett, 148, 153
Briggs, 178
Brinkhurst, 71, 136
Brown, E. S., 69, 70
Brown, L. A., 158, 159
Brown, M. E., 273
Brundin, 176

Bryan, 228
Bullough, 166
Buscemi, 176
Butcher, 177, 180
Buxton, 71, 83, 84, 86

Carpenter, 263
Chimits, 88
Christophers, 164
Claridge, 49
Claus, 199, 230-2
Clawson, 148
Clemens, 159
Coker, 88
Colless, 86
Corbet, 69
Couégnas, 248
Creaser, 148
Crisp, D. J., 163
Crisp, D. T., 147, 255
Croghan, 92, 242
Crow, 285

Dahm, 156
Dallinger, 166
D'Ancona, 211
Darwin, 4, 102
David, 285
Davies, 142
Davis, 158, 164, 165
de Beaufort, 210
Decksbach, 241
Degerbøl, 147
Diver, 4, 103, 105
Dodds, 180
Donaldson, 153
Dorier, 122, 124, 127
Duffield, 113
Dunn, 83, 86
Dussart, 138, 139
Dyke, 178
Dymond, 100

Ebeling, 261
Ebling, 91
Eckel, 139
Edmondson, 130, 254
Ehrenberg, 45

Einsele, 94, 111, 112
Ellis, A. E., 67
Elson, 148, 150
Elton, 1, 2, 4, 61, 102, 103, 131
Emerson, 141
Engelhardt, 122
Ewer, 202
Eyles, 71

Fabricius, 88
Fauré-Fremiet, 93, 94
Ferguson, 154
Fernando, 70, 105
Fish, 155
Fisher, 150
Fleming, 207
Foerster, 270
Forsman, 215
Foster, 153
Fox, 180, 192, 193, 198, 202, 204, 235, 260, 262
Fraser, 80
Frey, 11
Frömming, 251
Fry, 11, 148-51, 153, 184-7
Fryer, 104, 105, 106, 112, 235

Gaarder, 263
Gajevskaja, 95
Gameson, 178, 179, 180
Gardner, A. E., 80
Gardner, R. S., 285
Garside, 154, 155
Geijskes, 251
Geldiay, 128
Gessner, 135, 239, 240
Gibson, 153, 187
Gilbert, 101, 105
Gilchrist, 202
Gillett, 134
Gillies, 79
Gislén, 66, 68
Gledhill, 20, 27, 61, 102
Godwin, 74, 75
Gorham, 139
Graham, J. F., 108
Graham, J. M., 150, 151, 184-7
Greenbank, 176
Grenier, 122
Grinnell, 102
Griffith, 178, 179, 180

Hall, 134
Halsband, 168, 172, 173, 251
Hamilton, 250
Hardin, 98, 100-2, 104, 106
Hardy, 72
Harker, 122, 132
Harnisch, 184, 201, 202, 204, 209, 262, 264
Harris, 79, 249
Hart, 150, 153, 185, 186
Hartley, 284, 285
Heal, 147
Hecht, 85
Hedgpeth, 92, 237
Hedwig, 94
Hendelberg, 246
Hepburn, 154
Herbert, 188
Hers, 201
Hetherington, 93
Hinton, 163
Hirst, 153
Hisaw, 180
Hobart, 101, 105
Hoestlandt, 217, 218
Holmquist, 73, 164, 180
Hopkins, 83, 84, 86
Howland, 78
Hubault, 127, 177, 181
Hubendick, 247, 251
Hudson, 86
Huet, 132, 279
Huffaker, 158
Hunter, 67, 168, 250
Huntsman, 148
Hurlbut, 158
Hutchinson, 46, 75, 94, 105, 106, 138, 176, 238, 239, 241, 262, 264
Hynes, 20, 23-7, 34, 47, 67, 75, 97, 98, 112, 129, 131, 134, 135, 160, 177, 181, 215, 249, 270

Illies, 22, 23, 27, 35, 68, 69, 101, 102, 103, 112

Jackson, 70
Jacobi, 242
Jakob, 94
Jenkin, 138, 240
Jewell, 75, 250, 254, 261
Jobling, 85, 86
Jónasson, 183, 198, 272
Jones, 24, 27, 188, 263
Johnson, 207

Kaiser, 67, 68
Kalleberg, 3, 99
Kimmins, 47
Kinne, 7, 160, 215, 232-4
Kitching, 91, 167, 221, 222
Kligler, 72, 84
Knowles, 178
Korringa, 165
Kramer, 158
Kreuzer, 76, 133
Krog, 147
Krogh, 209, 218, 226, 227

Lack, 3, 79, 101, 103, 111, 113
Lamborn, 83
Lang, 60
Lansbury, 68, 145
Laurie, E. M. O., 176
Laurie, R. D., 263
Le Cren, 112, 267, 268
Ledebur, 198
Leeson, 71
Lefèvre, 94
Leitch, 202
Leloup, 99
Leston, 144, 145
Lewis, 164
Lindberg, 212, 213
Lindeman, 274
Lindroth, 183, 187
Linduska, 129
Lloyd, Ll., 107, 108
Lloyd, R., 262
Lockwood, 228
Lotka, 100
Longfield, 69
Longwell, 177
Lumbye, 181, 182, 200
Lund, 3, 95, 256-8, 260, 261
Lundblad, 61, 92

Macan, 9, 20, 22, 23, 27, 35, 49, 50-4, 75, 102, 106, 112, 122, 127, 128 132, 140-2, 146, 212, 249, 255, 256, 258, 260, 279, 280, 281
Macfadyen, 4, 16, 249
Mackereth, J. C., 20, 112, 127, 132
Mackereth, F. J. H., 178, 256, 258, 261
McGavock, 204
McLeese, 5, 155, 201
Mann, K. H., 47, 48, 60, 75, 189-92, 204, 247
Margalef, 87
Margineanu, 211
Marshall, 216
Mason, 147
Mathiesen, 272
Matonickin, 147
Mer, 72
Merriam, 141
Miller, 2, 4
Mills, 263, 264
Milne, 72
Monard, 101, 105
Moon, 46-9, 68, 75, 76, 135
Moore, J. A., 160, 161
Moore, J. E., 239
Moore, N. W., 69, 80
Moore, W. H., 279
Morawa, 181
Mortimer, 175, 279
Muirhead-Thomson, 83-5, 139, 156, 157, 158
Müller-Liebenau, 45, 46
Mundie, 73, 279, 280
Munro, 88

Naylor, 215
Nielsen, 116, 117, 144, 165
Nilsson, 99
Nisbet, 94
Noland, 93
Nygaard, 248, 271

Ockelmann, 181-3, 198
Odum, H. T., 4, 272
Odum, E. P., 4
O'Gower, 85, 86
Orton, 164

Pacaud, 75, 92, 93, 157, 188, 189, 250
Palmén, 158
Panikkar, 234
Pantin, 215, 219, 251
Park, T., 95, 96, 141
Park, O., 141
Pattée, 183
Pause, 202
Pearsall, 62, 255
Pearson, 284, 285
Pennington, 3, 110, 112
Pentelow, 177, 180, 273
Percival, 79, 118, 128, 129, 133
Pesta, 138, 139
Petersen, 57, 58
Petit, 211
Phear, 202
Philipson, 122
Philippson, 122
Picken, 131
Pitt, 154
Pleskot, 79, 102, 132, 142, 143, 168
Popham, 131
Potts, 224
Pratt, R., 95
Pratt, D. M., 162
Provost, 71

Quennerstedt, 135

Ramage, 260, 262
Ramsay, 236
Rao, 87
Ravera, 183
Rawson, 150, 165, 176, 239
Reich, 95
Remane, 7, 209, 211, 213, 221
Reuter, 139
Reynoldson, 47, 48, 60, 75, 76, 101, 105, 108-11, 113, 247, 248, 250, 251
Richard, 69, 70
Richards, 113
Ricker, 270

Rodhe, 271
Roe, 283, 285
Rollinat, 147
Roos, 70
Rose, 113
Russell, 87
Ruttan, 176
Ruttner, 11, 138, 176
Ruttner-Kolisko, 135
Ryther, 95

Sachlan, 139
Saunders, 263, 264
Savage, 89
Sawyer, 80
Schachter, 74, 211
Schäperclaus, 204
Schiemenz, 81
Schlieper, 7, 157, 168-73, 198-200, 208, 209, 217-20, 222, 223, 225, 226, 251
Schmeing-Engberding, 155
Schmidt, 141
Schmitz, 27, 139, 211, 279
Schultz, 89
Scott, 120-3, 130
Segerstråle, 209, 211, 214, 215, 216, 219
Senior-White, 87
Shaw, 224, 228, 229
Shelubsky, 95
Shepard, 187
Simmonds, 192, 193, 198
Simpson, 283, 285
Slinn, 215
Sloane, 91
Smith, C. D., 142
Smith, M., 147
Smyly, 55, 56, 250
Sonneborn, 95
Solomon, 109, 110
Southgate, 210
Southward, 163
Spooner, 103, 105, 214, 215
Staddon, 49
Starmühlner, 147
Steinböck, 156, 157
Steiner, 81
Strenger, 129
Stuart, 72, 88
Sullivan, 150
Sverdrup, 207

Tait, 155
Tansley, 1, 2, 250
Taylor, A. E. R., 204
Taylor, E. W., 13
Theodor, 84
Thienemann, 4, 67, 68, 73, 74, 75, 92, 101, 106, 109, 141, 147, 180, 181, 198, 250, 261
Thorpe, 242
Treherne, 236
Tucker, D. S., 249, 250
Tucker, D. W., 216

Ullyott, 98

Vaas, 139
Vaillant, 122, 124, 127
Verrier, 69, 79
Viets, 47
Vivier, 100
Vollenweider, 109, 130, 183
Volterra, 100
Vorstman, 162

Walker, 150
Walshe, 192, 193, 201-3
Walton, 88
Warwick, 67
Washbourn, 192
Wautier, 183
Weber, 70
Weerekoon, 72, 73
Welch, 276
Wesenberg-Lund, 80
Whitehead, 79, 118, 128, 129, 133
Whitney, 156, 177
Wiebe, 204
Wigglesworth, 182, 235, 236
Wikgren, 223-9
Williams, 69, 102, 103
Williamson, 47, 49
Wilson, 261
Wingfield, 180, 193, 198
Woker, 6
Woodbury, 204
Woodley, 180
Wuhrmann, 6
Wundsch, 95

Zahner, 81, 82, 197
Zwicky, 182

Subject Index

Accessory growth factors, 261
Acclimation (*see* Acclimatization)
Acclimatization, oxygen, to, 188, 190, 200
  salinity, to, 216, 219, 236
  temperature, to, 148-50, 154-6, 160, 165-7
Adaptation, life in a current, to, 11, 119, 120
  oxygen, low, to, 188
  temperature high, to, 166, 167
Algae, decomposition of, 176
  food, as, 24, 92, 93, 110, 130
  outbursts and decline of, 3, 95
  oxygen and, 176, 177, 180, 204
  production by, 271, 272
  substratum for Protozoa, as, 131
  toxins produced by, 95
Altitude, 20, 23, 141
Ammonia, 83, 228, 229, 262
Athalassohaline waters, 237-43

Bala, L., 135
Baltic Sea, 209, 211-14, 216-20
Barriers, land, 66, 68, 74-6
Bear, L., 262
Birge-Ekman grab, 36, 280
Biocoenosis, 8, 36, 62
Biocoenotics, first law of, 75, 101, 106
Biotope, 8
Brackish water (*see also* mixohaline), 13, 207

C^{14}, 271-2
Calcium, 13, 27, 89, 228, 246-52, 262
  carbonate, deposition of, 251
  concentration of in: Esrom L., 36; Hodson's Tarn, 50; some saline lakes, 239; Store Grib Lake, 57; Windermere, 46
  occurrence of: Crustacea and, 75, 76, 248-50; Hirudinea and, 247; Mollusca and, 48, 246, 247, 251, 257-260; Platyhelminthes and, 27, 247, 248; Porifera and, 250; Rotifera and, 254
  resistance to unfavourable conditions, and, 172, 219, 251

Canals, 67
Cannibalism, 113
Carbonate, 89, 237, 239, 240, 251
Catastrophes, 61, 74, 104, 110
Chemical properties of water, 12-14
Chironomidae, reasons for neglect of, 15
Chloride, 208, 242, 262
  concentration: in body fluids, 219, 228; in waters, 237, 239, 257-9; limiting, 226-8, 236;
  rate of absorption of, 227-9
Cladocera, requirements of, 55-7, 92-4, 158, 159, 182, 188, 189, 262
Classification of streams, 27
Coleoptera, flight of, 70
Communities, definition of, 4, 16, 276
  difference between animal and plant, 1-3, 15, 16
  lakes and ponds, of, 36-62
  Protozoa, of, 93, 131
  running water, of, 16-36
  structure of, 277
Competition, 68, 91, 95-107, 147
  between: *Gammarus* spp., 97, 98, 215; Hirudinea, 60; planarians, 27, 98, 99; *Salmo* spp., 99
  limitation of range and, 7, 22, 163, 165, 167
Competitive exclusion principle, 102-7
Contractile vacuoles, 220-2
Copper, 262
Corixidae, Baltic, of, 212, 213
  calcium and occurrence of, 249
  flight of, 69, 70
  succession in, 255
Cosmopolitan species, 66, 67
Crustacea, occurrence of and calcium, 248-50
Culicidae, flight of, 71-2
  oviposition of, 83-7
Current speed, *see* flow

Decomposition, 12, 176, 177
Density of water, 12
Desiccation (*see* Drying up)
Development rate 152-4, 158-62
Diapause, 165, 166
Dispersal, 66
Dominance in plants, 15, 16

Drying up, 66
ponds, of, 61, 133, 134
streams, of, 20, 134, 135

Ecological Society, 100, 102, 103
Elements found in tissues, 260
Eltonian pyramid, 2, 113
Emergence traps, 22, 279
Emigration, 8
Energy, 4
solar, 3, 267
Ennerdale, L., 47, 48, 73, 256, 261
Ephemeroptera, flight of, 69
oviposition of, 79, 80
Ephydrid larvae, 237, 238, 240, 242
Esrom, L., 57, 62, 267
fauna of, 36-46, 48, 256
temperature of, 138
production in, 272
Erken L., production in, 271, 272
Evaporation, cooling by, 12, 140, 141

Feeding rate of: and temperature, 157, 158; and oxygen concentration, 203
Fen, 15
Flamingoes, diet of, 240
Flight, 69-72
Flow, 11, 28, 35, 116-33, 279
distribution and, 78, 82, 120-7
experiments on effect of, 122-7
measurement of, 118
oxygen and, 180, 192-8
size of objects moved, 117
spawning and, 88
territories of *Salmo* and, 99
Food, 2, 50, 91, 156, 168, 261, 263, 267
carried by current, 119, 120, 130, 133
control: of numbers, and, 3, 104, 110-13; of range, and, 56, 163, 251, 263
effect of on stream communities, 20, 22, 35
efficiency of conversion of, 273
Salmo, of, 99
sandy substratum, on, 116
source of in streams, 24, 131
specific requirements, 92-4
Ford Wood Beck: fauna of, 16-23, 27, 34, 44; methods used in survey of, 276
Fugitive species, 105
Fure L., 73, 131, 256
production in, 272

Gause's hypothesis, 100-6
Gribsø (Grib L.) (*see* Store Grib L.)

Habitat, definition of, 8
Haemoglobin, 184, 202-4
Hatching of *Aëdes* eggs, 134
Heat, absorption of, 11
Hibernation requirements: of *Anopheles*, 146, 162; of *Gerris*, 136
Hirnant, Afon, fauna of, 23-7
production in, 270
Hirudinea, calcium and distribution of, 247
effect of substratum on, 44
oxygen consumption of, 189-92
Hodson's Tarn, fauna of, 50-6, 61
methods used in survey of, 276-8
Horokiwi Stream, 268-70
Hot Lake, 239, 240
Hot springs, fauna of, 147
Humic substances, 60, 261
Hydrachnellae, irregular distribution of, 61
Hypersaline waters, 237

Ice, formation of, 12
Ice Age, 27, 67, 73, 97, 163, 235
Inland waters, peculiar features of, 11
Interaction of factors, 7, 85, 156, 262
Introductions, 67, 68, 100, 181
Isolation, a feature of fresh waters, 13
of Britain, 97
of Plecoptera, 27, 28, 69
by time of flight, 102
of marine invaders, 235
Ijsselmeer, 209, 213, 214

Lake District, English, 16, 117, 120, 144, 247, 249, 260
Lakes, fauna of, 36-50
serial arrangement of, 48, 62
Lead, 262, 263
Lethal levels, interaction of, 5-7
Level, water, rise and fall of, 135
Light, reproduction and, 166
penetration of, through ice and snow, 176
Light traps, 69, 70
Lismore I., 250
Little Manitou L., 238-40
Lomond, L., 72, 250, 251
Lunz, 142
Lyngby, L., 15
production in, 272

Magnesium, 89, 228, 239, 254, 258, 259, 262
Malaria, 72, 92, 100
Merriam's life zones, 141
Methods, 276-85
Migration, 8, 66
rate, 69, 70
Mixohaline water, 207-36
fauna of, 209-17
physiological problems of marine invaders of, 217-35

Mixohaline water—*cont.*
physiological problems of freshwater invaders of, 235, 236
Mölle, R., fauna of, 27, 28
Mollusca, good conditions for, 48,
occurrence of, and calcium, 247, 251; and other chemical factors, 257-60
Monard's hypothesis, 101-6
Mortality, 104, 113, 268-70
Mundie pyramid, 279

Naivasha L., 240
Nakuru L., 239, 240
Net-spinning, 11, 22, 50, 119, 133, 197
relation to current, 122
Net, error of, 281, 282
Niche, 2
Numbers, 2-4
control of, 109-14
on different substrata, 129
Nyasa, L., speciation of fish in, 104-6, 112, 235

Odonata, flight of, 69
oviposition of, 80-3
Oil wells, fauna of, 183, 242
Organic matter, attractive to mosquitoes, 83, 84, 86, 87
decomposition of, 176, 177, 273
distribution and, 75, 254, 255, 260, 261
Osmotic concentration of body fluids, 217-24, 226, 235, 242
Oviposition, 7, 57, 79-89, 126, 150, 156, 160
Oxygen, 175-206, 215, 278, 279
concentration of: catastrophic drop in, 61, 74, 107, 108, 110; effect on amount consumed, 169-71; effect on respiraation, 169-71; fluctuation of, 175-80; in lakes, 12, 47, 57, 176; low, as limiting factor, 35, 45, 56, 60, 61, 82, 92, 93, 130, 180, 181, 185-7; 189-92; running water in, 11, 177-80; temperature, and, 168, 175
consumption of, salinity and, 198-201, 219, 232
temperature, and, 150, 151, 153, 168
excess, 204
incipient limiting point, 185-7, 190-3, 198, 202
indicator of productivity, as, 271, 273
lethal level of, 7, 185-8, 193, 195-7
level of no excess activity, 185-7
life without, 11, 201-4
measurement of, by Fox's flagellate method, 180, 235
requirements of animals, 168, 181-3, 192

Parasites, 113
Permeability, 199, 200, 217, 220, 222, 225-7, 229, 242
pH, effect of: on range, 93, 254, 263, 264; on respiration, 169-71; on uptake of ions, 228, 229, 262; of Three Dubs Tarn, 50
Physical properties of water, 11, 12
Pipits, 79
Platyhelminthes, calcium and occurrence of, 247, 248
Plecoptera, flight of, 27, 28, 69
speciation in, 27, 28
Poisons produced by organisms, 91, 94, 95, 261
inorganic, 262, 263
Pollution, 34, 92, 93, 177, 178, 181
Porifera, occurrence of, 250, 254, 261
Potassium, absorption of, 230
body fluids, in, 219, 228
frog-spawning, and, 89
natural waters, in, 239, 258, 259
resorption of, 220, 236
toxicity of, 262
Predation, 23, 35, 55, 78, 91, 92, 98, 108, 112, 132, 133, 152, 165
Production, 4, 60, 62, 256, 267-75, 277
Protozoa communities, 93, 131

Rice-fields, 84, 86
Rheidol, R., 27, 263
Rotifera, occurrence of, 254

Salinity (*see also* mixohaline water), 13, 73, 147, 183, 207-43
effect: on oxygen consumption, 170, 171, 198-201; on rate of respiration, 170, 171
lethal level of, 7
methods of expressing, 207, 208
Sampler, Birge-Ekman, 280, 281
saw-cylinder, 281-3
shovel, 280
Surber, 280
tray, 280
Sampling, statistics of, 283-5
Scale Tarn, 55, 56
Scientific names, changes of, 8-10
Sewage works (*see* trickling filters)
Shelter, 112
Silt, 255
Silver Springs, production in, 272
Slope, 22
Sodium, absorption of, 228-30
body-fluids, in, 219, 228-30, 236, 242
frog spawning and, 89
in natural waters, 237, 239, 240, 258, 259
toxicity of, 262
Soil, aquatic animals in, 135

Speciation in fish, 104-6
  in Plecoptera, 27, 28, 69
Springs, fauna of, 98, 144, 251
  hot, fauna of, 147
Statistics, 283-5
Store Grib L., fauna of, 57-62
  production in, 272
Stratification in lakes, 12
Substratum, 55, 128-33, 279
  hard, 35, 44, 45
  influence: of water movement on, 116; of bedrock on, 117
  sampling of, 280-3
  sandy, 27
  silted, 255
  stony, 50
  territory in *Salmo*, and, 3, 99
  trickling filters, in, 107-10
  vegetable, 46, 56
Sulphate, 89, 237, 241, 257-9
Susaa, R., fauna of, 28-35, 131
Swarming, 69, 87
Synonyms, method of indicating, 9

Tanganyika, L., 94, 138, 262
Tees, R., 177
Tegid, L., 135
Temperature, 138-74, 180-97, 200, 219, 251
  adaptation to high, 166, 167
  competition and, 22, 98
  distribution and, 20, 22, 35, 45, 60, 130, 131, 141-8, 230, 247, 251
  experimental observations on, 148-61
  infection and, 155
  lakes, of, 47, 138, 240
  lethal level of, 7, 100, 148-50, 154, 156, 159, 161, 163-5, 172
  lethal level of, reduced by salts, 172, 173
  limiting factor, as 161-7, 234
  measurement of, 278, 279
  mode of action of, 167-72
  optimum, 108, 150, 152, 158, 161, 162, 273
  oviposition and, 83-5, 88, 150, 161
  oxygen consumption and, 150, 151, 153, 168, 169, 173
  oxygen in solution and, 168
  rate of development and, 152-4, 158-62, 233, 273
  rate of feeding and, 157, 158, 273
  reproduction and, 111, 160, 164-6, 215
  selection, 148, 150, 154, 155
  Store Grib L., of, 57
  streams of, 139, 140
Temporary water, 133-5, 163, 215, 216
Territory, 3, 105
  of *Agrion*, 82
  of *Salmo*, 99, 112
Thermistor, 278
Three Dubs Tarn, 50, 53, 54
Transport, 66-77
  active, 68-72
  passive, 67, 68
  water currents, by, 72
  wind, by, 66, 68, 72
Trichoptera, larvae undescribed, 4, 62
Trickling filters, fauna of, 107-9, 114
Trophic levels, 267, 274
Tso Kar L., 239
Turbidity, 135
Turbulence, 271

Van L., 239, 240
Vitamins, 261

Waves, 116, 255
Whelpside Ghyll, 21, 23
Wicken Fen, 103
Wind, transport by, 66, 68, 72
Windermere, algae in, 95, 261
  fauna of, 46-50, 67, 135, 136, 142, 250, 255, 256
  temperature of, 138
  molluscs of, 246
Wise Een Tarn, 55, 56

Zenker's organ, 228
Zinc, 262, 263
Zones, life, Merriam's, 141
Zuyder Zee (*see* Ijsselmeer)

DATE DUE

Strangely, however, there has been no biography of him in English, and little even in German. The reason is probably that he left very meager information about himself. A biographer would necessarily have to reconstruct his life out of the details gleaned from documents and records of a turbulent and confusing era. For Josel's life coincided with the Protestant Revolt, with its concomitant social, economic and political upheavals in Central Europe. Only a historian of the greatest erudition, and one gifted with a peerless ability to interpret facts and reconstruct an era, could undertake a task fraught with so many difficulties.

Dr. Selma Stern was just the person to bring Josel to life. An incomparable researcher into the German later Middle Ages and modern times, she knew where to look for material and how to interpret what she found. She possesses the gift of a constructive historical imagination. The volume restores for us the heroic figure of a man who, like Mordecai of old, "sought the good of his people and spoke peace to all his seed."

Selma Stern had written a number of historical works in Germany before her exile to the United States. In Cincinnati, she was the archivist of the American Jewish Archives. Upon her retirement, Dr. Stern established herself in Switzerland, where she has been working on an extensive documentary history of the Jews in Prussia, of which four volumes have already appeared. For the Jewish Publication Society, Dr. Stern has written a book of historical fiction called *The Spirit Returneth,* and a series of biographical sketches called *The Court Jew.* Like these others, the book on Josel, here presented in English translation, is characterized by authoritative and skillful presentation.

the Jews. Clearly, Josel was a man of extraordinary stature.

Jacket design by Richard Palmer